Information Processing in Social Insects

Edited by: C. Detrain
J. L. Deneubourg
J. M. Pasteels

Birkhäuser Verlag
Basel · Boston · Berlin

Editors:

Dr. Claire Detrain
Dr. Jean Louis Deneubourg
Prof. Dr. Jacques M. Pasteels
Université Libre de Bruxelles
50 Av. F.D. Roosevelt
C.P.160/12
B-1050 Bruxelles
Belgium

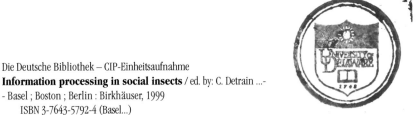

Die Deutsche Bibliothek – CIP-Einheitsaufnahme
Information processing in social insects / ed. by: C. Detrain ...-
- Basel ; Boston ; Berlin : Birkhäuser, 1999
   ISBN 3-7643-5792-4 (Basel...)
   ISBN 0-8176-5792-4 (Boston)

Library of Congress Cataloging-in-Publication Data
Information processing in social insects / edited by: C. Detrain, J. L. Deneubourg, J. M. Pasteels.
   p. cm
   Includes bibliographical references.
   ISBN 3-7643-5792-4 (hardcover : alk. paper). — ISBN 0-8176-5792-4 (hardcover : alk. paper)
   1. Insects--Behavior.   2. Insect societies.   3. Animal communication.
   I. Detrain, C. (Claire), 1963– .   II. Deneubourg, J. L.   III. Pasteels, Jacques M.
QL496.I1383   1999
595.715—dc21   99-25668
                                                              CIP

© 1999 Birkhäuser Verlag, PO Box 133, CH-4010 Basel, Switzerland
Printed on acid-free paper produced from chlorine-free pulp. TCF ∞
Cover illustration by A.C. Mailleux, based on Hölldobler (1976), *Science* 192: 912–914.
Cover design: Gröflin Graphic Design, Basel (www.groeflin.ch)
Printed in Germany
ISBN 3-7643-5792-4
ISBN 0-8176-5792-4

9 8 7 6 5 4 3 2 1

Table of contents

Part 2 Role and control of behavioral thresholds

Part 3 The individual at the core of information management

Part 4 Amplification of information and emergence of collective patterns

List of contributors

Carl Anderson, Department of Zoology, Duke University, Box 90325, Durham, NC 27708-0325, USA, E-mail: carl@duke.edu

Samuel N. Beshers, Department of Entomology, 320 Morill Hall, 505 S. Goodwin, University of Illinois, Urbana, IL 61801, USA
E-mail: beshers@life.uiuc.edu

Guy Beugnon, Laboratoire d'Ethologie et Psychologie Animale, UMR CNRS N°5550, Université Paul Sabatier, 118, route de Narbonne, F-31062 Toulouse Cedex 4, France, E-mail: beugnon@cict.fr

Eric Bonabeau, Santa Fe Institute, 1399 Hyde Park Road, Santa Fe, NM 87501, USA

Nicholas F. Britton, School of Mathematical Sciences and Centre for Mathematical Biology, Claverton Dawn, University of Bath, Bath BA2 7AY, UK
E-mail: N.F.Britton@maths.bath.ac.uk

Scott Camazine, Penn State University, Department of Entomology, University Park, PA 16802, USA

Deby L. Cassill, Department of Entomology, University of Arizona, Tucson, AZ 85719, USA

Blaine J. Cole, Division of Evolutionary Biology and Ecology, Department of Biology, University of Houston, Houston, TX 77204-5513
E-mail: bcole@uh.edu

James T. Costa, Department of Biology, Western Carolina University, Cullowhee, NC 28723, USA

Jean-Louis Deneubourg, Centre d'étude des phénomènes non-linéaires et des systèmes complexes, Unité d'écologie comportementale théorique, CP 231, Université Libre de Bruxelles, Boulevard du triomphe, B-1050 Bruxelles, Belgium, E-mail: jldeneub@ulb.ac.be

Claire Detrain, Laboratoire de biologie animale et cellulaire CP.160/12, Université Libre de Bruxelles, Avenue F. Roosevelt 50, B-1050 Bruxelles, Belgium, E-mail: cdetrain@ulb.ac.be

Claudia Dreller, Department of Entomology, University of California, Davis, CA 95616, USA, E-mail: 101336.2741@compuserve.com

Christine Errard, LEPCO, Université de Tours, Faculté des Sciences, Parc de Grandmont, F-37200 Tours, France, E-mail: errard@univ-tours.fr

Terrence D. Fitzgerald, Department of Biological Sciences, State University of New York, College at Cortland, Cortland, NY 13045, USA
E-mail: FITZGERALD@CORTLAND.EDU

Vincent J.L. Fourcassié, Laboratoire d'Ethologie et Psychologie Animale, UMR CNRS N°5550, Université Paul Sabatier, 118, route de Narbonne, F-31062 Toulouse Cedex 4, France, E-mail: fourcass@cict.fr

Nigel R. Franks, School of Biology and Biochemistry and Centre for Mathematical Biology, Claverton Down, University of Bath, Bath, BA2 7AY, UK
E-mail: N.R.Franks@bath.ac.uk

Dominique Fresneau, LEEC-CNRS, Université Paris Nord, Av JB Clément, F-93430 Villetaneuse, France, E-mail: fresneau@leec.univ-paris13.fr

Deborah M. Gordon, Dept of Biological Sciences, Stanford University, Stanford, CA 94305-5020, USA, E-mail: gordon@ants.stanford.edu

Abraham Hefetz, Department of Zoology, Georges S. Wise Faculty of Life Science, Tel Aviv University, Ramat Aviv 69978, Israel,
E-mail: hefetz@post.tau-ac.il

Zhi-Yong Huang, Department of Entomology, 505 S. Goodwin Ave., University of Illinois, Urbana, IL 61801, USA

Robert L. Jeanne, Department of Entomology, University of Wisconsin, 1630 Linden Drive, Madison, WI 53706, USA,
E-mail: jeanne@entomology.wisc.edu

Alain Lenoir, LEPCO, Université de Tours, Faculté des Sciences, Parc de Grandmont, F-37200 Tours, France, E-mail lenoir@univ-tours.fr

Jay E. Mittenthal, Department of Cell and Structural Biology, 505 S. Goodwin Ave., University of Illinois, Urbana, IL 61801, USA, E-mail: mitten@uiuc.edu

Robin F.A. Moritz, Institut für Zoologie, Martin Luther Universität Halle/Wittenberg, Kröllwitzerstr. 44, D-06099 Halle/Saale, Germany
E-mail: r.moritz@zoologie.uni-halle.de

Robert E. Page Jr., Department of Entomology, University of California, Davis, CA 95616, USA, E-mail: repage@ucdavis.edu

Jacques M. Pasteels, Laboratoire de biologie animale et cellulaire CP.160/12, Université Libre de Bruxelles, Avenue F. Roosevelt 50, B-1050 Bruxelles, Belgium

Francis L. W. Ratnieks, Department of Animal and Plant Sciences, University of Sheffield, Sheffield S10 2TN, UK, E-mail: F.Ratnieks@Sheffield.ac.uk

Gene E. Robinson, Department of Entomology, 320 Morrill Hall, 505 S. Goodwin Ave., University of Illinois, 505 S. Goodwin Ave., Urbana, IL 61801, USA
E-mail: generobi@uiuc.edu

Simon K. Robson, Department of Zoology and Tropical Ecology, James Cook University, Townsville, 4811, Australia, E-mail: Simon.Robson@jcu.edu.au

Bernhard Ronacher, Institute of Biology, Humboldt-University Berlin, Invalidenstrasse 43, D-10115 Berlin, Germany

Bertrand Schatz, Laboratoire d'Ethologie et Psychologie Animale, UMR CNRS N°5550, Université Paul Sabatier, 118, route de Narbonne, F-31062 Toulouse Cedex, France

Tim R. Stickland, School of Mathematical Sciences and Centre for Mathematical Biology, South Building, University of Bath, Bath BA2 7AY, UK

Guy Theraulaz, CNRS-URA 5550, Laboratoire d'Ethologie et de Psychologie Animale, Université Paul Sabatier, 118 route de Narbonne, F-31062 Toulouse, France

Franc I. Trampus, Division of Evolutionary Biology and Ecology, Department of Biology, University of Houston, Houston, TX 77204-5513
E-mail: ftrampus@bayou.uh.edu

James F.A. Traniello, Department of Biology, Boston University, 5 Cummington St., Boston, MA 02215, USA, E-mail: jft@bio.bu.edu

Walter R. Tschinkel, Department of Biological Science, Florida State University, Tallahassee, Florida 32306-4370, USA, E-mail: tschinkel@bio.fsu.edu

P. Kirk Visscher, Department of Entomology, University of California, Riverside, CA 92521, USA

Rüdiger Wehner, Department of Zoology, University of Zürich, Winterthurerstrasse 190, CH-8057 Zürich, Switzerland

Foreword

Claire Detrain, Jean-Louis Deneubourg and Jacques Pasteels

Studies on insects have been pioneering in major fields of modern biology. In the 1970 s, research on pheromonal communication in insects gave birth to the discipline of chemical ecology and provided a scientific frame to extend this approach to other animal groups. In the 1980 s, the theory of kin selection, which was initially formulated by Hamilton to explain the rise of eusociality in insects, exploded into a field of research on its own and found applications in the understanding of community structures including vertebrate ones. In the same manner, recent studies, which decipher the collective behaviour of insect societies, might be now setting the stage for the elucidation of information processing in animals.

Classically, problem solving is assumed to rely on the knowledge of a central unit which must take decisions and collect all pertinent information. However, an alternative method is extensively used in nature: problems can be collectively solved through the behaviour of individuals, which interact with each other and with the environment. The management of information, which is a major issue of animal behaviour, is interesting to study in a social life context, as it raises additional questions about conflict-cooperation trade-offs. Insect societies have proven particularly open to experimental analysis: one can easily assemble or disassemble them and place them in controllable situations in the laboratory. They provide a unique research opportunity to explore some questions such as the link (or the independence) between behavioural algorithms and their physiological implementation. They offer a complete range of contrasts in complexity between the individual and collective levels and therefore are useful for discriminating the contribution of individual capacities in the emergence of efficient global responses. Ultimately, they should enable researchers to replace intuition about behavioural simplicity and complexity by real scientific quantification of these concepts.

Authors of this book, who are at the leading edge of the field, provide extensive and often thought-provoking reviews of current knowledge. This book

C. Detrain et al. (eds) Information Processing in Social Insects
© 1999, Birkhäuser Verlag Basel/Switzerland

encompasses different and complementary approaches in a synergy between theoreticians and experimentalists. The first section describes how information is shared between nestmates and stresses on the regulation of task allocation. The influence of colony size on interaction patterns is one of the major topics discussed here. The second section focuses on the role of behavioural thresholds in the management of information and in the emergence of polyethism. Several contributions untangle environmental, development and genetic factors which shape behavioural responses to stimuli and which ultimately influence division of labour in insect societies. The third section puts the stress on the complexity of information management at the individual level. The individual perception of the environment and the role of key individuals in the development of social behaviour are some of the questions raised in this section. In contrast to the previous approach, contributions of the fourth section refer to some simplicity of both decision and information management by the individual. Local rules and amplifying phenomena are a common theme of these contributions, which show their structuring role in the emergence of decisions at the colony level.

The impressive amount of research dealing with the function and adaptive value of collective behaviour in insects might have led researchers to conclude that there is nothing really new to know about information processing in insect societies. This book clearly demonstrates that the individual and collective management of information as well as its integration into a coherent system are questions still under debate.

Acknowledgments

Referees have given generously of their time to provide insightful comments on the manuscripts. We are endebted to Anne-Catherine Mailleux who illustrated the cover page of this book and to Bert Hölldobler who gave permission to use photographs of *Myrmecocystus* ants. The editing of this book owes much to the workshop organised by the "Fondation Les Treilles". Beside the authors, we benefited greatly from interactions with the following colleagues who participated in debates: Serge Aron, Michael Krieger, Arnaud Lioni, Philippe Rasse, Fabrice Saffre and Thierry Van Damme. Many thanks go to Mrs. A. Postel-Vinay, to Mrs C. Bachy and to the staff of the "Fondation Les Treilles" for making our stay so enjoyable for all of us in a delightful environment.

Part 1 Group size and information flow inside the colony

Group size, productivity, and information flow in social wasps

Robert L. Jeanne

Summary

Social wasp species segregate into two behavioral groups: independent founders, in which queens found new colonies independent of workers, and swarm founders, in which new colonies are founded by swarms comprising several queens accompanied by many workers. Recent work on *Polybia occidentalis*, a swarm founder, indicates that productivity (measured as nest size and grams of brood reared) per adult per day increases with founding swarm size. I consider four mechanisms than can account for this pattern, and adduce *evidence supporting two of them. First, large colonies appear to allocate a larger proportion of their worker population to foraging for resources. Second, foragers in large colonies transfer their materials to nest workers more efficiently than in small colonies. I suggest that differences in the stochastic properties of small versus large groups lead to shorter queuing delays and a greater ability of large colonies to keep the size of interacting worker groups in balance in the face of perturbations and changing conditions. Finally, I argue that the mode of social organization seen in the swarm founders works most efficiently for large groups, while the simpler organization of independent founders works most efficiently for very small groups.

One of the more intriguing attributes of insect societies is the variability of their size. Typical mature colony size for a species usually falls within fairly narrow limits, although it may vary in response to local environmental conditions and climatic differences across the species' range. When we compare species, however, we find a much wider range of colony sizes. Across the four major groups of social insects—termites, ants, bees, and wasps—colony size spans seven orders of magnitude, from only a handful of individuals for some species of

C. Detrain et al. (eds) Information Processing in Social Insects

wasps and bees to millions in some of the termites, ants, and wasps, and even tens of millions in the African driver ant, *Anomma wilverthi* [1, 2].

We have a surprisingly rudimentary understanding of the causes and consequences of such differences in the sizes of these social groups. Is there an optimal colony size for each species, a size at which colony fitness is highest, and if so, what factors influence it? Why, for example, do colonies of honey bees typically contain tens of thousands of workers, instead of hundreds or millions? What is the interplay between the size of the colony and its social organization? Are there correlations between colony size and the organization of material and information flow within the colony? Does the colony size achieved by a species constrain that species to certain forms of social organization? Conversely, do certain types of social organization work better for colonies of some sizes than others?

Species-typical colony size is likely to be a balance struck in response to numerous forces impinging on colonies, and these forces no doubt include both intrinsic and extrinsic factors. In this chapter I focus on intrinsic factors, citing recent developments in the study of social wasps (Vespidae). I review recent work on the interaction between colony size, colony productivity, and mode of organization of work in one of the two groups of social wasps and investigate possible mechanisms behind the resulting pattern. Then I compare it to the alternative mode of organizing work found in the other group of wasps, and suggest that each mode may be best adapted to maximize colony productivity within the range of colony sizes characterizing each group.

Two groups of social wasps

Social vespids segregate into two groups according to the type of colony founding they exhibit [3, 4]. The independent founders are species in which reproductive females found colonies alone or in small groups, but independent of any sterile workers (Fig. 1a, b). Into this group fall the subfamily Stenogastrinae, most of the subfamily Vespinae (hornets and yellowjackets), and the polistine genera *Polistes*, *Mischocyttarus*, *Parapolybia*, *Belonogaster*, and most species of *Ropalidia*. The species in the second group initiate colonies by means of a swarm of workers and queens (Fig. 1c). Swarm founding characterizes the vespine genus *Provespa*, the ropalidiine genera *Polybioides* and *Ropalidia* (in part), and all of the 21 genera in the polistine tribe Epiponini. A cladogram of the genera of

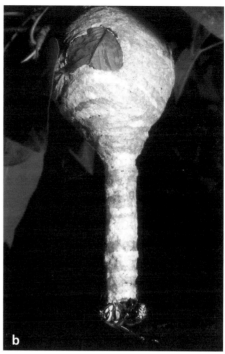

Figure 1 Colony founding in two groups of social wasps, independent founders (a and b) and swarm founders (c)

(a) A founding queen of Polistes pacificus is shown on her nest before any workers have eclosed (Costa Rica). (b) A queen of the bald-faced hornet, Dolichovespula maculata, is on the tip of the entrance tube of her nest (New Hampshire). (c) A swarm of Polybia occidentalis has settled on the twig, where a fraction of the workers have begun construction of the first comb of the nest (center), while the remaining workers and the numerous queens form an inactive cluster a few cm up the twig (Brazil). Photographs by the author.

social vespids is shown in Figure 2. No intermediates are known; species for which we know anything about founding can clearly be assigned to one or the other group. Thus the two behavioral groups appear to be distinct enough for us to treat them as discrete.

I focus on the swarm founders and independent founders within the subfamily Polistinae (Fig. 2). Together these comprise over 900 of the approximately 1000 species of social Vespidae worldwide. Besides the mode of colony founding, there are other differences that distinguish the independent from the swarm founders. Most striking are differences in colony size (Fig. 3). Independent-

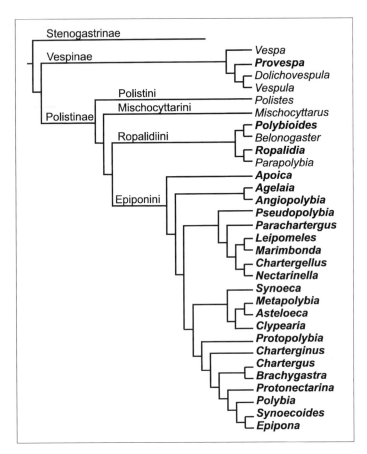

Figure 2 Cladogram of social Vespidae, showing the genera of Vespinae and Polistinae

Genera containing swarm-founding species are shown in boldface type, independent-founding genera are shown in regular type. Cladogram based on [5, 6].

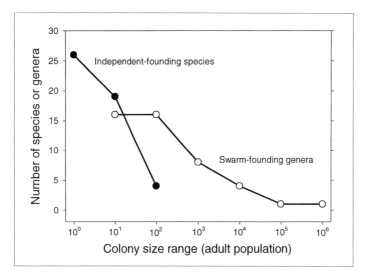

Figure 3 Distribution of colony sizes for IF and SF polistines
Values represent numbers of polistine taxa falling within each size category. Note that the x-axis is logarithmic; thus values for 10^0, for example, include colonies of 1–9 individuals, 10^1 colonies of 11–99, and so on. Only colonies collected without regard to stage of development are included; data from studies focusing on the founding stage are omitted. Minimum and maximum colony size reported for each species or genus defines the range included in the figure. A given taxon may be included in more than one size category. Data for independent-founding species cover 42 species in 5 genera [7–20]. Data for swarm founders include 20 genera (after [4]).

founding polistines have small colonies, typically in the range of up to 10^2 individuals at maturity (Fig. 4a). Reproductive dominance is rudimentary, largely involving direct physical attacks and threat displays by the single queen [21–23]. The swarm founders, on the other hand, are characterized by larger colony sizes, ranging from 10^1 to 10^6 adults, among which 0.1–20% may be queens (Fig. 4b). Little is known about reproductive dominance in this group, but it appears to involve pheromones and behavioral displays [24, 25], with physical dominance interactions rarely in evidence, except when competing queens fight [26]. Although the swarm founders are less speciose than the independent founders, they are vastly more dominant ecologically (sensu Wilson [27]), individuals of swarm-founding species outnumbering independent founders by 100 or 1000 to 1 at some Neotropical sites [4].

Figure 4 Mature colonies of (a) an independent founder, Polistes fuscatus (Wisconsin), and (b) a swarm founder, Polybia occidentalis (Costa Rica)
The Polistes *colony has reached the reproductive stage, as evidenced by the presence of males (with curled antennal tips), yet contains only a few dozen individuals. The nest lacks any covering envelope. The enclosed* Polybia *nest contains five or six brood combs and several hundred workers and egg-laying queens. Photographs by the author.*

A number of representative independent founders, particularly in the genera *Polistes, Mischocyttarus,* and *Ropalidia,* have been well studied. Although the diversity of these genera is greatest in the tropics, their ranges extend far into temperate regions, where they commonly nest in disturbed areas and even on buildings, making them easily accessible to many biologists. Furthermore, their colonies comprise a modest number of adults on a single, uncovered comb, usually in semiexposed places, virtually inviting detailed study of their behavior. As a measure of the extent and significance of research on these wasps, a number of important principles of insect sociobiology were discovered through work on them [1].

By comparison, the biology of swarm founders is relatively poorly known. One reason is that they occur only in the tropics. For the majority of biologists,

Figure 4b

this necessitates costly travel to remote field sites to study them. Moreover, unlike the independent founders, the nests of most swarm founders are multilayered structures covered with an envelope (Fig. 4b). These features, combined with the hundreds or thousands of often small and aggressive wasps in a typical colony, have made them less attractive to study than the independent founders. The challenges they pose to sociobiological theory, however, have stimulated a small but increasing number of biologists to find ways to look into the details of their social lives. Modern genetic techniques such as the polymerase chain reaction (PCR) and improved video-recording technology are among the recent advances that have encouraged the quest.

The case of *Polybia occidentalis*

Background biology

Because *Polybia occidentalis* is the most common epiponine in many locations in Central and South America, my students and I have selected it as a paradigm of the swarm-founding syndrome. In Guanacaste, Costa Rica, where we do our fieldwork, colonies grow throughout the wet season (May–November), then give off reproductive swarms toward the end of the year [28; R.L. Jeanne, personal observation]. Reproductive swarms typically contain 75–350 adults, up to 10% of which are young queens [28]. Even as the last members of the swarm are arriving at the newly selected nest site, typically a twig of a tree or shrub, the first-arriving workers are already beginning construction of the nest (Fig. 1c). The structure, completed within a week or so, typically consists of 3–5 combs containing 1–3000 brood cells. The queens, meanwhile, begin laying eggs in the cells. The first worker offspring begin to eclose about 30 days after founding, and the colony grows during the course of the wet season, eventually, barring accident or unusual weather, reaching a population of 750–1500 or more. As the colony grows, the number of queens is reduced to one or a few due to queen-queen or worker-queen fighting, a process called cyclical oligogyny [25, 28, 29].

Division of labor in *P. occidentalis* is characterized by a marked age polyethism [30, 31]. The average worker begins adult life working in the nest at brood care. At an age of a week or so she begins to spend more time on the outside of the nest, where she unloads foragers and works on nest construction and maintenance. At the age of approximately 3 weeks she switches to foraging and defense. Specialization on either nest work or foraging is fairly rigid, resulting in task partitioning [32]. Task partitioning means that the collection of material in the field and its handling or utilization at the nest are performed in sequence by two different age subcastes of workers. Thus, foragers hand off their loads to receivers at the nest, then return to the field for another load. Virtually complete task partitioning characterizes all four of the materials foraged by *P. occidentalis*: nest material, water, prey, and nectar [33, 34] (Fig. 5a), in contrast to the weak task partitioning found in the independent founders (Fig. 5b). Furthermore, foragers tend to show a strong fidelity to one of the four materials [35, 36].

In addition to normal reproductive swarms produced at the end of the growing season, a swarm will form if the colony is forced to abandon its nest by pre-

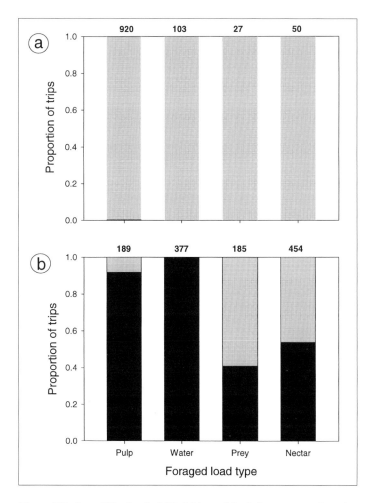

Figure 5 Task partitioning in (a) **Polybia occidentalis,** *a swarm founder,*
and (b) **Polistes instabilis,** *an independent founder*
Bars represent the 4 types of foraged loads. Each bar is divided into the
proportion of observed forager trips for which the forager herself utilized at least part of her load at the nest (black) versus transferred the
entire load to a nest worker (gray). In Polybia, only pulp foragers were
ever seen to utilize part of a load (thin black base to "pulp" bar), and
then only rarely. Numbers above bars give sample sizes. Data from [33,
35]; R.L. Jeanne, unpublished data.

dation or accidental destruction. Absconding swarms may occur at any time of
year and are often larger than reproductive swarms because they comprise the
entire adult population. The emigration of the swarm to a new site and the con-

struction of a new nest occur just as in reproductive swarms. We have worked exclusively with absconding swarms, which are easily triggered by dismantling the nest. Causing absconding is a way of resetting the colony's developmental stage to zero. This has made it possible to study the effects of colony size while controlling for the effects of stage of colony development. In this way we have been able to study the development of newly founded colonies across a size range of 24–1500 adults.

Large colonies produce more *per capita* than small ones

One of our more surprising discoveries using this technique is the relationship between colony size and productivity. We allowed 21 swarms to initiate nests and develop for 25 days, then we collected each and measured its adult population and total productivity to that point. Productivity was measured in two ways: total number of brood cells constructed by the swarm and total dry weight of brood (eggs, larvae, and pupae) reared by the 25th day. Because eggs take 30 days or more to produce an adult, the entire brood-rearing effort of the colony was represented in the mass of the brood we removed from the cells and weighed. For both measures, we found that the larger the founding swarm, the greater the *per capita* (per adult female—queens and workers—in the colony) output of the colony [37] (Fig. 6). To take two examples from the 1984 data shown in Figure 6a, a colony of 278 adults produced only 3.6 cells and 10.4 mg of brood per adult, while one of 719 adults produced 6.3 cells and 19.4 mg of brood per adult. This pattern of rising *per capita* productivity with swarm size in *P. occidentalis* is surprising because it is the precise opposite of the pattern previously reported for most social insects, including *Polybia*, albeit based on data collected for other purposes [38].

Another manifestation of the effect of colony size on output in *P. occidentalis* comes from a detailed analysis of rates of performance of a fixed amount of work. Working with nest construction behavior, an operation that takes place primarily on the outside of the nest where all the component acts can be observed, I estimated the mean number of worker-minutes required to collect and add to the nest an average foraged load of pulp (662 µg). For small colonies (mean size: 32 females) this worked out to 35.4 worker-minutes, nearly double the 20.1 worker-minutes required by large colonies (mean size: 512 females) [39].

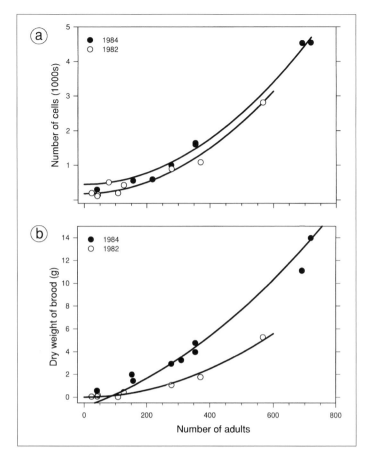

Figure 6 Productivity of 25-day-old P. occidentalis colonies
(a) Number of cells in the nest built by the swarm as a function of number of adults in the colony at the end of 25 days of growth. Regression equations: $cells_{1982} = 182.4 + 0.0082A^2$; $cells_{1984} = 449.5 + 0.0082A^2$; where A = number of adults. (b) Dry weight of brood (grams) in the nest at day 25 as a function of the number of adults in the colony at the end of 25 days. Regression equations: $Wt_{1982} = 0.0000155A^2$; $Wt_{1984} = -2.554 + 0.00926A + 0.0000155A^2 + 19.211p(Q)$, where $p(Q)$ = proportion of queens in colony (for the regression line shown, the value is set to the mean of 0.0896). Modified from [37]. (See [37] for a discussion of two outlier colonies omitted from these figures.)

This result raises the question of what the proximate causes of the pattern may be. Although we have only preliminary answers to this question, there is good evidence that at least some of the explanation has to do with how colony size influences the flow of materials and information through the colony. The pattern translates to behavior very simply: the increase in *per capita* productivity of nest cells and brood biomass with colony size means that larger colonies bring in more materials per adult per day than do smaller colonies. There are four possible ways, none mutually exclusive, this could come about. Each is taken up in turn below, with consideration of the evidence for and against it.

How do they do it? Four hypotheses on mechanism

One way larger colonies could bring in more wood pulp and prey *per capita* is if their foragers carry larger loads of material back to the nest (Fig. 7, hypothesis 1). For this to be the whole explanation would mean that foragers in the larger of the two colonies mentioned above would have had to bring in loads nearly twice the size of those foraged by the smaller colony. Observations of returning foragers on colonies of different sizes suggest that this is not the case (R.L. Jeanne, unpublished observations), so this hypothesis seems unlikely to be true.

The alternative to larger forager loads is a higher number of successful foraging trips per adult per day in larger colonies (Fig. 7). There are two logically possible ways for this to come about. The first is that larger colonies may allocate a higher percent of their worker population to foraging for pulp and prey (Fig. 7, hypothesis 2). If this were going on, it would indicate that colonies of different sizes were making different life history decisions about how to allocate resources to maintenance, growth, and reproduction. There are as yet no data to indicate that the percent of foraging workers is greater in larger colonies than in smaller ones, but there is evidence that life history strategies may vary with colony size. Karen London, a student in my laboratory, studied the defensive response of *P. occidentalis* as a function of colony size. She measured the number of wasp attacks on a target in response to simulated vertebrate attack (standardized tapping) on the nest on the 25th day of colony development for 20 colonies. She found that the smaller the colony, the higher the number of attacks *per capita* (K. London, personal communication). This suggests that small colonies allocate proportionately more worker effort to maintenance, including defense, and less to growth and reproduction, including foraging, than do large colonies.

The second way for larger colonies to achieve a higher *per capita* rate of foraging is for each forager to make more round trips per day. There are in turn two

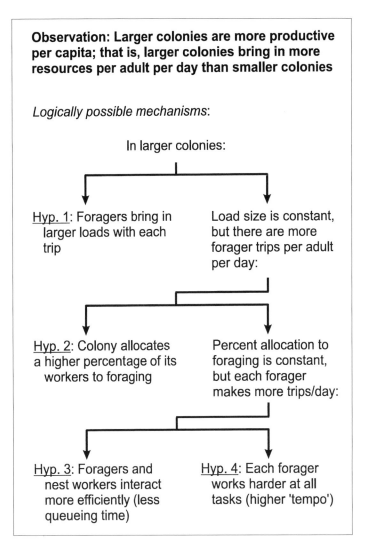

Observation: Larger colonies are more productive per capita; that is, larger colonies bring in more resources per adult per day than smaller colonies

Logically possible mechanisms:

In larger colonies:

Hyp. 1: Foragers bring in larger loads with each trip

Load size is constant, but there are more forager trips per adult per day:

Hyp. 2: Colony allocates a higher percentage of its workers to foraging

Percent allocation to foraging is constant, but each forager makes more trips/day:

Hyp. 3: Foragers and nest workers interact more efficiently (less queueing time)

Hyp. 4: Each forager works harder at all tasks (higher 'tempo')

Figure 7 Branching logical tree of possible explanations of the pattern seen in Figure 6

ways this can happen. The first is for foragers to transfer their loads more quickly to receivers, thus reducing the time they must spend on the nest before taking off on their next trip (Fig. 7, hypothesis 3). The second way is for foragers in larger colonies to work at a higher "tempo" in the sense of Oster and Wilson [40], that is, to perform all their movements more quickly than foragers in smaller colonies (Fig. 7, hypothesis 4).

Before considering the evidence bearing on these last two hypotheses, I shall need to explain more about how the work of nest construction is organized in *P. occidentalis*. Nest construction is a more complex social activity than food foraging because it involves three (rather than just two) interacting groups of workers—water foragers, wood pulp foragers, and builders. As indicated above, each of these workers tends to specialize at its task. The round of tasks performed by each type of specialist, and its interaction with each of the others, is diagrammed in Figure 8. Each specialist cycles through the acts comprising her task as shown

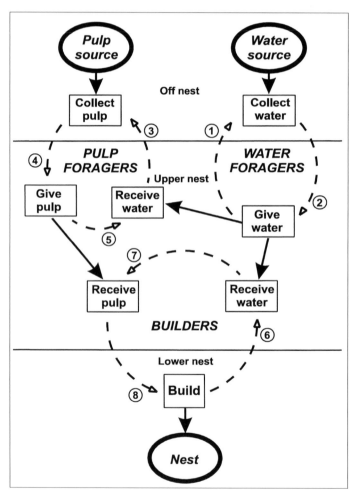

Figure 8 Activity cycles for water foragers, pulp foragers, and builders engaged in nest construction
Polybia occidentalis. *After [39].*

by the broken arrows. A water forager imbibes water at a source away from the nest, filling the crop (act 1). Returning to the front of the nest, she searches for willing receivers, which may be both pulp foragers and builders, and regurgitates her load to them (act 2). A pulp forager leaves the nest with water in her crop, flies to a dead twig or a fence post, and uses the water to moisten the surface as she scrapes up a wad of the softened fibers with her mandibles (act 3). Upon returning to the nest with her pulp load, she moves about the landing area on the front of the nest seeking waiting builders to unload to (act 4). After unloading all her pulp, the forager seeks water from a water forager (or a builder) on the nest (act 5), then takes off on her next trip. A builder first receives water from a water forager (or another builder) (act 6), then seeks pulp from a pulp forager (or another builder) (act 7). She uses her water to further moisten the pulp before subdividing it with other builders and adding it to the nest (act 8). Builders frequently interrupt act 8 to give or receive partial pulp loads from other builders. Thus, of the 8 acts, only two (acts 1 and 3) do not involve interactions with nestmates.

Now let's return to hypotheses 3 and 4. Hypothesis 3, the transfer efficiency hypothesis, predicts that only the six nest construction tasks involving material transfer will require less time in larger colonies than in small ones. It predicts no difference for acts not involving material transfer, that is, acts 1 and 3. Hypothesis 4, the tempo hypothesis, on the other hand, predicts that workers in large colonies will perform all eight tasks more rapidly than workers in small colonies because they perform all their movements at a higher rate. As seen in Table 1, all six of the acts involving transfer of water or pulp took significantly longer in small colonies than in large, often by wide margins. On the other hand, there were smaller differences between large and small colonies in the duration of water and pulp foraging, the two acts that do not involve transfers of material from one individual to another. These results lend tentative support to hypothesis 3 but not to hypothesis 4.

Keeping builder and forager numbers in balance

What affects the speed with which donors and receivers find each other? Foragers and builders do, in fact, spend a finite amount of time searching for a receiver or waiting for a donor. The efficiency of the whole operation—rate of work accomplished per worker-minute—will be maximized if the waiting times of donor and

Table 1. Mean duration (sec) of tasks comprising nest construction in Polybia occidentalis in small and large colonies

Act (no.)	Small colonies			Large colonies				
	Mean	95% CI	n	Mean	95% CI	n	Ratio	P
Water foragers								
Time in field/trip (1)	41.3	38.9–43.8	295	42.9	41.5–44.4	467	.96	.8
Find receivers, unload (2)	45.4	41.1–50.2	283	34.6	32.7–36.6	465	1.3	.001
Pulp foragers		Tasks involving material transfers						
Time in field/trip (3)	219.0	206.3–232.5	240	188.8	181.8–194.8	414	1.2	.001
Find receivers, unload (4)	16.1	14.1–18.4	330	6.7	6.3–7.1	477	2.4	.001
Find donors, get water (5)	44.4	38.3–51.6	207	27.7	24.5–31.2	270	1.6	.001
Builders								
Find donors, get water (6)	116.6	78.4–173.5	49	37.7	21.9–65.1	34	3.1	.01
Find donors, get pulp (7)	69.9	60.8–80.4	129	44.3	36.6–53.8	42	1.6	.001
Build, adjust load size (8)	96.6	87.6–112.2	218	61.9	53.3–71.9	87	1.6	.001

Act numbers refer to the numbers used in Figure 8. "95% CI" = 95% confidence intervals. Mean size of colonies: small = 32 females (n = 4), large = 512 (n = 3). Durations for small and large colonies were compared by t-test on log-transformed data. Data from [39].

receiver groups are equal [41], and this can come about only if the numbers of workers in each of the work groups are in balance with one another. If there are too many pulp foragers, for example, pulp loads will arrive too fast for the builders and water foragers to keep up with them, and the pulp foragers will experience long queuing times on the nest. On the other hand, if there are too few water foragers, say, both pulp foragers and builders will experience delays in obtaining enough water before moving on to their next tasks. Thus, just the right number of workers engaged in each task—water foraging, pulp foraging, and building—will keep material flowing through the system while minimizing the time lost overall to waiting in queue.

For the sake of illustration, it is possible to compute a crude estimate of the relative numbers of workers involved. Using mean values obtained during several days of observation on three large colonies (average size = 512 adults) [39], we can calculate that 15 pulp foragers, 7 water foragers, and 60 builders constitute an in-balance workforce under the average conditions and work rates in effect over the period during which the data were collected. The actual numbers of workers engaged in each task will vary depending on the average cycling time for each type of worker and on such extrinsic factors as relative humidity (see below). Furthermore, work rates of individual workers vary tremendously. Typically there are 1 or 2 "elites", foragers that become fixated on one material and make long series of rapid round trips separated by only a few seconds on the nest to unload. Other foragers make less frequent trips, often separated by long breaks. Elites comprise only 10–50% of the active foragers, yet are responsible for bringing in 80–95% of the material [42].

Can the colony keep the numbers in each work group in approximate balance with one another under changing conditions? Two lines of evidence suggest that it can. The first comes from the colony's response to naturally changing environmental conditions. The amount of water required by a pulp forager to collect a load of pulp and for builders to add it to the nest is sensitive to relative humidity (RH). The drier the air, the more quickly collected pulp dries out and the more water is required to keep it moist enough to work with. During the wet season in Guanacaste, when a colony begins its construction workday at around dawn, RH is typically in the high-90% range. Then, as the temperature rises toward noon RH drops to 65–70%. Preliminary data indicate that the ratio of water trips to pulp trips increases from 0.4 at 95% RH to 1.6 at 65% RH (R.L. Jeanne, unpublished data). Thus, colonies engaged in nest construction are able to track these changes throughout the morning and continually adjust the ratio of water-foraging trips to pulp-foraging trips to compensate, meanwhile keeping queuing delays to a minimum.

The second line of evidence is more direct, and comes from a series of perturbation experiments. By artificially upsetting the balance in the flow of materials during repair of the nest, I was able to show that the colony can adjust the overall rate of work by each group to accommodate the perturbation, restoring the balance and bringing waiting times back down [43]. For example, after the majority of the pulp foragers were removed from a nest, the rate of pulp inflow dropped temporarily, causing a slowdown in building. Within 20–45 min, however, new pulp foragers were pressed into service, returning the rate of the operation to close to its premanipulation level. Likewise, supplying water directly to the nest caused a significant drop in water-foraging rate within 20 min, due to the dropping out of some foragers and the slowing down of others. Other manipulations showed a similar ability of the colony to respond appropriately to increases and decreases in rates of input of pulp and water into the nest so as to return the construction effort to previous levels. Furthermore, workers appear to make the necessary response by using information that flows from the nest through the builders to the foragers (Fig. 9). Only the builders obtain information about the nest damage directly from the nest; foragers rarely stray from the landing area on the front of the nest (Fig. 4b). Feedback from the nest damage and among the builders determines the magnitude of the overall construction effort. The appro-

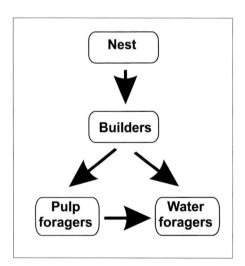

Figure 9 Paths of information flow during nest construction in P. occidentalis *After [43].*

20

priate rate of pulp-foraging is set by feedback from builders to pulp foragers, while water foragers similarly adjust their group's activity level in response to feedback they receive from both of these groups [43].

The adjustments made to bring numbers back into balance following a change in RH or a perturbation are, of course, the result of decisions made by individuals whether to start, stop, or change their work rate. Builders make those decisions using information they receive from the nest and from each other; foragers appear to receive information while searching for and transferring to receivers. The nature of the information is unknown, but it likely takes the form of cues, such as the length of the wait to unload. The decision rule for a pulp forager, for example, could take a form such as, "If I have to wait >30 s for a receiver to take my load, then I will quit foraging." Seeley and Tovey [44] have shown for honey bees that the waiting time to unload is an accurate indicator of relative rates of foraging and receiving. Because the whole operation appears to be regulated through interactions among the workers, nest construction in *P. occidentalis* is an example of a self-organizing system [45].

Stochasticity is the hobgoblin of small colonies

The only way queuing time can be eliminated from an activity such as nest construction is if the system is deterministic, that is, if the numbers of specialists in the three groups are in exact balance with each other and there is no stochasticity in time of arrival into each queue. This would be virtually impossible to achieve in a biological system. Stochasticity is a fact of life for *P. occidentalis*. There are at least three ways it has the potential to have a greater influence on small colonies than on large ones.

First, because workers are discrete units, it will be difficult to achieve an optimal balance of numbers of workers among the three work groups when the total number of active participants is small. In the example given above, 15 pulp and 7 water foragers were in balance with 60 builders, for a total work force of 82. If this were scaled down to a force of 10, maintaining the same optimal balance would require 1.8 pulp foragers, 0.9 water foragers, and 7.3 builders, hardly a realizable mix. Rounding to 2 pulp and 1 water forager and 7 builders, for example, would increase queuing times for the foragers. As Anderson and Ratnieks [41] point out, deviations from the optimum ratio greatly increase mean waiting time.

Second, as group size decreases, the relative effect of random variation in forager arrival times increases, causing an increase in mean waiting time [41, 46]. That is, as worker group size decreases, stochastic effects alone have an increasing negative effect on the efficiency of the operation. This is a possible explanation of the observation that mean and variance of waiting times are larger in small colonies of *P. occidentalis* than in large ones [39] (Tab. 1).

Third, increased stochasticity in small groups may decrease the efficiency of group activity in yet another way. As suggested above, foragers and builders may use an increase in waiting time to detect a drop in the level of demand for their task and to decide whether to continue working or not. According to information theory, the threshold for signal detection is equal to the standard deviation of the noise (waiting time) [47]. Assuming that waiting times are randomly distributed, their variance increases with the mean [48]. Thus, compared to her counterpart in a large colony, a forager in a small colony would have to experience a waiting time that was a greater increase over the mean before she would recognize it as a signal indicating that waiting times had actually increased and that demand for her material had dropped. To take a hypothetical example, suppose a forager acts according to a decision rule that says "Quit or switch to another task if waiting time to unload exceeds y seconds." If the standard deviation of waiting time for foragers in a large colony is ± 3 s, the mean waiting time, $\bar{x}$, must increase by at least 3 s for the rise to be detectable. If the set point in the forager's decision rule were 6 s (two standard deviations above the mean), she would quit if she experienced a waiting time of $y > (\bar{x} + 6)$, a rare event. In a small colony, in contrast, both the mean, $\bar{x}$, and the variance of waiting time are greater. If standard deviation were ± 5 s, a forager would frequently experience waiting times of 6 s above the mean, even though the mean waiting time had not changed. If her decision rule had the same set point of $y > (\bar{x} + 6)$, this could lead her to quit working or to switch tasks prematurely. At the global level, this effect would make it more difficult for a small colony to tune the number of individuals active in each work group precisely enough to maintain an optimal balance for efficient operation. The result would be that work group size would be less stable in smaller colonies, because workers would switch on and off more frequently in response to random fluctuations in waiting times. Indeed, workers of *P. occidentalis* engaged in nest construction switch among pulp foraging, water foraging, and building tasks at higher rates in small colonies than in large ones [39].

In summary, the evidence so far provides the most support for the hypothesis that at least some of the rise in *per capita* productivity with colony size in *P. occidentalis* is an effect of the more efficient integration of foraging and building tasks in large colonies than in small ones (hypothesis 3). Recent data on defensive behavior in small versus large colonies provides reason to suspect that differential allocation of colony resources to growth versus maintenance (hypothesis 2) may also contribute to the observed pattern, but direct evidence for this is still lacking. Finally, there is less reason to believe that forager load sizes (hypothesis 1) or tempo (hypothesis 4) vary with colony size and contribute to the effect.

It seems established that the type of organization of work seen in *P. occidentalis*—task partitioning with the integration of tasks of nest workers and foragers through material and information flow—works better the larger the size of the group. Whether this type of organization characterizes all swarm-founding species remains to be seen. It will be especially interesting to investigate whether species with very small colony sizes, such as *P. catillifex*, whose colonies often contain fewer than 50 adults [13], have a system much like *P. occidentalis*, or one that is modified to function better for smaller groups. At the other extreme, does *Agelaia vicina*, whose colonies may contain in excess of a million workers [49], have an even more specialized system of organization? Study of a diversity of taxa is needed to test whether the *P. occidentalis* form of organization of material and information flow is robust and stable across a wide range of colony sizes.

A comparison of the swarm-founding and independent-founding systems of organization

The *P. occidentalis* system stands in contrast to that seen in the independent-founding wasps such as *Polistes* and *Mischocyttarus*. The fundamental difference is that the independent founders exhibit much less complete task partitioning. The virtually complete task partitioning seen in *P. occidentalis* (Fig. 5a) means that during an activity such as nest construction information flows from the nest to the foragers indirectly, by way of the builders. Builders gather information about demand for material from the nest then pass it on via still-unidentified cues to pulp and water foragers while the latter are on the nest (Fig. 10). Independent founders, by comparison, exhibit a much reduced degree of task partitioning (Fig. 5b), especially with respect to pulp and water [33, 50]. This means that the

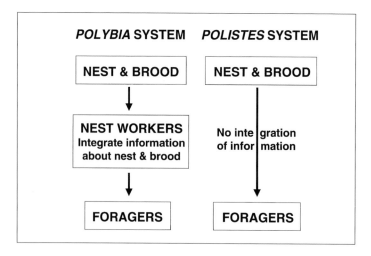

Figure 10 Indirect information flow of the Polybia *system compared with the direct information flow of the* Polistes *system*

path of information flow from needs on the nest to foragers is more direct (Fig. 10). While there is evidence for some species of *Polistes* that the queen stimulates activity on the part of foragers [22, 51], there is no evidence that this stimulus contains information about what materials are needed. It appears instead that a worker inspects the nest, determining for herself that food or nest material is needed [52]. She then departs, collects the material, and returns to apply it herself where needed. This direct route of information flow from nest/brood to forager holds even if the forager occasionally shares or gives her load to a nestmate, because the forager was stimulated to forage by her direct perception of the need for the material. In the task-partitioning, indirect information flow system of *P. occidentalis*, in contrast, the builder subcaste can be viewed as an information-gathering and integrating group that informs foragers about requirements at the nest. As Seeley has elegantly shown for food storer and nectar-foraging honey bees, information flows from the group (builders in this case) to the individual (forager) via cues that sum to convey the colonywide need for material [52, 53].

Because the direct information flow system of independent founders lacks task partitioning, there are no queuing delays. Unlike in *Polybia*, a returning *Polistes* forager can move directly to applying her load of pulp to the nest or to feeding a hungry larva without having to wait for a receiver. Although in some cases loads are shared with nestmates, these transfers occur en route to direct utilization by the forager. The most critical difference is that, unlike in *Polybia*, the *Polistes* forager is not blocked from further action until the transfer occurs.

The absence of queuing delays in independent founders suggests that the efficiency of the system will not increase with group size. In fact, empirical evidence suggests it may decrease. Several investigators have measured colony productivity (typically number of cells in the nest at the time of eclosion of the first worker) as a function of founding group size (Fig. 11). Although the results of these studies generally show a downward trend of *per capita* output with increasing group size, it is not at all clear that the causes have anything to do with how work and information are organized in these colonies. The drop could well be due instead to the disruptive effects of reproductive competition among cofoundresses, for example. On the other hand, reproductive competition should have a smaller influence after worker offspring eclose and evidence from postemergence colonies of *Polistes chinensis antennalis* suggests that the biomass of reproductives produced per worker declines with increasing number of workers

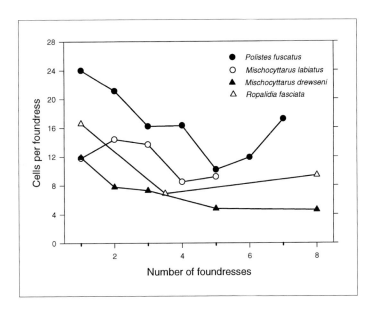

Figure 11 Productivity of preemergence colonies of independent-founding species
Y-axis values are the number of cells in the comb at the time the first offspring eclosed divided by the number of co-founding females. Data from [8, 55–57].

produced [58]. But whether this pattern can be attributed to effects of group size on the efficiency of colony organization remains to be seen.

There are at least two ways that the *Polybia* system of organization can confer efficiency gains on large groups, compared with the *Polistes* system. First, by specializing to a greater extent than *Polistes* workers, *Polybia* workers probably improve their performance efficiency [39, 59, 60]. *Polybia* pulp and water foragers visit their sources frequently and can track changes in their quality more closely than if they made less frequent visits to their sources. In addition, the very rapid cycling of some *Polybia* foragers, the "elites" mentioned above, bring in large amounts of material in a short time [42]. This type of concentrated, rapid-fire foraging is seldom seen in independent-founding species.

The second way the *Polybia* system confers an advantage on large groups is by maximizing material handling efficiency [32]. By specializing, foragers collect loads that appear to be limited by what they can carry and not by what they could handle back at the nest if they were to utilize the material themselves. *P. occidentalis* pulp foragers, for example, bring in loads that are more than six times larger than builders can work with. Maximizing carried load by specializing on foraging means that fewer trips must be flown to collect a gram of material, reducing both the energy expended and the exposure to predation per unit of material returned [39].

It is tempting to speculate that the direct information flow system of independent founders is more efficient for groups in the range of 2–50 or so, while the indirect flow system is more efficient for groups larger than this. Given the increasing inefficiencies in material and information flow experienced by very small *P. occidentalis* colonies, it seems likely that the indirect information flow system would not work as well as direct flow in groups of less than 50 or so. Unfortunately, how *per capita* productivity of swarm founders compares with that of independent founders has not yet been investigated. To make such a comparison, data on productivity would have to be standardized to a form such as grams of brood produced per gram of adults per day. Even then, in making such cross-specific comparisons one must be alert to other differences between the species that could influence colony productivity. These could include differences in life history strategy, food availability, susceptibility to predators and parasites, and development rate of immatures.

Although it is unlikely that the type of organization of material and information flow will fully explain colony size at the species level in wasps, it is possible that the full complement of benefits of the *Polybia* system may have enabled, at least in part, the impressive expansion of colony size seen in the swarm-founders as a group and the tremendous ecological dominance of these wasps over the independent founders in tropical regions. At the very least, it is clear that colony organization, information flow, productivity, and colony size are intimately linked.

Acknowledgements

I gratefully acknowledge Sean O'Donnell, Karen London, Andy Bouwma, Eric Bonabeau, Mary Jane West-Eberhard, and Rick Jenison for valuable discussion of some of the ideas contained herein. Mary Jane West-Eberhard suggested ways to improve the manuscript. I thank Kandis Elliot for digital enhancement of the photographs used in Figures 1 and 4. Research was supported by the College of Agricultural and Life Sciences, University of Wisconsin, Madison, and by National Science Foundation grants BNS-8112744, BNS-8517519, and IBN-9514010.

References

1   Wilson EO (1971) *The insect societies.* Harvard University Press, Cambridge, MA

2   Raignier A, van Boven JKA (1955) Etude taxonomique, biologique et biométrique des *Dorylus* du sous-genre *Anomma* (Hymenoptera Formicidae). *Ann Mus R Congo Belg* 2: 1–359

3   Jeanne RL (1980) Evolution of social behavior in the Vespidae. *Annu Rev Entomol* 25: 371–396

4   Jeanne RL (1991) The swarm-founding Polistinae. *In*: KG Ross, RW Matthews (eds): *The social behavior of wasps.* Cornell University Press, Ithaca, New York, 191–231

5   Carpenter JM (1991) Phylogenetic relationships and the origin of social behavior in the Vespidae. *In*: KG Ross, RW Matthews (eds): *The social behavior of wasps*, Cornell University Press, Ithaca, New York, 7–32

6   Wenzel JW, Carpenter JM (1994) Comparing methods: adaptive traits and tests of adaptation. *In*: P Eggleton, R Vane-Wright (eds): *Phylogenetics and ecology.* Linnean Society London, 79–101

7   Gadagkar R (1991) *Belonogaster, Mischocyttarus, Parapolybia*, and independent-founding *Ropalidia. In*: KG Ross, RW Matthews (eds): *The social behavior of wasps.* Cornell University Press, Ithaca, New York, 149–190

8 Jeanne RL (1972) Social biology of the Neotropical wasp *Mischocyttarus drewseni*. *Bull Mus Comp Zool* 144: 63–150

9 Keeping MG (1989) *Social biology and colony dynamics of the polistine wasp* Belonogaster petiolata *(Hymenoptera: Vespidae)*. PhD dissertation, University of the Witwatersrand

10 Pardi L, Marino Piccioli MT (1970) Studi sulla biologia di *Belonogaster* (Hymenoptera, Vespidae) 2. Differenziamento castale incipiente in *B. griseus* (Fab.). *Monit Zool Ital NS 3*, 11: 235–265

11 Pickering J (1980) *Sex ratio, social behavior and ecology in* Polistes *(Hymenoptera, Vespidae),* Pachysomoides *(Hymenoptera, Ichneumonidae) and* Plasmodium *(Protozoa, Haemosporidia)*. PhD dissertation, Harvard University

12 Rau P (1933) *The jungle bees and wasps of Barro Colorado Island (with notes on other insects)*. Phil Rau, Kirkwood, MI

13 Richards OW, Richards MJ (1951) Observation on the social wasps of South America (Hymenoptera Vespidae). *Trans R Entomol Soc Lond* 102: 1–170

14 Richards OW (1978) *The social wasps of the Americas, excluding the Vespinae*. British Museum (Natural History), London

15 Rodrigues VM (1968) *Estudo sôbre as vespas sociais do Brasil (Hymenoptera-Vespidae)*. PhD dissertation, Universidade de Campinas, Brazil

16 Snelling RR (1953) Notes on the hibernation and nesting of the wasp *Mischocyttarus flavitarsis* (de Saussure). *J Kans Entomol Soc* 26: 143–145

17 Wenzel JW (1987) *Ropalidia formosa*, a nearly solitary paper wasp from Madagascar (Hymenoptera: Vespidae). *J Kans Ent Soc* 60: 549–556

18 Wenzel JW (1992) Extreme queen-worker dimorphism in *Ropalidia ignobilis*, a small-colony wasp (Hymenoptera: Vespidae). *Insect Soc* 39: 31–43

19 Yamane S (1972) Life cycle and nest architecture of *Polistes* wasps in the Okushiri Island, northern Japan (Hymenoptera, Vespidae). *J Fac Sci, Hokkaido Univ, ser 6, Zool* 18: 440–459

20 Yamane S (1980) *Social biology of the* Parapolybia *wasps in Taiwan*. PhD dissertation, Hokkaido University

21 Pardi L (1948) Dominance order in *Polistes* wasps. *Physiol Zool* 21: 1–13

22 Reeve HK (1991) *Polistes*. In: KG Ross, RW Matthews (eds): *The social behavior of wasps*. Cornell University Press, Ithaca, New York, 99–148

23 Röseler P-F (1991) Reproductive competition during colony establishment. In: KG Ross, RW Matthews (eds): *The social behavior of wasps*. Cornell University Press, Ithaca, New York, 309–335

24 West-Eberhard MJ (1977) The establishment of reproductive dominance in social wasp colonies. *Proc 8th Int Cong Int Union Study Soc Insects*, 223–227

25 West-Eberhard MJ (1978) Temporary queens in *Metapolybia* wasps: nonreproductive helpers without altruism? *Science* 200: 441–443

26 Forsyth AB (1975) Usurpation and dominance behavior in the polygynous social wasp *Metapolybia cingulata* (Hymenoptera: Vespidae: Polybiini). *Psyche* 82: 299–303

27 Wilson EO (1990) *Success and dominance in ecosystems: The case of the social insects*. Ecology Institute, Oldendorf/Luhe, Germany

28 Forsyth AB (1978) *Studies on the behavioral ecology of polygynous social wasps*. PhD dissertation, Harvard University

29 Queller DC, Negrón-Sotomayor JA, Strassmann JE, Hughes CR (1991) Queen number and genetic relatedness in a neotropical wasp, *Polybia occidentalis*. *Behav Ecol* 4: 7–13

30 Jeanne RL, Downing HA, Post DC (1988)

Age polyethism and individual variation in *Polybia occidentalis,* an advanced eusocial wasp. *In*: RL Jeanne (ed): *Interindividual behavioral variability in social insects.* Westview Press, Boulder, CO, 323–357

31 Jeanne RL, Williams NM, Yandell BS (1992) Age polyethism and defense in a tropical social wasp. *J Insect Behav* 5: 211–227

32 Jeanne RL (1986b) The evolution of the organization of work in social insects. *Monit Zool Ital NS* 20: 119–133

33 Jeanne RL (1991) Polyethism. *In*: KG Ross, RW Matthews (eds): *The social behavior of wasps.* Cornell University Press, Ithaca, New York, 389–425

34 Hunt JH, Jeanne RL, Baker I, Grogan DE (1987) Nutrient dynamics of a swarm-founding social wasp species, *Polybia occidentalis* (Hymenoptera: Vespidae). *Ethology* 75: 291–305

35 O'Donnell S, Jeanne RL (1990) Forager specialization and the control of nest repair in *Polybia occidentalis* Olivier (Hymenoptera: Vespidae). *Behav Ecol Sociobiol* 27: 359–364

36 O'Donnell S, Jeanne RL (1995) Worker lipid stores decrease with outside-nest task performance in wasps: implications for the evolution of age polyethism. *Experientia* 51: 749–752

37 Jeanne RL, Nordheim EV (1996) Productivity in a social wasp: per capita output increases with swarm size. *Behav Ecol* 7: 43–48

38 Michener CD (1964) Reproductive efficiency in relation to colony size in hymenopterous societies. *Insect Soc* 11: 317–341

39 Jeanne RL (1986) The organization of work in *Polybia occidentalis:* the costs and benefits of specialization in a social wasp. *Behav Ecol Sociobiol* 19: 333–341

40 Oster GF, Wilson EO (1978) *Caste and ecology in the social insects.* Princeton University Press, Princeton, NJ

41 Anderson C, Ratnieks FLW (1999) Effect of colony size on the efficiency of task partitioning in insect societies. *In*: JM Pasteels, JL Deneubourg, C Detrain (eds): *Information processing in social insects.* Birkhäuser, Basel

42 Jeanne RL (1987) Do water foragers pace nest construction activity in *Polybia occidentalis? In*: JM Pasteels, JL Deneubourg (eds): *From individual to collective behavior in social insects.* Birkhäuser, Basel, 241–251

43 Jeanne RL (1996) Regulation of nest construction behaviour in *Polybia occidentalis. Anim Behav* 52: 473–488

44 Seeley TD, Tovey CA (1994) Why search time to find a food-storer bee accurately indicates the relative rates of nectar collecting and nectar processing in honey bee colonies. *Anim Behav* 47: 311–315

45 Bonabeau E, Theraulaz G, Deneubourg JL, Aron S, Camazine S (1997) Self-organization in social insects. *Trends Ecol Evol* 12: 188–193

46 Wenzel JW, Pickering J (1991) Cooperative foraging, productivity, and the central limit theorem. *Proc Natl Acad Sci USA* 88: 36–38

47 Green DM, Swets JA (1966) *Signal detection theory and psychophysics.* Wiley, New York

48 Freund JE (1971) *Mathematical statistics,* 2nd edn. Prentice-Hall, Englewood Cliffs, NJ

49 Zucchi R, Sakagami SF, Noll FB, Mechi MR, Mateus S, Baio MV, Shima SN (1995) *Agelaia vicina,* a swarm-founding polistine with the largest colony size among wasps and bees (Hymenoptera: Vespidae). *J New York Entomol S* 103: 129–137

50 Post DC, Jeanne RL, Erickson EH Jr (1988) Variation in behavior among workers of the primitively social wasp *Polistes fuscatus variatus. In*: RL Jeanne (ed):

Interindividual behavioral variability in social insects. Westview Press, Boulder, CO, 283–321

51 Reeve HK, Gamboa GJ (1983) Colony activity integration in primitively eusocial wasps: the role of the queen (*Polistes fuscatus*, Hymenoptera: Vespidae). *Behav Ecol Sociobiol* 13: 63–74

52 O'Donnell S (1998) Effects of experimental forager removals on division of labour in the primitively eusocial wasp *Polistes instabilis* (Hymenoptera: Vespidae). *Behaviour* 135: 173–193

53 Seeley TD (1989) Social foraging in honey bees: how nectar foragers assess their colony's nutritional status. *Behav Ecol Sociobiol* 24: 181–199

54 Seeley TD (1989) The honey bee colony as a superorganism. *Amer Sci* 77: 546–553

55 Ito Y (1979) Colony development and social structure in a subtropical paper wasp, *Ropalidia fasciata* (F.) (Hymenoptera: Vespidae). *Res Pop Ecol* 27: 333–349

56 Litte M (1981) Social biology of the polistine wasp *Mischocyttarus labiatus*: survival in a Colombian rain forest. *Smithson Contr Zool* 327: 1–27

57 West-Eberhard MJ (1969) The social biology of polistine wasps. *Misc Publ Mus Zool Univ Mich* 140: 1–101

58 Suzuki T (1981) Flesh intake and production of offspring in colonies of *Polistes chinensis antennalis* (Hymenoptera, Vespidae) II. Flesh intake and production of reproductives. *Kontyû* 49: 283–301

59 Seeley TD (1982) Adaptive significance of the age polyethism schedule in honeybee colonies. *Behav Ecol Sociobiol* 11: 287–293

60 O'Donnell S, Jeanne RL (1992) Forager success increases with experience in *Polybia occidentalis* (Hymenoptera: Vespidae). *Insect Soc* 39: 451–454

Task partitioning in foraging: general principles, efficiency and information reliability of queueing delays

Carl Anderson and Francis L. W. Ratnieks

Summary

The collection and handling of colony resources such as food, water, and nest construction material is often divided into subtasks in which the material is passed from one worker to another. This is known as task partitioning. Transfer between workers may be direct with the material passed between individuals, or indirect, with the material left in a cache or dump for others to collect. In this chapter, we review the various forms of task partitioning found in insect societies. If tasks are partitioned with direct transfer of material between foragers and receivers, queueing delays will occur as individuals search or wait for a transfer partner. A computer simulation of direct transfer of material between two groups of workers is used to study the effects of colony size and worker allocation on the magnitude and variance of these delays. These delays are important in colony organisation because they can be used in recruitment. Two mechanisms, multiple transfer and averaging delays over multiple trips, by which workers can enhance the information content of these delays are discussed.

Introduction

Within a typical insect society specific tasks are performed simultaneously by many individuals, and individuals are specialised to perform specific tasks [1]. This parallel-series organisation is hypothesised to enhance colony performance both by increasing the reliability that tasks are carried out and by efficiency gains due to demographic factors [2] and behavioural, physiological, and morphological specialisations of individuals [3].

C. Detrain et al. (eds) Information Processing in Social Insects
© 1999, Birkhäuser Verlag Basel/Switzerland

Two major features characterise the organisation of work in insect societies: (i) division of labour—individuals or groups of individuals vary in the tasks they perform over periods of a few days to their whole life [1–3]; (ii) task partitioning—two or more individuals contribute sequentially to a particular task or piece of work, such as the collection, retrieval, and storage of one load of forage [2, 4].

The extent to which tasks are partitioned varies depending on task, species, and colony size [2, 4]. In the case of foraging for food or nest material, there are two main types of organisation. Foragers may collect material, and on returning to the nest use, store, or distribute the material themselves. Thus, in the honey bee, each pollen forager deposits her own pollen pellets into a cell [5]. However, foragers may also transfer the material collected to other individuals. In the honey bee, nectar and water collected by foragers are transferred to receiver bees who store the nectar or distribute the water [5, 6]. Nectar and water transfer in the honey bee take place in the nest at the end of each foraging trip. But transfer can also occur away from the nest. Leaf cutter ants frequently transfer leaves or leaf fragments to other workers to cut into smaller pieces or to carry back to the nest (Fig. 1; [7, 8]). Transfer of forage material may be either direct, that is, directly passed between individuals as in the case of nectar transfer in the honey bee, or indirect, that is, placed in a dump or cache from where it is retrieved by other individuals.

Task partitioning in the handling of forage has both benefits and costs. Potential benefits include enhanced performance through greater ergonomic or demographic efficiency [2]. Potential costs include the time wasted in transfer between individuals, loss of material during transfer, and ergonomic inefficiencies caused by nonoptimal proportions of foragers and receivers. In the case of indirect transfer, time costs include the time taken to load or unload at the dump site and time taken to locate the dump site. Direct transfer incurs an additional time cost—queueing delays incurred while waiting to be served by a forager or receiver. A queue of foragers or receivers waiting for transfer partners will occur, depending on which group is in excess.

Because the arrival rates of foragers and receivers at the transfer area will vary stochastically, direct transfer incurs a queueing delay even when the relative number of foragers to receivers is optimal. That is, the numbers of foragers and receivers are such that both can collect or process an equal amount of material per unit time. Queueing delays due to stochastic fluctuations need not occur with indirect transfer, because the dump site serves to uncouple the effect of short-

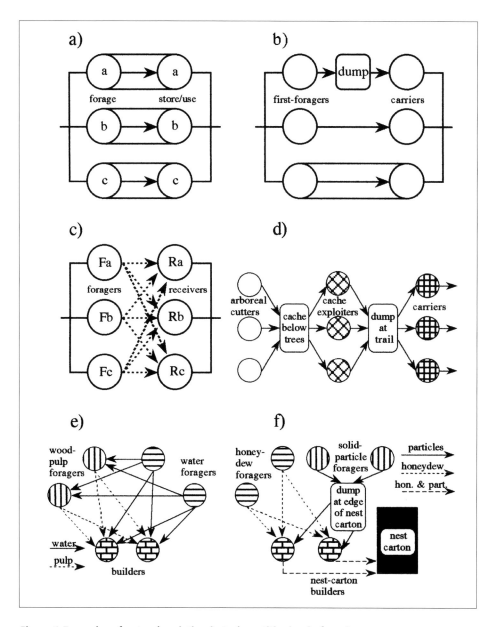

Figure 1 Examples of natural variation in task partitioning in foraging
(a) No transfer, each individual forages and stores or uses material; (b) one set of "first for-agers", indirect, direct or no transfer to "carrier" receivers; (c) one set of foragers, direct trans-fer to receivers; (d) one set of "arboreal cutters", indirect transfer to "cache exploiters"; indi-rect transfer to "carriers"; (e) two set of foragers, both transfer directly to receivers, transfer from one set of foragers to the other; (f) two sets of foragers, one transfers directly to "builder" receivers and one indirectly.

term fluctuations in the arrival rates of foragers and receivers. [Zero queueing delay occurs under two biologically unrealistic conditions when the relative numbers of foragers and receivers are optimal: (i) infinite colony size; (ii) simultaneous arrival of pairs of foragers and receivers at the transfer area.] Because stochastic fluctuations in the arrival rates of foragers and receivers will be proportionately greater in smaller colonies, foragers and receivers in smaller colonies will experience greater mean queueing delays ([2, 9]; C. Anderson and F.L.W. Ratnieks, unpublished observations). Greater stochasticity in the smaller colonies also causes greater variation in the queueing delays experienced by individual foragers and receivers ([9]; F.L.W. Ratnieks and C. Anderson, unpublished observations). This variance effect may be of great biological importance because delays can be used as sources of information for adjusting the relative numbers or work rate of foragers and receivers. Honey bee foragers that experience short delays waiting to be unloaded by a receiver are more likely to perform waggle dances, thereby recruiting additional foragers [5, 10]. Honey bee foragers that experience long delays are more likely to perform tremble dances, thereby recruiting additional receivers.

This chapter provides both an overview of task partitioning in foraging and a more detailed examination of the effect of colony size on the mean and variance of queueing delays, and is organised as follows. Section one presents examples of task partitioning in foraging with the intention of showing a representative range of observed patterns rather than a complete survey. Section two focuses on results from our computer simulations of task partitioning ([9]; C. Anderson and F.L.W. Ratnieks; F.L.W. Ratnieks and C. Anderson, unpublished observations). In section 2.1, we describe the effect of "colony size", actually the number of foragers plus receivers, on the mean queueing delays experienced by foragers and receivers in the transfer area, the dynamics of the queues that develop (section 2.2), and the effect of nonoptimal proportions of foragers and receivers on queueing delay (section 2.3). Finally, in section 2.4 we describe the variation in queueing delays experienced by individual foragers and receivers, and the usefulness of these delays in informing individuals about the relative work capacity of the colony's foragers versus receivers. This information can be used by individuals to determine whether the relative numbers of foragers versus receivers is suboptimal and thereby permit recruitment in a way that moves the colony closer to the optimum.

Examples of task partitioning in foraging

Figure 1 shows some of the diversity in patterns of task partitioning in foraging. No transfer (Fig. 1a) is common and may be the sole means of handling foraged material, as in bumble bees which transfer neither pollen nor nectar [11]. Of the four materials collected by honey bee foragers (nectar, pollen, water, propolis), only pollen is handled in this way. Nectar, water, and propolis are all directly transferred to receivers inside the nest (Fig. 1c) [5, 6, 11, 12]. Nectar also appears to be transferred in meliponine bees ([13]; F.L.W. Ratnieks, unpublished observations).

Foraged material may sometimes, but not always, be transferred [14]. In *Vespula* wasp colonies, nectar is directly transferred in large colonies but not in small colonies [15, 16]. *Vespula* colonies grow from a single queen and zero workers in spring to as many as 5000 workers at the end of the ergonomic phase of their annual life cycle [17], suggesting that this changeover is an adaptive response to changing colony size. Figure 1b shows the case of the leaf cutter ant *Atta cephalotes* [8], in which direct, indirect, and no transfer occur simultaneously. The first foragers at a new patch do not carry their leaf fragments back to the nest but only as far as the main trail, where they are usually either dropped or transferred to other "carrier" workers who take them back to the nest [8]. First foragers sometimes take their leaf fragment all the way to the nest themselves. The proportion dropping or transferring their load decreased from 89% to 25% from the first to the fifth 10-min period following the discovery of an artificial food patch of bread crumbs. It is suggested that first foragers transfer their loads in order to return to the newly established food patch more quickly and easily, given that the new patch is tenuously connected to the main trail by a pheromone trail laid by just a few ants. Transfer of material between nestmates at the foraging area occurs in other species. Minor workers of *Oecophylla longinoda*, the African weaver ant, transfer honeydew, gathered from scale insects, directly to major workers for transport back to the nest ([18]; Figure 13-38 of [19]). Also, in *Ectatomma ruidum* and *E. quadridens* "stingers" directly and indirectly transfer killed insect prey to "transporters" [20]. This is similar to the termite-killing hunters of *Leptogenys ocellifera* [21] and *Megaponera foetens* [22, 23]. It is reported that *Messor* harvester ants will readily transfer seeds to "newcomers" when returning with material on the trail [24]. More complex examples of task partitioning involving three groups of workers also occur. In *A. sexdens* (Fig. 1d) leaf fragments are transferred indirectly between "arboreal cutters" and "cache exploiters", and between "cache exploiters" and "carriers" [7]. The cutters climb trees and cut leaves. The leaves fall to the ground, which acts as a cache or dump site. Here they are cut up by the exploiters, who take the pieces to the foraging trail from where they are taken to the nest by carriers [7]. Cutter workers are smaller than carriers. Hubbell et al. [8] report that dropping leaves from trees sometimes occurs in Costa Rican *A. cephalotes*.

Additional examples of task partitioning involving three groups of workers occur in foraging for nest-construction materials in the wasp *Polybia occidentalis* (Fig. 1e) and the ant *Lasius fuliginosus* (Fig. 1f). Note that in comparison to *A. sexdens* (Fig. 1d), these examples are not linear in their task partitioning. In *P. occidentalis* there are two groups of foragers collecting building materials, wood pulp and water. On returning to the nest, foragers transfer wood pulp and water directly to receivers, who then build nest carton and comb [16]. At the nest, water foragers also transfer water to pulp foragers who use the water to soften the wood they are collecting. In *L. fuliginosus* there are also two independent sets of foragers, collecting honeydew from aphids and particles of soil or organic matter [19]. The honeydew is directly transferred to builders. Particles are indirectly transferred after being dumped near the nest carton within the nest cavity.

The above examples serve to introduce the great diversity that exists not only in the form of partitioned task but also taxonomically as we have presented examples from ants, bees, wasps, and termites. Within a single species different materials may be transferred or not (e.g. nectar versus pollen in the honey bee). Within a single species (e.g. honeydew versus particles in *L. fuliginosus*) or even a single type of material (leaf fragments in *A. cephalotes* and nectar in *Vespula* wasps), there may be both direct and indirect transfer. The two most complex systems, involving two cycles of foragers interlocking with one of receivers, both occur for building materials. This complexity almost certainly reflects the nature of building, which in both cases requires builders simultaneously to use two raw materials. Conversely, food materials such as pollen and nectar need not be combined before storage or use, and the task partitioning is essentially a "bucket-brigade" mechanism [8]. This mode of task partitioning has also been dubbed "chain transport" [25, 26] and "relay principle" [27].

Task partitioning in foraging: results of computer simulations

The results in this section were obtained using a stochastic simulation program written in C ([9]; C. Anderson and F.L.W. Ratnieks, unpublished observations). In the simulation, there are two groups of workers—foragers and receivers (Fig. 2). Foragers and receivers pair up in the transfer area of the nest, and zero time is wasted searching for a partner if a partner is free in the transfer area. If there is no partner available in the transfer area, then the forager of receiver queues for a partner. For example, if receivers are queueing for foragers, then foragers arriving in the transfer area begin unloading with zero delay, and *vice versa*. All workers follow the same queueing discipline, either "serve in random order" (SIRO) or "first come, first served" (FCFS). Because FCFS is unlikely to be used by worker insects, as opposed to British humans, all results presented are

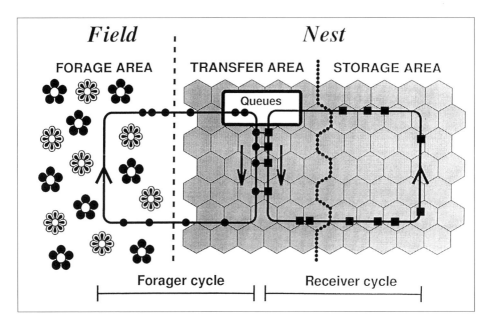

Figure 2. Graphic representation of task partitioning as modelled by computer simulation
There are two groups of workers, foragers (circles) that collect material and receivers (squares) that store or use the material. Foragers and receivers pair up for direct transfer of material. Foragers and receivers do not switch tasks or alter their work rate. Queues form when either foragers or receivers have no available partner in the transfer area. The arrival rates of foragers and receivers in the transfer area fluctuate because storage trips and foraging trips vary in duration.

for SIRO. Queueing discipline has no effect on the mean queueing delay, but the variance in queueing delay experienced by individuals is lower with FCFS.

The capacity of foragers and receivers is symmetrical, in that one receiver can exactly unload one forager. In most of our simulations one receiver always unloads one forager. In section 2.4.2, however, we explore the effects of multiple transfer in which every forager unloads to a specified number of receivers, with each receiver unloading a corresponding number of foragers. In particular, we evaluate the hypothesis that multiple transfer is a mechanism for increasing the reliability of the information provided by queueing delays. Multiple transfer appears to be a normal part of task partitioning in the honey bee. Kirchner and Lindauer [28] found that nectar foragers collecting syrup from a feeder unload to an average of 2.3 receivers, with a range of 1–5 (W.H. Kirchner, personal communications). Multiple transfer also occurs in pulp-foraging *P. occidentalis* wasps (see above; [16]). In this case, although improved information may occur, the primary reason appears to be the ability of a forager to collect more pulp than a single builder can handle.

In our simulation the durations of foraging trips, unloading trips, and transfers were sampled at random from distributions of different mean, variance, and type (i.e. normal, exponential, gamma, uniform, triangular). Unless otherwise reported, the results below were obtained using a standard parameter set in which both forage and storage trips had mean and variance of 500 time units, transfers had mean and variance of 50 units, and all distributions were normal. (The type of distribution had little effect on the results; increased variance in trip durations increased mean queueing delays, with the increase being greater in smaller colonies [9].) When multiple transfer was modelled, the total time of all transfers had a mean of 50 time units, with each partial transfer taking $50/N$ time units on average, where N is the number of transfers per trip. Our parameters were not chosen to correspond to any particular species or situation, but are a reasonable representation of the honey bee. The simulation was run for colony sizes (number of foragers + receivers) from 10 to 10 000.

In conducting our simulations, we were normally interested in the mean or the variance of the queueing delays. The former is relevant to the ergonomic cost of task partitioning to the whole colony, while the latter is relevant to the information content of the queueing delays experienced by individuals and the use of these delays in recruitment. To explore a realistic range of colony conditions, our simulations were run both at the optimum proportion of foragers to receivers, that is, when the forager and receiver groups have equal capacity for collecting or receiving material, and at deviations from this optimum. Given that stochastic variations in the arrival rates of foragers and receivers at the transfer area will be relatively greater in smaller colonies, we ran the simulations across a range of colony sizes to determine more precisely the relationship between colony size and the mean and variance effects. Results for a colony size of two, that is, one forager plus one receiver, were obtained analytically [9].

Effect of colony size

The amount of time wasted queueing for a partner in the transfer area declines exponentially (Fig. 3), although this is not obvious to the eye because of the logarithmic horizontal axis, with the mean delay declining from approximately 13

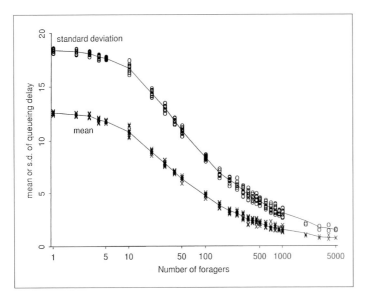

Figure 3. Mean and standard deviation of queueing delay as a function of number of foragers as obtained from the simulation model
These results were obtained at the optimum proportion of foragers to receivers (1:1) and apply to both receivers and foragers because the work capacity and mean trip duration of foragers and receivers were equal. (Trip durations: mean, 500 time units; variance, 500; shape, normal; transfer durations: 50, 50, normal.)

time units to 12, 6, 2, 1 units as colony size increases from 2 to 10, 100, 1000, 10 000. Half of the reduction in queueing delay occurs as colony size increases from 2 to 100. These results apply for both foragers and receivers because of the symmetrical nature of the model and because the relative numbers of foragers and receivers are equal, that is optimal.

The queueing delay in a colony with 10 000 foragers plus receivers, for example a large honey bee colony, is clearly insignificant. But even in a much smaller colony queueing delay may still be relatively unimportant. Thus, from purely ergonomic considerations the queueing delay is likely to impose only a small cost to the colony in all but the smallest colonies and is unlikely to impose, by itself, a major barrier to the evolution of task partitioning in foraging. However, queueing delay is not the only cost in task partitioning. Additional costs include searching costs incurred by foragers and receivers in locating a partner already present in the transfer area, and any loss of material during transfer. These costs were not explored in our simulation, and in any case are unlikely to be greatly affected by colony size, as is the queueing delay. If these costs are not affected by colony size, then it is a reasonably strong conclusion that task partitioning will not incur much greater costs to smaller colonies.

Figure 3 shows how colony size affects the standard deviation of the queueing delays experienced by foragers (and, by symmetry, receivers) under SIRO. Because all the foragers in our simulation are identical, this variation, which is determined for all delays of all workers in a particular simulation, is the same as that experienced by an individual, subject to the usual sampling effects. Clearly, there is a great drop in the variation in queueing delays as colony size increases.

Queue dynamics

In larger colonies queues are longer and fluctuate more rapidly ([9]; C. Anderson and F.L.W. Ratnieks, unpublished observations). In a colony of 1000 workers queues of foragers and receivers alternate dozens of times across 1000 time units and contain up to 12 workers. In a colony of 100 workers queues alternate about 10 times in 1000 time units and contain up to 4 individuals. In a colony of 10 workers queues alternate only a few times per 1000 time units and contain only one worker. However, the percentage of workers that are queueing is smaller in larger colonies, which is why the mean queueing delay (above) is greater in smaller colonies. Because queues of foragers and receivers alternate more rapidly than the mean trip duration, it is unlikely that a forager or receiver will be part of the same queue on the next return to the transfer area. In large colonies it is unlikely that an individual will rejoin the same queue during a single trip to the transfer area when multiple transfer occurs. The importance of this is that the queueing delays experienced will be independent between transfers on consecutive trips, and may be partly independent even within a single trip with multiple transfer.

Effect of nonoptimal proportions of foragers and receivers

The above results were obtained when the numbers of foragers and receivers were exactly sufficient to balance each others' work capacity. Because the mean trip times of foragers and receivers are equal in our simulation, this is when the number of foragers equals the numbers of receivers. However, the numbers of foragers and receivers could easily deviate from the optimum. For example, if nectar becomes more or less easy to collect [5, 6], then the number of foragers will become greater or lower than the work potential of the receivers. Similarly, death of foragers reduces the total work capacity of the foragers. In fact, when task partitioning occurs, it is unlikely that the system is ever exactly at the optimum.

How do nonoptimal proportions of foragers versus receivers affect mean queueing delay? In an infinitely large colony, workers of whichever group, forager or receiver, is present in excess would experience queueing delays, with workers of the other group experiencing zero delays. This "deterministic" case is given by the solid and dotted lines in Figure 4. The deterministic solution deviates from the stochastic model close to the mean foraging duration, 500 time units, at which foragers and receivers have equal work capacity as a group. In this zone both foragers and receivers experience queueing delays. In a colony size of 1000 the deterministic model only deviates from the simulation model for mean foraging durations between approximately 490 and 510 time units. If the mean foraging duration exceeds approximately 510 time units, that is, if forage becomes more difficult to collect, but receiving trips continue to have a mean duration of 500, then only foragers queue. If the mean foraging duration is less than approximately 490 time units, then only receivers queue. Between 490 and 510, both groups have nonzero mean queueing durations. As colony size reduces, the zone in which both groups have nonzero queues increases. In other words, as

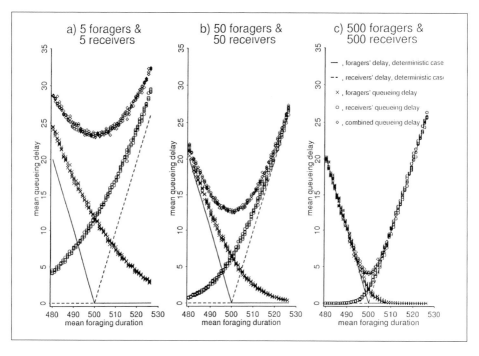

Figure 4. Effect of non-optimal proportions of foragers versus receivers on mean queueing delay as a function of colony size

When the foraging trip duration is 500 time units, the total work capacities of the forager and receiver groups are equal. Greater foraging trip durations lower the total amount of forage collected per unit time, leading to increased queueing delays of receivers, and vice versa. legend: Solid line, queueing delay of foragers (deterministic model); dashed line, queueing delay of receivers (deterministic model); crosses, queueing delay of foragers (stochastic model); circles, queueing delay of receivers (stochastic model); diamonds, combined queueing delay of foragers and receivers (stochastic model).

colony size decreases, the deterministic case fails to represent the actual situation over a wider range of nonoptimal work capacities.

The slope of the curve of queueing durations of foragers and receivers combined is lower in the zone around 500 time units for smaller colonies, but the absolute value is greater. This means that smaller colonies suffer a greater queueing cost (i.e. delay experienced by both foragers and receivers), but this cost is less sensitive to changes in the total work capacity of either foragers or receivers. This is because the greater importance of stochastic effects in smaller colonies reduces the impact of nonoptimality in the relative work capacities of the forager and receiver groups.

Information content of queueing delays

The information content of queueing delays

In a deterministic situation the queueing delay provides extremely reliable information about the work capacity of foragers versus receivers. Whichever group experiences a queueing delay has excess work capacity. The delay is directly proportional to the excess and has no variation (Fig. 4).

However, in a real insect colony things will not be so simple, and the information that can be extracted from queueing delays will be subject to various limitations. One limitation is that queueing delays and the time spent searching for a partner will probably not be distinguished, so that the variance in total delay will be the variance in queueing delay plus the variance in searching duration. Thus, a worker in the group that does not have excess work capacity may still experience delays that could be attributed to queueing even if the delay actually arises from searching.

As noted above, our simulation does not include searching duration. More relevant to our simulation is the effect of colony size on the variance in queueing delay. For a wide range of parameter values around the optimum, depending on colony size (Fig. 4), both groups experience nonzero queueing delays. Thus, simple rules such as "if delayed then recruit more partners" or "if not delayed then recruit more workers of own group" will often result in a worker making recruiting errors, that is, recruiting more workers into a group that already has excess work capacity, for example, recruiting more foragers when the work capacity of the foragers already exceeds that of the receivers. (However, a number of studies have shown that simple rules like these can lead to complex adaptive behaviour [29], such as the "threshold model" [30–32] and the "foraging for work" model [33, 34].) Even if the net result of recruitment behaviours still causes the colony to move towards the optimum proportion of receivers versus workers, this would introduce an error cost. For example, to redress a shortage of foragers, it would be less costly to have 10 receivers switch to being foragers and 0 foragers switch to being receivers than 50 receivers and 40 foragers switching, even though the net effects are identical.

Examination of the queueing delays experienced by individual foragers and receivers over 10 trips at colony conditions that are optimal (foraging trip duration 500 time units), near optimal (e.g. foraging trip duration 510 time units;

work capacity of receivers 2% in excess of foragers), and far from optimal (e.g. foraging trip duration 600 time units; work capacity of receivers 20% in excess of foragers) shows that although the near-optimal situation cannot reliably be distinguished from the optimal situation, based on the queueing delay from a single trip, the far-from-optimal situation can be ([9]; F.L.W. Ratnieks and C. Anderson, unpublished observations). In particular, at the far-from-optimal situation a large proportion of the queueing delays experienced by the group not in excess are long, and the delays experienced by the group in excess are always zero. As colony size increases, a greater proportion of the delays experienced by the underworked group are small. Even if additional delays due to searching occur, only the in-excess workers will experience long delays.

Thus, long queueing delays provide reliable information, but with an interesting asymmetry in the use these delays can be put to. A long delay is a reliable indication that the forager or receiver experiencing the delay is in the group in excess. However, a short delay does not reliably tell a forager or a receiver arriving at the transfer area that she is in the group not in excess. Because queueing is SIRO a worker in the excess group may still experience a zero or short queueing delay.

There will be times when colony and external conditions are such that more foragers should be recruited, and other times when more receivers should be recruited. This means that if recruitment decisions are made exclusively by one group of workers, as in the honey bee [5] in which foragers recruit both receivers, then recruitment errors are unavoidable if only a single queueing delay is used as information. Recruitment errors can be avoided if each group only recruits partners. That is, foragers experiencing long delays recruit receivers, and receivers experiencing long delays recruit foragers. (By the same argument, negative feedback should be directed at one's own group, not the partner group.) This type of dual-control positive feedback may not be used by the honey bee, because each honey bee forager is in possession of additional important information unavailable to receivers, such as the distance to her food patch, how long it takes to collect the nectar, and the weather. The only food patch information shared by forager and receiver is the sugar concentration of the nectar. Thus control by foragers may be a better strategy than dual control. Foragers can also modulate total foraging effort, including the provision of negative feedback, by individually making decisions about the tempo of their own foraging, and may even temporarily abandon foraging if their flower patch becomes sufficiently unrewarding [5].

Increasing the information content of queueing delays

Individual workers, and honey bee foragers in particular, could benefit from obtaining more information than that available from a single queueing delay. One obvious method is to take an average of several queueing delays. Statistical theory predicts that as the number of trips averaged over increases, the mean remains constant but the variation decreases in inverse proportion to the square root of the sample size. That is, averaging over four trips will halve the standard deviation. Figure 5 shows that, as expected, the standard deviation of queueing delays decreases in this way when multiple consecutive delays are averaged. Although all colonies in Figure 5 experience the same proportional decrease in standard deviation, it is the smaller colonies that experience the greatest absolute decrease, which is presumably of greater importance when considering information quality. Averaging the queueing delay across multiple trips is almost certainly a workable strategy when foraging trips are short in relation to the timescale over which colony and external conditions change. In both the honey

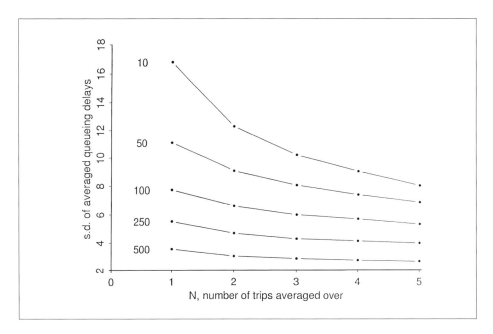

Figure 5. Effect of colony size and averaging queueing delays over one to five consecutive trips to the transfer area on the standard deviation of a large sample of mean queueing delays of individual workers.

bee and *P. occidentalis* wasps, this kind of averaging is likely to be realistic given that many foraging trips can be made per day. In *P. occidentalis* the foraging trips are very short, ca. 40 s for water and 3–4 min for pulp [16], and the external conditions (availability of wood or water) are unlikely to change much. In the honey bee, nectar-foraging trips are much longer, from ca. 20 min to several hours [12], and nectar availability changes on a daily basis [5, 35].

A second method of obtaining more information from queueing delays is to queue multiple times per trip to the transfer area. As mentioned previously, multiple transfer occurs in the honey bee [28]. There is no strong alternative hypotheses for why honey bee foragers transfer to multiple receivers. Possibly, receivers take less than a full crop load from foragers to facilitate the evaporation of excess water, a behaviour involving repeated movements of the mouthparts to expose the nectar to the air, which is often carried out before the nectar is deposited in a cell [5, 36]. Results from our simulation show that, as expected, multiple transfer greatly reduces the variation in queueing delay experienced in a single trip. Unexpectedly and most interestingly, our results show that multiple transfer reduces variation in queueing delays more than expected by averaging effects alone—standard deviation is more than halved with four transfers per trip—and at very little cost in terms of increased total queueing delay (Fig. 6). It seems that multiple transfer decreases individual queueing delays by making the group of workers not in excess more available. (Although we did not model searching durations and their probable combined increase when multiple transfer occurs, it is likely that this "increased availability" factor would prevent the total searching duration from increasing in direct proportion to number of transfers per trip.) In that multiple transfer appears to give an important benefit (more reliable information about colony conditions) at low cost (increase in total queueing delay), the hypothesis that multiple transfer in the honey bee is a strategy to increase the reliability of information arising from queueing delays is supported.

Discussion

The value of modelling biological processes is usually not in the model itself, but in the testable hypotheses and additional insights into the biology that are gained. The simulation model used to generate the results described in this chapter is a

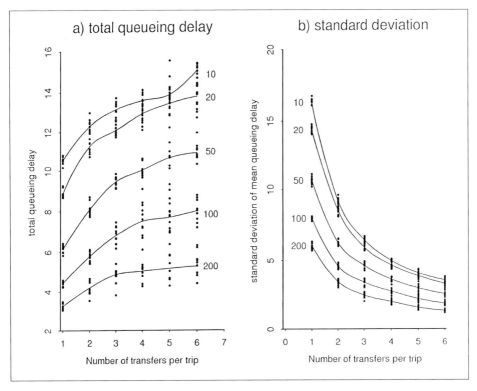

Figure 6. Effect of multiple transfers per trip to the transfer area on queueing delay as a function of colony size (number of foragers; 1:1 ratio foragers to receivers)
(a) Total queueing delay (= mean queueing delay per transfer multiplied by number of transfers per trip); (b) standard deviation of a large sample of mean queueing delays per individual per trip to the transfer area.

great simplification of task partitioning in foraging in insect societies. Nevertheless, we are optimistic that it has produced results of value.

In terms of testable hypotheses, one strong prediction is that multiple transfer and averaging over multiple trips are effective mechanisms for improving the reliability of the information content of queueing delays. Given that queueing delays are used by honey bee foragers in modulating recruitment, a worthwhile experimental study would be to collect data to determine whether foragers respond to some weighted average of queueing delays experienced in the current and previous trips to the nest [28]. In addition, alternative hypotheses for multiple transfer should be critically tested, such as whether receivers take only small amounts of nectar to facilitate evaporation [5, 36] or the addition of enzymes.

Increased insight has been provided into several facets of task partitioning and recruitment behaviour. Foremost is the relationship between colony size and queueing delay, particularly that queueing delays rapidly diminish as colony size rises. Also of importance is the realisation that queueing delays can be reliably used by individual workers to detect that they belong to the group in excess, but not that they belong to the group not in shortage, when the relative work capacities of the forager and receiver groups deviates by 20% from the optimum. One result was very unexpected: that multiple transfer results in only a small increase in total queueing delay.

A question amenable to modelling that may well be worth pursuing in the future is the fine tuning of recruitment systems, such as the thresholds at which behaviour changes from nonrecruitment to recruitment. This chapter explicitly raises the concept of recruitment errors, but does not investigate the role of these errors in an optimisation model in which benefits in colony performance can be achieved though recruitment, by adjusting the relative numbers of foragers to receivers closer to the optimum, but where recruitment errors have costs. In the honey bee, recruitment of foragers is undoubtedly costly because most recruits never find the flower patch to which they were directed by the waggle dance of the recruiting forager [5]. Recruitment also takes time and energy in the performance of the dance itself. In contrast, unnecessary recruitment of receivers may have a low cost because the newly recruited receivers do not then embark on risky and energetically costly flights away from the nest.

Task partitioning itself has several advantages as an avenue of research. Descriptive and natural history studies have great value in increasing our knowledge of an important and fascinating aspect of insect social life. Additionally, task partitioning in foraging provides excellent model systems [5, 7, 16] for experimental studies of the ergonomics of insect societies, a field that has traditionally focused on caste and division of labour [3]. Finally, the simulation model we have explored is very general. Specific models could be made for alternative systems, such as the three interacting groups of workers in *P. occidentalis*, pulp foragers, water foragers and builders, for actual parameters measured in actual societies, or for use in conjunction with experimental studies of real societies.

Acknowledgements

Paul Blackwell and Chris Cannings provided valuable advice about the simulation model, and Tom Seeley and Bob Jeanne helped us to keep in mind the biology of real honey bees and wasps. We also thank the editors of this book for their encouragement. C.A. acknowledges support from a BBSRC special research studentship.

References

1   Robinson GE (1992) Regulation of division of labor in insect societies. *Annu Rev Ecol Syst* 37: 637–665

2   Jeanne RL (1986) The evolution of the organization of work in social insects. *Monitore Zool Ital* (N S) 20: 119–133

3   Oster GF, Wilson EO (1978) *Caste and ecology in the social insects*. Harvard University Press, Cambridge, MA

4   Jeanne RL (1991) Polyethism. *In*: KG Ross, RW Matthews (eds): *The social biology of wasps*. Cornell University Press, New York, 389–425

5   Seeley TD (1995) *The wisdom of the hive*. Harvard University Press, Cambridge, MA

6   Lindauer M (1961) *Communication among social bees*. Harvard University Press, Cambridge, MA

7   Fowler HH, Robinson SW (1979) Foraging by *Atta sexdens*: seasonal patterns, caste, and efficiency. *Ecol Entomol* 4: 239–247

8   Hubbell SP, Johnson LK, Stanislav E, Wilson B, Fowler H (1980) Foraging by bucket-brigade in leaf-cutter ants. *Biotropica* 12: 210–213

9   Anderson C (1998) *The organisation of foraging in insect societies*. PhD thesis, School of Mathematics, University of Sheffield, UK

10 Seeley TD (1992) The tremble dance of the honey bee: message and meanings. *Behav Ecol Sociobiol* 34: 51–62

11 Michener CD (1974) *The social behavior of the bees: a comparative study*. Harvard University Press, Cambridge, MA

12 Ribbands RR (1953) *The behaviour and social life of honeybees*. Bee Research Association, London

13 Sommeijer MJ, De Rooy GA, Punt WA, de Bruijn LLM (1983) A comparative study of foraging behavior and pollen resources of various stingless bees (Hym., Meliponinae) and honeybees (Hym., Apinae) in Trinidad, West Indies. *Apidologie* 14(3): 205–224

14 Jeanne RL (1991) Polyethism. *In*: KG Ross, RW Matthews (eds): *The social biology of wasps*. Cornell University Press, Ithaca, NY, 389–425

15 Akre RD, Garnett WB, MacDonald JF, Greene A, Landolt P (1976) Behavior and colony development of *Vespula pensylvanica* and *V. atropilosa* (Hymenoptera: Vespidae). *J Kans Entomol Soc* 49: 63–84

16 Jeanne RL (1986) The organization of work in *Polybia occidentalis*: costs and benefits of specialization in a social wasp. *Behav Ecol Sociobiol* 19: 333–341

17 Greene A (1991) *Dolichovespula* and *Vespula*. *In*: KG Ross, RW Matthews (eds): *The social biology of wasps*. Cornell

University Press, New York, 263–308

18 Hölldobler B (1984) The wonderfully diverse ways of the ant. *Nat Geog M* 165(6): 778–813

19 Hölldobler B, Wilson EO (1990) *The ants*. The Belknap Press of Harvard University Press, Cambridge, MA

20 Schatz B, Lachaud JP, Beugnon G (1996) Polyethism within hunters of the ponerine ant, *Ectatomma ruidum* Roger (Formicidae, Ponerinae) *Insect Soc* 43(2): 111–118

21 Maschwitz U, Mühlenberg (1975) Zur Jagdstrategie einiger orientalischer *Leptogenys*-Arten (Formicidae: Ponerinae). *Oecologia* 20: 65–83

22 Longhurst C, Johnson RA, Wood TG (1978) Predation by *Megaponera foetans* (Fabr.) (Hymenoptera: Formicidae) on termites in the Nigerian Southern Guinea Savanna. *Oecologica* 32: 101–107

23 Brian MV (1983) *Social insects. Ecology and behavioural ecology*. Chapman and Hall, London

24 Sudd JH (1965) Transport of prey by ants. *Behaviour* 25(3–4): 234–271

25 Stäger R (1935) Über Verkehrs- und Transportverhältnisse auf den Strassen der Waldameise. *Rev Suisse Zool* 42(3): 459–460

26 Dobrzanska J (1966) The control of territory by *Lasius fuliginosus* Latr. *Acta Biol Exp (Warsaw)* 26(2): 193–213

27 Goetsch W (1934) Untersuchungen über die Zusammenarbeit im Ameisenstaat. *Zs Morph Ökol Tierre* 28: 319–401

28 Kirchner WH, Lindauer M (1994) The causes of the tremble dance of the honeybee, *Apis mellifera*. *Behav Ecol Sociobiol* 35: 303–308

29 Bonabeau E, Theraulaz G, Denebourg JL, Aron S, Camazine S (1997) Self-organization in social insects. *Trends Ecol Evol* 12(5): 188–193

30 Calabi P (1988) Behavioural flexibility in Hymenoptera: a re-examination of the concept of caste. *In*: JC Trager (ed): *Advances in myrmecology*. Bill Press, Leiden, 237–258

31 Robinson GE (1987) Modulation of alarm pheromone perception in the honey bee: evidence for division of labour based on hormonally regulated response thresholds. *J Comp Physiol* 160: 613–619

32 Bonabeau E, Theraulaz G, Deneubourg JL (1996) Quantitative study of the fixed threshold model for the regulation of division of labour in insect societies. *Proc R Soc Lond B* 263: 1565–1569

33 Tofts C (1993) Algorithms for task allocation in ants (a study of temporal polyethism: theory). *Bull Math Biol* 55: 891–918

34 Franks NR, Tofts C (1994) Foraging for work: how tasks allocate workers. *Anim Behav* 48: 470–472

35 Visscher PK, Seeley TD (1982) Foraging strategy of honeybee colonies in a temperate deciduous forest. *Ecology* 63: 1790–1801

36 Winston ML (1987) The biology of the honey bee. Harvard University Press, Cambridge, MA

Interaction patterns and task allocation in ant colonies

Deborah M. Gordon

Summary

Social insect colonies must accomplish many tasks, such as foraging, tending brood, constructing a nest, and so on. Task allocation is the process that adjusts the numbers of workers engaged in each task. This chapter discusses how information from other individuals is used in task decisions, and in particular, how workers use the pattern of interactions they experience, rather than the content of messages received. Empirical studies of harvester ants led to a mathematical model of task allocation in which environmental stimuli and interaction patterns both influence an individual's task. I outline the main results from this model, and describe recent empirical work that begins to examine how interaction patterns contribute to task allocation in harvester ants.

A colony's organization determines how the colony gets things done. Colonies must accomplish many tasks, such as foraging, tending brood, constructing a nest, and so on. Task allocation is the process that adjusts the numbers of workers engaged in each task. Such numbers change as external conditions and the needs of the colony vary. Colony organization operates, at a finer scale than that of task allocation, to determine how each task is accomplished. The organization of some tasks, such as recruitment to food, nest construction, and brood transport, has been well studied in some social insects. This chapter is about the broader scale of colony organization, that of task allocation.

One way to understand task allocation is to figure out how the behavior of individuals comes together to produce the behavior of colonies. The basic mystery of colony organization in social insects is that there is no one in charge. No individual is able to assess the global needs of the colony, or to count how many workers are engaged in each task and decide how many should be allocated differently. Instead, the capacities of individuals are fairly limited. Each worker

C. Detrain et al. (eds) Information Processing in Social Insects
© 1999, Birkhäuser Verlag Basel/Switzerland

must make fairly simple decisions. There is abundant evidence, throughout physics, the social sciences, and biology, that such simple decisions by individuals can lead to predictable patterns in the behavior of groups [e.g. 1]. In the study of social insects, most of this evidence is concerned with the organization of particular tasks. For example, a tendency of individuals to move in the direction of certain chemicals can lead to the formation of foraging trails [2]. Evidence is growing that task allocation can be explained in a similar way, as the consequence of simple decisions by individuals.

Most early work on task allocation, especially in ants, was concerned with division of labor. Task allocation has two components: which individual performs which task, and how many individuals perform a task at any instant. Division of labor characterizes the first component, specifying which individuals perform a task. The notion of division of labor was introduced by the economist Adam Smith in the 18th century, to describe a workplace in which individuals are specialized, such that each one performs the same task over and over; the labor is "divided" among the individuals. The most extreme form, conceptually, is that each individual performs only its assigned task: that is, whenever worker x is active, it always performs task i. The second component of task allocation is what determines whether, at any instant, an individual is active and performs its task. Recent work suggests that complete specialization is rare in social insects: workers usually change tasks, both as they grow older and, in the short term, as conditions vary (reviewed in [3, 4]). Thus the study of task allocation has to resolve two questions: what determines an individual's task, and what determines whether, at a particular instant, that individual is active.

What influences which task a worker performs and whether the worker is active? Clearly workers react to their immediate environments, because task allocation changes as conditions vary. When more food is available, more foragers appear; when the nest is damaged, more nest builders become active; when predators are nearby, workers remain inactive inside the nest. The conditions a worker encounters as it performs its task must affect the probabilities that it continues that task, becomes inactive, or takes up a different task.

But each worker's task decisions are based on more than its own independent assessment of the environment. Workers also use signals from each other. Chemical signals are an obvious example. A worker exposed to alarm pheromone may stop performing its task and begin to circle around and perhaps to attack. The worker is responding to an interaction, in the form of transfer of an airborne chemical signal, with another worker. In the same sense, people who rush out of a building when someone shouts "Fire!" are responding to an interaction with the shouter.

An individual worker's task decisions thus depend on information from two sources: the environment, and other individuals. Information from other individuals may be packaged in chemical, mechanical, or visual messages, which can be understood by analogy to human language. For example, a chemical might say "Disturbance!" or "I belong to this colony" or "There is a queen here" or "I walked here". However, specific chemical messages cannot underlie all task decisions. There would have to be a distinct chemical message to represent all the possible task transitions, such as "start doing task i", "stop doing task j"; "change from task i to task j", for all tasks i and j, and a worker would have to be capable of using them all. This is asking social insect workers to use a very large vocabulary. But beyond the requirement for an impossibly large vocabulary, there is another improbable aspect of the one-message-per-task-decision scenario. How would workers know which command to give? For one ant to tell another, "stop doing task j", that ant would have to make some judgement about whether task j is needed, and this requires a global assessment of colony needs.

This chapter discusses how information from other individuals is used in task decisions. I focus on how workers use the pattern of interactions they experience, rather than the content of messages received. Various aspects of interaction pattern have been shown to be important: the rate of interactions, that is, the number per unit time; the interval between interactions; or the total number of interactions. The common feature of these results is that each interaction that an individual experiences contributes to a pattern, and the pattern itself influences the individual's task decisions.

"Interaction" covers many different kinds of behavior. An interaction with a time delay occurs when one individual responds to a chemical signal deposited on a substrate by another individual, such as a trail pheromone or scent mark. Or interactions may be temporally constrained; for example, an interaction occurs immediately when one individual touches another with its antennae. Daily and short-term temporal patterns of interactions, often in the form of brief antennal contact, occur in many species of ants [5–7]. For most species, we do not know the function of such interaction patterns.

An interaction may not transmit any message besides the incidence of interaction itself. For example, ants may use the rate at which they contact others as an indication of density. In this case the message one ant gets when it touches another is simply that it met another ant. This appears to occur in the ant *Lasius fuliginosus* [7]. In laboratory experiments, ants were kept in groups either of 35 or 75 ants. To each host group, we introduced 15 ants of another colony. The rate of contact with non-nestmates depended on the relative proportions of nestmates and non-nestmates. The 15 non-nestmates circulated freely, so each of the 35 ants would tend to encounter one of the 15 non-nestmates at a higher rate than would each of the 75. If the host ants' response depended on the total amount of alien pheromone introduced by the 15 non-nestmates, then both the 35-ant and 75-ant host groups should have had the same response. However, the 35-ant groups responded differently from the 75-ant groups. This suggests that ants were responding to the rate of interaction with non-nestmates, that is, to the proportion of non-nestmates present. The cue to non-nestmate density may be the incidence of interaction.

Another example of a response to interaction pattern is in the work of Seeley and colleagues on the organization of foraging, in particular the allocation of bees to nectar foraging [8–10]. This work is concerned, not with the question of which bee does which task, but the second component of task allocation, how a bee decides at a given instant whether to pursue its task actively. Whether a nectar forager pursues its task actively, leaving the nest to gather nectar, depends on the time that elapses between its arrival and its interaction with a nectar unloader. When there is less nectar to unload, more unloaders are available, and the rate at which foragers encounter unloaders increases. When there is more nectar to unload, unloaders are more likely to be busy elsewhere, unloading nectar, and the rate at which foragers encounter unloaders decreases. The time a forager has to wait for an unloader is directly, negatively proportional to this encounter rate. In this case, the unloader does not give the forager a message "Go out and forage some more" or "Remain inactive". Instead, the pattern of interaction provides a cue that influences the nectar forager's task decision. The probability a nectar forager will be active at any instant is, of course, also influenced by cues related to the forager's environment, such as the quality of the food sources it has recently experienced, and their distances from the hive. It is also, of course, influenced by the unique physiology of that forager [11]. The link between interaction pattern and the forager's decisions allows the colony to regulate its nectar intake according to colony need, just as the link between nectar quality and the forager's decisions allows the colony to regulate its nectar intake according to food quality.

Interaction patterns and group size

There are vast numbers of social insect species that have never been studied, and those that we have studied differ greatly. There is no reason to expect that every species will use the same features of interaction patterns, or that cues based on interaction patterns will have exactly the same function in every species. Instead, it seems likely that many different features of interaction patterns are used, and in many ways. In general, though, as we learn more about colony organization and more about other species, should we expect to discover more individual decisions based on interaction patterns?

I think that individual decision rules based on interaction patterns will prove to be widespread in social insects. Interaction patterns are an obvious device for transmitting the effects of group size to individuals without requiring individuals to count. There is a basic spatial relation between group size and interaction pattern. If some objects are jostling around in an enclosed space, the number of times they will bump into each other, and the interval that elapses between encounters, will depend on the number or density of objects present. The pattern of encounters each individual experiences depends on the size of the group, so that each individual's experience tracks group size without any need for the individual to assess the overall size of the group.

Interaction patterns provide a way for individuals to track group size, and group or colony size is crucial to colony organization. A colony's size determines in many ways its needs and capacities. In many species, colonies grow larger as they grow older. Colony size influences how large a space the colony needs for a nest, how much food the colony requires, and how it competes with its neighbors. Colony size also affects the composition of task groups within colonies, and thus the resources available to perform each task; larger colonies have more foragers, and so on. Task group size, the numbers performing a task, determines how much of the task gets done. Of course, individuals may differ in the speed and quality of their work, but in general more workers on a task means more of that task is accomplished.

The main problem that colony organization must solve is to find ways that simple individuals, in the aggregate, can do what is necessary for colonies to live, grow, and reproduce. Because group size influences so strongly what colonies need, and what they can do, it seems likely that decision rules will appear that use simple cues which reflect group size. Interaction patterns can translate group size, which is difficult for individuals to assess, into simpler cues.

Group size and interaction patterns in harvester ants

My research on interaction patterns began from a surprising finding involving colony size in harvester ants (*Pogonomyrmex barbatus*), a species of seed-eating ants I study in the desert of southeastern Arizona. Perturbation experiments showed that the numbers engaged in one task, such as foraging, depended on the numbers engaged in another task, such as nest maintenance work [12]. These relations among distinct worker groups depend on colony size [13].

The ants clearly used environmental cues in task decisions: when more food became available, more ants foraged. When I introduced foreign objects on the nest mound, more ants appeared to clear them away. This was expected; if colonies did not adjust numbers engaged in certain tasks to current conditions, they would be unable to track their environments or respond to disturbances.

The experiments revealed a less obvious feature of task allocation, the relation of different task groups. It was conceivable that task groups could be independent of each other; for example, foragers might decide whether to forage regardless of events affecting only nest maintenance workers. The experiments showed that this is not the case. Further work with marked individuals [14] showed one way that numbers in one task affect numbers in another: ants switch tasks. When more foragers are needed, ants previously engaged in other tasks switch to foraging. This switching increases the numbers foraging, but decreases the numbers available to perform the tasks the new foragers came from. However, task switching does not occur in all directions; ants that have switched from nest maintenance to another task do not switch back to nest maintenance. Instead, when more nest maintenance workers are needed, they are recruited from ants previously working only inside the nest.

Task switching did not account for all of the interactions between task groups. In some cases, a change in numbers performing one task causes the workers in a different task group to change their active-inactive status. For example, experiments that increase the numbers of nest maintenance workers recruited from inside the nest led to a decrease in the number of workers foraging outside the nest. This is not because foragers switch to nest maintenance; switching does not occur in this direction. Instead, the foragers remain inactive inside the nest while other ants are recruited to perform nest maintenance. This may reflect a cycle in nest maintenance and foraging that seems to be common during the summer rains: after a storm, there is an increase in nest maintenance work to repair the damage done by flooding, and then follows a burst of foraging activity, as ants go out to collect newly exposed seeds, and nest maintenance subsides. It may be that perturbation experiments trigger the usual response to this cycle; if it's a day when intensive nest maintenance work is required, then foraging is low priority and the foragers tend to remain inside the nest.

These results show that task decisions by individuals, which task to perform and whether to be active, must be influenced by interactions among workers. Ants engaged in different tasks outside the nest are spatially segregated from each other (which is what made it possible to design experiments that directly affected only the workers engaged in one task). But workers of all tasks performed outside the nest mix as they come in and out of the nest, with many opportunities to interact.

The most surprising result of these experiments was an effect of colony age, which is related to colony size. Harvester ant colonies grow larger as they grow older, being founded by a single queen and reaching a stable size of about 10 000 workers by the time they are 5 years old and begin to reproduce [15, 16]. Colonies live for 15–20 years, until the queen dies [17]. Young, 2-year-old colonies, with 2–4000 workers, responded differently to perturbations than colonies 5 years or older with about 10 000 workers [13]. Older colonies were more consistent in response to repeated experiments, and their reactions to perturbations were more likely to emphasize foraging at the expense of other tasks. Since ants live at most a year [18], this is not due to the experience of older ants. The simplest explanation is an effect of colony size. This led me to consider what kind of decision rules individuals might use that could lead to different outcomes in a large colony and a small one.

Decision rules based on interaction patterns seem a good candidate to explain how colony size affects task allocation. Ants in a small colony might use the same rules as ants in a large one to decide which task to perform, and whether to perform it actively. But colony size could influence interaction patterns. In a small colony, ants might experience fewer interactions, or a lower interaction rate, or a longer interval between interactions, than ants in a large colony. Thus in the aggregate, smaller colonies might tend to show different outcomes, in allocation of workers to various tasks, than larger ones.

Questions about interaction patterns

Investigating the relation of group size, task allocation, and interaction patterns raises a set of related questions. Some are empirical: What determines interaction rates in ants? How do such patterns depend on group size? Do species vary in interaction patterns? Do ants use interaction patterns in task decisions?

Some questions are theoretical: How do interaction patterns depend on the ways that ants move around in space? Can we predict how spatial behavior will affect interaction rate? What is the simplest possible decision rule based on interaction patterns that could still give the observed features of task allocation? This exercise in model making generates another set of empirical questions: How well does the behavior of real ants match that of the model?

Many of these questions have been addressed elsewhere, including species-specific encounter patterns [7], the effect of group size on encounter pattern [7, 19], the relation between spatial behavior and interaction pattern [19, 20], and models involving interaction patterns [21, 22], which are reviewed in [4, 23]. Here I will discuss some theoretical and empirical work that asks: Do ants use interaction patterns in making task decisions?

A model of task allocation

In one model of task allocation, an individual's decision is based on two kinds of stimuli: the rate of interaction with other workers, and the state of the environment relevant to a particular task. The combination of interaction rate and environmental stimuli is one feature of the model that makes it more realistic than previous ones that incorporate only interactions (e.g. [21]) or those that incorporate only environmental cues. There is overwhelming evidence that such decisions are affected by cues from the environment. If this were not the case, social insect colonies would not respond to changes in environmental conditions by reallocating workers to various tasks. Yet we know that this occurs frequently; worker allocation is adjusted in response to changes in food supply [8], predation

[24, 25], and nest condition [12, 25]. Thus the cues individuals use in decisions that generate task allocation must involve both interaction with others and environmental stimuli. Both factors, interactions and environmental cues, are components of other models of task allocation. For example, response to environmental stimuli, or to the extent to which other workers have successfully accomplished a task, is a component of the models of Jeanne [25, 26]. Numbers of interactions, rate of interaction or time elapsed since the last interaction are a component of several recent models of the organization of certain tasks (e.g. [2, 10]).

How the model works

The model of Pacala et al. [22] predicts the numbers allocated to each of Q tasks. The vector representing the numbers engaged in each of Q tasks is $\underline{X} = \{ X_1, X_2, ... X_Q \}$. An individual's state is defined in the following way: either it is actively engaged in one of the Q tasks, or else it is inactive. If it is actively engaged in one of the tasks, it is either successful or unsuccessful. In the model, an individual can detect the task of another individual that it meets; empirical evidence concerning this capacity in harvester ants is discussed below.

Success is represented as s_i, the probability that an individual engaged in task i is in the successful state. This depends on current environmental conditions; for example, few foragers will be successful when there is little food available. In the model, an individual switches from task i to task j when she is unsuccessful at task i and encounters another ant that is successful at task j. It is well known that social insect workers can detect the success of other workers at some tasks, such as foraging. For example, harvester ant foragers are more likely to leave the nest to forage when the rate of successful returning foragers is high [17]. In the model, unsucccesful individuals engaged in task i will quit task i, and become inactive, at a rate q_i. This quitting rate depends in part on the environment, which determines how likely an individual is to be successful or unsuccessful. The model assumes a unique rate q_i for each task. Differences among tasks in the quitting rate might depend on the cost of performing a task. A high-cost task might not be worthwhile to pursue when workers engaged in it cannot succeed. For example, harvester ants obtain water by metabolizing the lipids in the seeds they collect. Foragers are using water when they travel in the hot desert sun, and if they are unsuccessful at collecting seeds, they may be losing more water than they gain. At this point, they might quit foraging and remain inactive inside the nest until more food becomes available. However, ants working inside the nest to enlarge a chamber may not be using as much energy. Suppose they hit a rocky patch of soil in which few individuals can succeed in scraping away at the wall of the chamber. These unsuccessful nest workers may continue longer before they quit than would unsuccessful foragers. A high-cost task, less worthwhile to pursue when workers engaged in it cannot succeed, might have a higher q.

The rate at which an individual engaged in task i encounters individuals engaged in task j is proportional to D_{ij}, the average local density (numbers per unit area) of task-j individuals in the vicinity of a task-i individual. D_{ij} is averaged over all individuals engaged in task i. Local densities depend on the movement patterns of individuals. These may vary from task to task, and the resulting interaction rates among workers engaged in different tasks could vary accordingly in ways that might be complicated to predict [20]. To imagine how this might work, one can think of interactions as simple collisions (though in fact there could be a time delay so that the two participants in the interaction are in the same place at different times). One extreme in the process that determines collision rate would be when all individuals mix completely, and D_{ij} is proportional to X_j/N. That is, the higher the proportion of task-j individuals, the more likely any task-i individual is to meet one. The other extreme would occur when individuals engaged in different tasks are spatially segregated. Then D_{ij} would be proportional to X_j. A task-i individual that happened to fall in among the task-j individuals would tend to experience more contacts, the more task-j individuals there were. In fact, the processes that determine interaction rates probably fall between these two extremes. Such processes may differ among tasks, since tasks differ in use of space.

The basic model is

$$\frac{dX_i}{dt} = -X_i (1 - s_i) q + [\Sigma X_i s_i D_{ij} (1 - s_j) - \Sigma X_j s_j D_{ji} (1 - s_i) + (N - \Sigma X_j) D_{Ii} s_i]$$

where all sums are over all Q tasks, and D_{Ii} is the average local density of task-i individuals around an inactive individual.

The change in numbers engaged in task i (the left-hand side) depends on four terms: (i) individuals that are unsuccessful at task i and quit, thereby becoming inactive; (ii) individuals from other tasks that encounter successful task-i ants and switch tasks to perform task i; (iii) individuals that are unsuccessful at task i, encounter successful individuals from another task, and switch from task i to the other task; and (iv) inactive individuals that encounter successful task-i individuals, and switch to perform task i.

Investigating the dynamics of this model led to the following results.

Optimal task allocation

First, if success rates are related to fitness, then the model leads to some optimality predictions. The predicted optimum distribution of workers among tasks will depend on whether we add up the fitnesses of each individual in the colony, such as the net amount of resource brought into the colony by each individual, or instead use some measure of the fitness of the whole colony, such as the total amount of resource brought into the colony given the colony's costs in obtaining these resources. If the former, then the optimal outcome of the model is the ideal free distribution: no individual should perform a task if it could have higher success at some other task. In the latter case, when colony fitness is maximized, then all tasks performed actively at any moment must yield equal marginal benefits, and the benefit of performing each task must at least equal the cost of performing it.

Stability of task allocation

The second result is that the dynamics of the model lead to equilibria that resemble the optimal outcomes described above. The conditions when the allocation of workers into various tasks is stable depend on the way in which the performance of each task is assumed to modify the environment, and thus on the ensuing success rate of individuals engaged in that task. Three possible relations of task performance and success rates were considered:

(1) As *per capita* returns from task i increase, so does s_i, the proportion of task-i workers that are successful; s_i is a monotonically increasing function of the *per capita* returns from task i. In addition, the *per capita* returns are independent of X_i. An example of this is when foragers are in an environment with very abundant food: each individual returning with food is a successful for-

ager, and no matter how many individuals forage, the amount of food available does not decrease.

(2) As *per capita* returns from task *i* increase, so does s_i, the proportion of task-*i* workers that are successful; as in case (1), s_i is a monotonically increasing function of the *per capita* returns from task *i*. But the *per capita* returns are a decreasing function of X_i. For example, suppose food is scarce. Initially the more foragers are active, the more food will be retrieved, and the more foragers will be successful—but eventually the foragers will deplete the food, and the returns in food will go down as the numbers foraging go up.

(3) Individuals somehow interfere with each other, with the consequence that successful individuals that meet unsuccessful ones sometimes change state to become unsuccessful themselves. In this case, as in the previous one, *per capita* returns are a decreasing function of X_i. But unlike the previous two cases, s_i is a monotonically increasing function of the marginal rate of return to the group as a whole, not the *per capita* rate of return.

For (1) and (2) the stable equilibrium resembles the predictions for the optimal distribution of workers into tasks to maximize individual fitness, that is, individual returns. For (1) or (3) the stable equilibrium also resembles the predictions for the optimal distribution to maximize the fitness of the colony as a whole.

The equilibria are unstable when the success rate for any task not currently performed is greater than the success rate common to all tasks currently performed. Thus if the environment changes, causing the success rate of particular tasks to change, then the distribution of workers in various tasks will be altered so that all equally successful tasks are being performed.

Colony size and ability to track a changing environment

The third result is a prediction for how quickly task allocation can track a changing environment. Tracking speed depends on colony size. Everything else being equal, large colonies can track a changing environment more rapidly than small ones. This is because the larger the colony, the higher the rate of interaction. High interaction rate means a high rate of dissemination of information about the environment, as successful individuals interact with unsuccessful ones.

However, there is a disadvantage to large colony size. In a large colony, a successful individual might interact with large numbers of unsuccessful or inactive individuals that would then be recruited to the task of the successful one. As a consequence there might be more workers performing the task than the environment warrants. For example, suppose a successful forager comes back to a large colony and meets large numbers of unsuccessful or inactive workers inside the nest. This would lead many workers to switch to forage, possibly more than would be warranted by the amount of food available. In general, workers in large colonies might be more likely to be active in unprofitable environments.

If real colonies operate the way the model does, then one way they could overcome this disadvantage of large colony size would be to curtail interaction rates. The ant *Lasius fuliginosus* seems to do this [7]. We manipulated density experimentally. When density, number of ants per unit area, is high, ants curtail the rate at which they engage in antennal contact with others. The ants can avoid interaction before it occurs. We found that in this species one ant can perceive another at a distance of about a body length [7], and thus an ant can move out of the range of contact with an approaching ant.

Interaction patterns in task allocation

Recent empirical work shows that interaction patterns affect the task decisions of harvester ants [28]. Observers followed marked ants for about 20 min, and recorded the task of the focal ant and those of all ants it met. An encounter was defined as a brief contact by the antennae of the focal ant and any part of the body of another ant; the great majority of such contacts were between the antennae of the two ants. Whether an ant switched tasks, and whether it remained inactive or pursued a task actively, depended on its recent encounter history. There was a statistically significant association between the number and type of contacts a focal ant experienced and its subsequent tasks.

This study focussed on midden work, the sorting and piling of the colony refuse pile, or midden. Ants not engaged in midden work, that encountered high numbers of midden workers, were likely to begin midden work. Ants that were walking or standing around, apparently not enaged in any task, were likely to remain so if they met low numbers of midden workers.

In this laboratory study, interaction patterns provide positive feedback to perform midden work. In the field, midden material is removed by other insects. The absence of midden material to work with probably provides negative feedback. In general, however, there is no reason to suppose that interaction patterns generally provide positive or negative feedback; either one is equally probable a priori. There is, however, abundant reason to believe that interaction patterns alone do not determine task decisions. First, if an ant's decision whether to perform midden work depended only on its interaction with midden workers, our results suggest that once a few ants begin midden work, all ants will end up joining in. Second, if ants were to rely solely on interaction patterns, they would be unable to respond to any environmentally induced changes in their colony's need for midden work. Moreover we never observe all ants in a colony engaged in midden work. This means that some process of negative feedback, probably linked to the quantity of midden material present, influences worker decisions about midden work.

Harvester ants of different task groups differ in their cuticular hydrocarbon profiles [27]. Ants may use the cuticular hydrocarbon odour to distinguish the task of the ants they meet.

What feature of interaction pattern do the ants use?

We found a statistical effect of number of encounters on task decisions. We do not know which feature of an ant's interaction history, correlated with number, influences its behavior. We counted number of encounters in a fairly constant period of time. Thus numbers of encounters are positively correlated with rate of encounter, and negatively correlated with the interval between encounters. The effect of number may occur because ants respond to the rate, in numbers per unit time, or to the interval between encounters. We tested whether there was any effect of the sequence of encounters; did the probability an ant performs task i depend on whether its last encounter was with an ant of task i, its second-to-the-last encounter, and so on? We found no such effect.

The question of which aspect of its recent interaction history influences an ant's task decisions is an empirical one, but it is also an interesting theoretical question. What feature of the pattern should an ant use? There must be some time interval that is relevant. An ant's decision about its task this minute almost certainly does not depend on the whole pattern of every interaction it has experienced since the moment it eclosed. Our observations show that ants change task state several times, at least from from active to inactive or back, in 20 min. This suggests the relevant interval is on the scale of tens of minutes or less. Let us call this the ant's forgetting time. During this forgetting time, the ant might count interactions. It would then respond to the total number of interactions it has counted, as far back as it can remember. Another alternative is that the ant might calculate an interaction rate. That is, it would count number of interactions, for some interval less than or equal to its forgetting time, and divide this by the amount of time elapsed. A third alternative is that the ant tracks the interval between successive encounters, measuring somehow the time elapsed since the last time it met another ant of task i.

Asking which of these alternatives is more likely branches into two questions. The first is about the capacities of ants. Which alternative requires the least cognitive power? Measuring intervals seems to require the least calculation, and there is much work to suggest that many animals, including honey bees and ants, can assess time intervals (Fourcassié et al., this volume). However, the result that the last encounter rarely determines task decisions indicates that more than one encounter, and thus more than one interval, is involved. Tracking many intervals, or counting many encounters, or calculating a rate involving many encounters, may be equally difficult operations. At least, it is not obvious that one is much easier than the rest.

A second way to ask which of these alternatives is more likely is to ask what would be most effective for the colony. One consideration is that the larger the number of interactions involved in the decision, the less likely are decisions to be affected by sampling error. An ant moves around and encounters other ants. In the short term, who each ant meets is greatly affected by individual variation in movement. A larger sample may more accurately reflect the actual proportion of ants engaged in each task; if the sample were large enough that the ant met every other ant in the colony, its interaction record would be equivalent to the distribution of ants into each task.

Future work

Task allocation, how colonies change the allocation of workers in response to changing conditions and changing colony needs, has been studied in relatively few species of social insects. The basic empirical question for any species is how different tasks are related. How do numbers engaged in one task depend on numbers engaged in another? The next question is how individual decisions create shifts in task allocation by the colony. Under what conditions do individuals switch tasks, and which transitions are most likely? Under what conditions do individuals decide to become active, or to remain inactive? Once we know something about how task allocation operates in a colony, it is possible to construct models or hypotheses. The problem is then to determine what rules individuals use that, in the aggregate, produce colony behavior. Such rules will probably involve information from the environment and from interaction with other individuals. The ways that colony organization produces task allocation may be as various as the tens of thousands of social insect species. Or there may be common features, such as perhaps the use of interaction patterns, across broad taxonomic groups. Only empirical work on a variety of species will make it possible to discover the general principles of task allocation in social insects, and to understand the relation between task allocation, interaction patterns, and colony size.

References

1   Langton CG (1989) Artificial Life. *In*: CG Langton (ed): *Artificial Life*. SFI Studies in the Sciences of Complexity, vol 6. Addison-Wesley Publishing Company, Redwood City CA

2   Deneubourg JL, Goss S (1989) Collective patterns and decision-making *Ecol Ethol Evol* 1: 295–311

3   Gordon DM (1989) Caste and change in social insects. *In*: P Harvey, L Partridge (eds): *Oxford surveys in evolutionary biology*. Oxford University Press, Oxford

4   Gordon DM (1996) The organization of work in social insect colonies. *Nature* 380: 121–124

5   Franks NR, Bryant S, Griffiths R, Hemerik L (1990) Synchronization of the behaviour within nests of the ant *Leptothorax acervorum* (Fabricius). I. Discovering the phenomenon and its relation to the level of starvation. *Bull Math Biol* 52: 597–612

6   Cole BJ (1991) Short-term activity cycles in ants: generation of periodicity by worker interaction. *Amer Naturalist* 137: 244–259

7   Gordon DM, Paul REH, Thorpe K (1993) What is the function of encounter patterns in ant colonies? *Anim Behav* 45: 1083–1100

8   Seeley TD (1989) Social foraging in honey bees: how nectar foragers assess their

colony's nutritional status. *Behav Ecol Sociobiol* 24: 181–199

9   Seeley TD, Camazine S, Sneyd J (1991) Collective decision-making in honey bees: how colonies choose among nectar sources. *Behav Ecol Sociobiol* 28: 277–290

10 Seeley TD, Tovey CA (1994) Why search time to find a food-storer bee accurately indicates the relative rates of nectar collecting and nectar processing in honey bee colonies. *Anim Behav* 47: 311–316

11 Robinson GE, Page REJr, (1989b) Genetic basis for division of labor in an insect society. *In: The Genetics of Social Evolution.* Westview Press, Boulder, CO

12 Gordon DM (1986) The dynamics of the daily round of the harvester ant colony. *Anim Behav* 34: 1402–1419

13 Gordon DM (1987) Group-level dynamics in harvester ants: young colonies and the role of patrolling. *Anim Behav* 35: 833–843

14 Gordon DM (1989) Dynamics of task switching in harvester ants. *Anim Behav* 38: 194–204

15 Gordon DM (1992) How colony growth affects forager intrusion in neighboring harvester ant colonies. *Behav Ecol Sociobiol* 31: 417–427

16 Gordon DM (1995) The expandable network of ant exploration. *Anim Behav* 50: 995–1007

17 Gordon DM (1991) Behavioral flexibility and the foraging ecology of seed-eating ants. *Amer Naturalist* 138: 379–411

18 Gordon DM, Hölldobler B (1987) Worker longevity in harvester ants. *Psyche* 94: 341–46

19 Gordon DM (1995) The development of an ant colony's foraging range. *Anim Behav* 49: 649–659

20 Adler FR, Gordon DM (1992) Information collection and spread by networks of patrolling ants. *Amer Naturalist* 40: 373–400

21 Gordon DM, Goodwin B, Trainor LEH (1992) A parallel distributed model of ant colony behaviour. *J Theor Biol* 156: 293–307

22 Pacala SW, Gordon DM, Godfray HCJ (1996) Effects of social group size on information transfer and task allocation. *Evol Ecol* 10: 127–165

23 Gordon DM (1994) How social insect colonies respond to variable environments. *In*: LA Real (ed): *Behavioral mechanisms in evolutionary biology.* University of Chicago Press, Chicago

24 Nonacs P, Dill LM (1990) Mortality risk versus food quality trade-offs in a common currency: ant patch preferences. *Ecology* 71: 1886–1892

25 Jeanne RL (1986) The organization of work in *Polybia occidentalis:* the costs and benefits of specialization in a social wasp. *Behav Ecol Sociobiol* 19: 333–341

26 Jeanne RL (1996) Regulation of nest construction behaviour in *Polybia occidentalis. Anim Behav* 52: 473–488

27 Wagner D, Brown MJF, Broun P, Cuevas W, Moses LE, Chao DL, Gordon DM (1998) Task-related differences in the cuticular hydrocarbon composition of harvester ants, *Pogonomyrmex barbatus. J Chem Ecol* 24: 2021–2037

28 Gordon DM, Mendiabadi N (1999) Encounter rate and task allocation in harvester ants. *Behav Ecol Sociobiol* 45: 370–377

Information flow during social feeding in ant societies

Deby Lee Cassill and Walter R. Tschinkel

Summary

In fire ants, social feeding is regulated by two hungers, one among larvae, the other among workers. Workers donate to larvae or to workers hungrier than themselves, and solicit from workers more satiated than themselves. Food flows via a chain of demand initiated by hungry solicitors rather than a chain of transfer initiated by full donors.

Colony patterns of food distribution emerge from the rules by which individuals transfer food. The key elements are the rate and duration of individual food transfers. The rate of transfer is regulated by two groups: (i) a population of diverse workers coarse-tuned by size and to some degree by age, and fine-tuned by crop fullness and food preference to forage, store food, solicit from, or donate to other colony members; and (ii) a population of larvae who regulate their diet by soliciting at rates based upon individual food preference, midgut fullness, and body size. The duration of worker-worker trophallaxis is variable, generating an uneven distribution of crop fullness among workers. Variation in crop fullness contributes to the distribution of worker labor among social feeding tasks. The duration of worker-larva trophallaxis is fixed and brief. The brevity of larval feedings and the rate at which they solicit feedings result in a relatively even distribution of food per unit of larval size. This even distribution contributes to colony growth by diverting excess food to the incoming generation of larvae.

In total, social feeding is a decentralized homeostatic system composed of individuals constantly moving toward fullness. Relatively simple rules of thumb, foraging-for-work algorithms, food-preference templates, cue clouds, negative feedback and self-organizing processes contribute to the interactions between individuals and the distribution of individuals among feeding tasks inside and outside the nest.

C. Detrain et al. (eds) Information Processing in Social Insects
© 1999, Birkhäuser Verlag Basel/Switzerland

Introduction

One of the pressing issues of social insect research is the synthesis of colony organization from the actions of individuals. A number of hypotheses have emerged within the last 2 decades, each attempting to explain information flow among workers, moment to moment, in an ever-changing social environment. Although most describe colony organization as a decentralized process, each hypothesis offers a different mechanism regulating the flow of information. Mechanisms thought to regulate worker behavior and division of labor include polymorphism [1–3, pp. 298–354], learning [4], simple "rule-of-thumb" responses to local cues [5], "foraging-for-work" algorithms [6], random amplification (reviewed in [7, 8]), genetic differences in fixed-response thresholds (reviewed in [9, 10]), temporal/age polyethism (reviewed in [11, p. 400]), social context [12], and stigmergy (Grasse, 1979, cited in [7]). It is possible that, like the parable of the blind men describing different features of the same elephant (poet John G. Saxe, 1816–1887), each mechanism applies to a different feature or a different level of organization in colony life. The intent of this chapter is to review the mechanisms regulating the flow of food among colony members and the flow of colony members among social feeding tasks in an attempt to describe the flow of information therein.

Social feeding

In social feeding, food is the currency that links the majority of colony members and colony tasks to one another. For example, in fire ants, food links the colony to its environment. Food is acquired in quantities that far exceed the metabolic requirements of the individual and is transported from the environment to the nest by a fraction of colony members (W.R. Tschinkel, unpublished data), the foragers. Inside the nest, food links foragers to intermediaries, who store it, share it with still other workers, or feed it to the important end users, the queen and the larvae [13–15]. Food also links brood and soldiers who defend the concentrated supply of newly converted ant tissue from predators. Finally, metabolic waste products may link workers to their nest and to each other. Workers plaster the inside walls of the brood chambers and tunnels with their excretions, providing additional structural support (casual observation). Workers regularly step into the toilet area where larval excretions are deposited and groom themselves for extended periods of time (casual observation), possibly marking themselves with a common scent used in nestmate recognition. The last decade has seen a great deal of novel research revealing the mechanisms regulating the components of social feeding. A summary of this work follows.

Brood organization

In contrast to bees and wasps that rear larvae individually in small cells, ants rear larvae communally, piled together in relatively large chambers. Brood items are spatially organized in either of two configurations depending upon the species and the complexity of its nest. In founding colonies (*Solenopsis invicta*, casual observation) and in species that nest in a single chamber (*Leptothorax unifasciatus* [16]), brood items are arranged in concentric rings with eggs and microlarvae clustered at the center and surrounded, in progressively larger rings, by larger larvae and pupae. One mechanism regulating this arrangement could be passive, whereby the egg-laying queen occupies a central position with younger brood pushing older brood outward. However, passively generated concentric rings are often destroyed when the queen moves or when the colony migrates within the nest or to a new nest site. If brood organization is adaptive, then workers must be able to restore order after each migration. An active mechanism, with workers sorting brood based on several simple rules [17], could quickly and efficiently reorganize brood. Fire ant workers appear to have a general rule, "cluster all brood items together" followed by a more specific rule, "move like brood item next to like brood item." Fire ant workers sort brood item by repeatedly antennating first the brood item in its mandibles, then the surrounding brood items until a match (probably odor) is found. At this point, the worker gently places the brood item (egg, larva, prepupa, or pupae) next to the like brood item. The shape, size, and pilosity of fire ant eggs, larvae, and pupae determine their domain of care (defined as the amount of empty space around each brood item [17]). In all probability, these morphological differences in brood items contribute to the concentric arrangement of brood as smaller, more closely packed items naturally become centralized within larger items.

For species building complex nests, brood items are dispersed among many chambers depending upon the microclimate. Eggs and microlarvae are housed near the queen in humid chambers, older larvae in moderately warm, humid chambers, and pupae in hot, dry chambers [18, 19]. Workers could easily organize brood within and among brood chambers, operating with the original two rules and one or two additional rules for vertical sorting of brood depending upon the microclimate inside the nest. From the worker's point of view, the organization of brood emerges from each worker's response to local cues with a few simple rules of thumb. From the colony's point of view, the allocation of workers to the task of brood sorting is regulated by foraging for work algorithms [6] and negative feedback—fewer workers stop to engage in brood sorting as more brood item matches are completed.

Brood assessment

Once brood items are spatially organized, they are groomed and fed incessantly until the next disturbance or migration. Fire ant workers are tireless caregivers. Each brood item is assessed approximately 700 times per hour or once every 5 s [20]. For larvae, the assessment:feeding ratio is high—on average, larvae are contacted 70 times for every feeding they receive. Similarly, in *Pheidole dentata*, larval feedings occurred, on average, once every 50th behavioral act (2%) by minor workers [21]. Workers achieve constant care of brood by maintaining a high constant density of workers on the brood pile (85% coverage). Worker density remains constant regardless of brood satiation [22], worker:brood ratio or colony size [14]. Although warmer temperature slightly reduces the density of workers on the brood pile, workers move faster, thus maintaining a steady contact rate (D.L. Cassill, unpublished data). Potentially, workers regulate their density by monitoring the time between worker contact [23, 24]. From the colony's point of view, the allocation of workers to the task of brood assessment is regulated by negative feedback at the individual level—workers leave the brood pile when the time between contacts becomes too short.

Another feature of brood care is the great simplification of worker decision making when feeding larvae. During each trophallactic event, a larva is fed a tiny, fixed amount of food (~1.5 nl [25]) regardless of differences in larval size, hunger, orientation, or location on the brood pile. One outcome of a fixed volume is that workers do not have to assess how hungry or how large a larva is and adjust the amount of food they transfer accordingly. Instead, the worker feeding response is a simple, binary decision—feed or do not feed this larva. The brevity and constancy of worker-larva trophallaxis was found in other ant species as well, suggesting that small food increments are a common feature of larval feeding in ants.

A fourth feature of brood care is that workers frequently switch from one task to another and back again (D.L. Cassill, unpublished data). In theory, if workers were prone to staying on task, a single worker could feed hundreds of larvae each hour, even allowing time for assessment and crop refills [20]. In actuality the average worker feeds fewer than 30 larvae each hour [14]. Such erratic task switching by active workers may be the mechanism that ensures that a large number of workers are foraging-for-work, thus fine-tuning the allocation of workers among multiple tasks. We suggest that "foraging-for-work" algorithms be expanded to include information signals from other colony members as well as inanimate cues from the environment. After all, it is the intense assessment of brood by hundreds of wandering workers that produces the highly reliable brood care system, turning a probability of being fed, groomed, or moved into a certainty. If one worker fails to feed, groom, or move a brood item, another responds shortly thereafter.

Larval hunger

Meal volume is tightly regulated, not by workers, but by the larvae themselves who actively solicit food from workers [22]. Several lines of evidence suggest that the solicitation cue is a nonvolatile chemical—a pheromone or a metabolic waste product—rather than a behavioral or tactile cue. For a fourth-instar larva, a typical meal consists of hundreds of tiny feedings delivered over 8 to 12 h [22, 13]. The rate at which hungry larvae are fed is not affected by larval orientation, its location in the brood pile, or the hunger or size of adjacent larvae. Rather, larvae solicit, and are fed, at rates in proportion to their size and level of hunger. On average, each larva is fed at a similar rate per unit of larval volume such that all larvae are brought to fullness together. Nevertheless, larval appetites vary, resulting in different meal sizes that, in turn, may produce different adult sizes or castes.

We simulated larval hunger [20] to determine which feeding rules affected the patterns of meal size and nutritional mix among larvae. Varying the response thresholds among workers affects only the total time required to bring larvae to satiation but not the even distribution of food among larvae. When the rate of larval feeding is a function of larval size but not hunger, larvae are brought to fullness evenly over time, but are grossly overfed when food is abundant. When the rate of larval feeding is a function of larval hunger but not size, larvae are not overfed, but some larvae became full far sooner than others. The sum total of the patterns by which larvae solicit for food is that surplus food is retained over the short term inside worker crops. This phenomenon has potential implications for colony growth: surplus food stored in worker crops can act as a buffer against sporadic food availability, keeping a steady flow of food moving to the larvae. This allows more continuous larval feeding and growth, at least on a scale of a few days. Such buffering would also reduce the necessity for cannibalizing larvae to retrieve food during shortages. During periods of food deficit, hunger (the absence of food) is also distributed evenly over the larval population, thus reducing the possibility that some larvae receive no food and starve to death. Reduction of growth would then be spread evenly over all larvae allowing larval numbers to remain constant, offering clear advantages for colony growth when food supplies fluctuate.

Larvae not only regulate food volume but food quality as well [15]. Larvae have independent appetites for food, preferring concentrated rather than diluted solutions, and ingesting food at rates characteristic for that food type or food state, regardless of whether they are empty or full of other food types. Because meals are delivered in tiny morsels by hundreds of workers carrying different nutrients, unmixed, in their crops, each larva can fine-tune the nutritional mix of its diet. This ability provides larvae the potential for regulating their own development in competition with other larvae. Any extrinsic factor that regulates larval appetite (such as temperature or queen inhibitory pheromone; D.L. Cassill, unpublished data) could potentially regulate caste determination.

From the larva's point of view, meal quality is regulated by a food preference template (learned or innate); meal volume is regulated by negative feedback (less food is ingested with satiation). From the worker's point of view, the distribution of food among larvae results from their responding to larval hunger with simple rules of thumb rather than by a central guiding process. From the colony's point of view, the allocation of workers to larval feeding is regulated by negative feedback; as larvae become full, fewer workers are engaged in feeding them. The distribution of food among larvae, a colony-level pattern, emerges from the accumulation of thousands of workers responding independently to the solicitation signal of thousands of larvae at rates determined by their size and hunger. In the final analysis, colony hunger [26] does not exist. At least for larvae, it is an abstraction that generalizes the specific hunger of the individuals within the nest.

Worker motivation

We now focus our attention on the role of the worker in social feeding. Are all workers equally likely to forage, donate to workers, or feed larvae? In ants, young workers almost universally engage in brood and queen care, and move to general nest duties and finally to foraging as they age [8]. Polymorphism also plays a role in the distribution of workers among colony tasks [3, pp. 298–354]. In *Solenopsis invicta*, the tendency to feed larvae declines minimally and unevenly with age [14]. Worker size plays a larger role in motivating individuals to engage in different social feeding tasks. On average, medium-sized workers feed larvae most often, small workers groom larvae most often, and large workers recruit most strongly to food [27, 14].

Worker hunger plays a substantial role in motivating workers to tend brood [14]. Workers initiate feedings to hungry larvae in direct proportion to the volume of food in the workers' crops. Additionally, the type of food that workers ingest affects their feeding decisions. Workers carrying sucrose initially stay off the brood pile and donate crop contents to other workers, whereas workers carrying amino acids move directly to the brood pile to feed larvae [15]. This bifurcation of behavior based on the type of food being carried in their crops suggests a relatively sophisticated level of decision making by workers.

Once workers fill themselves, the timing of their last meal (volume and type) is erased [14], and they respond according to the current contents of their crops. Full workers do not actively push food on other workers. Rather, they advertise, actively by antennating others or passively by remaining stationary with their mandibles open in a stereotypic donor-display posture waiting for soliciting workers to contact them (D.L. Cassill, unpublished data). Howard and Tschinkel [26] claimed that full donors pushed food onto passive recipients. A review of their methods revealed this interpretation to be incorrect. Full foragers ingested less radiolabeled sugar water than did empty foragers. Therefore, less radiolabeled sugar water reached nestmates, not because they fed fewer workers but because their crop contents were diluted. Altogether, the physiological process initiating the flow of food into the colony appears to be hunger rather than full-ness.

Whereas hunger is evenly distributed among larvae, it is unevenly distributed among workers. The uneven distribution of hunger among workers is created by the highly variable frequency and duration of worker-worker trophallaxis [14]. There are significant differences in the quantity of food consumed by nurses, for-agers, and reserves [14, 28]. Nurses solicit when empty and donate to larvae or other workers when full. Scouts forage when empty and donate to other workers when full. Reserves solicit or are recruited to food sites when empty and donate or remain inactive when full. Likewise, different food types are unevenly distrib-uted among workers, with small volumes of amino acid solution reaching the most workers and larger volumes of sugar water reaching the fewest workers [29]. The skewed distribution of food volume and food type among workers has been reported for a number of other species (reviewed in [30]), suggesting that it is a fundamental feature of social feeding. The degree of individual worker hunger is an example of a single mechanism generating a distribution of labor from the colony's point of view and an alteration of behavior between foraging, soliciting, or donating from the worker's point of view.

Considerable variation in individual behavior occurs among workers that can-not be attributed to differences in size, age, or hunger [14]. Some nurses feed many larvae; some feed few larvae. Additionally, the same worker may be repelled by contact with larvae one moment (antennae jerk backwards and the worker immediately changes direction) and attracted to them (grooming or feed-ing) the next (D.L. Cassill, unpublished data). This variation in worker response poses a problem for the fixed response threshold hypothesis [9, 10] (see Beshers and Robinson, this volume, and Bonabeau and Theraulaz, this volume), in which workers are viewed as captive to their neurology, producing a stereotypic response when a cue is sufficiently strong enough to surpass a worker's percep-tion threshold. We propose an alternative, a cue cloud hypothesis in which work-ers are capable of perceiving cues regardless of their intensity, and actively choose to respond to one particular cue among a suite of perceived cues. The parameters that shape an individual's cue cloud may vary depending upon extrin-sic factors such as social context (cue clouds change moment to moment as workers wander about the nest or territory) and on intrinsic factors such as a worker's size, age or hunger. The internal state changes regulating worker motivation are unknown and offer an interesting opportunity for the field of social neurology and endocrinology.

Forager hunger

Foragers do not respond directly to the nutritional needs of larvae. When foragers and reserves are hungry, the presence or absence of larvae does not affect recruitment to food sites [14]. Starved nurses actively solicit food from reserves, which increases food solicitation by reserves and foraging activity by foragers [31]. Likewise, in the ant *Myrmica rubra*, hungry nurses cause foragers to ingest more food than they will if nurses are full—the influence is not reciprocal [32]. From the above, it is apparent that the mechanism by which larvae communicate their need for protein to foragers is indirect. Larvae act as a protein sink [33] and are fed protein by nurses, creating a protein hunger that reverberates from nurses to reserves to foragers who forage for protein outside the nest. Likewise, workers inside the nest act as a sugar sink, creating a hunger for sugar in the foragers which motivates them to leave the nest to forage for more sugar.

Foragers influence larval nutrition by selecting which food they transport back to the nest. Upon encountering food outside the nest, workers evaluate its quality, then return to the nest and recruit other workers by antennating a variable number of workers depending upon their enthusiasm for the food find (*Pheidole* [34]; *S. invicta*, D. L. Cassill, casual observation). When starved as a group, foragers and reserves consistently recruit to different food types at characteristic rates, recruiting twice as strongly to sucrose as to amino acids solutions. When a colony becomes satiated on one food type, such as sugar water, fewer workers will recruit to it again but will recruit strongly to amino acid solution and *vice versa* [14]. This ability to discriminate food types may be the cause of the idiosyncratic colony food preferences in fire ants observed by Glunn et al. [35]. The mechanism for food preference is thought to lie in the individual's ability to discriminate based upon some learned or innate understanding of food value rather than by comparison shopping among available food (*Pheidole* [34, 36] *Apis mellifera* [24, 37, 38]). Foragers may reject several food finds (casual observation) before ingesting and will ingest from only one food site before returning to the nest [15]. Additionally, foragers pass food on to a clique of workers in a chain reaction that usually involves one bout of solicitation before that worker becomes a donor. The advantage of small discrete feedings by workers carrying unmixed nutrients in their crops is that individual larvae can fine-tune the nutritional mix of their meal.

The self-organization (positive reinforcement from random amplification) of foraging trails has been well described for species that mass-recruit to food sites [39–41]. In *Pheidole pallidula*, differential response thresholds between castes are thought to affect which ants follow recruitment trails [42]. In fire ants, returning scouts advertise the food quality or novelty with high tempo movement and donor displays (D.L. Cassill and L.E. Chase, unpublished data). These behaviors regulate the number of recruits that leave the nest. Trail pheromone then guides recruits to the food source. Once at the food source, workers decide individually to reject and begin a random search elsewhere or to ingest and return to the nest. Initially, group movement along a trail is a positively reinforced process that can result in leading workers to a low-quality food source [43]. Ultimately, however, group movement along trails is regulated by negative feedback; as food resources run out, trail formation diminishes to that site (reviewed in [44]).

Conclusion

In the final analysis, social feeding is a decentralized, homeostatic process organized by hunger and bound by food exchange. Interactions are initiated by hungry individuals soliciting food from full donors in a chain of demand. The transactions of this chain of demand, based on individual hunger, an ability to discriminate food quality and novelty, plastic motivation levels, simple rules of thumb and foraging-for-work algorithms, are the foundation of the distribution of workers among social feeding tasks. No single worker possesses an overview of the nutritional status of either larvae or other workers. Rather, colony nutrition is an emergent property, the product of thousands of individuals (workers and larvae) independently adjusting their rates of ingestion based upon the food choices they encounter and their current appetites for that food.

Acknowledgements

We thank two anonymous reviewers who provided critical comments essential to the completion of this manuscript. We thank Keith Mason and Scott Powell for listening to and commenting on several emergent ideas as they crystallized. The work summarized herein was carried out under the support of the National Science Foundation, grants BSR-8920710 and IBN-9317853.

References

1 Oster GF, Wilson EO (1978) *Caste and Ecology in the Social Insects.* Princeton University Press, Princeton, NJ

2 Herbers JM, Cummingham M (1983) Social organization in *Leptothorax longispinosus* (Mayr). *Anim Behav* 31: 759–771

3 Hölldobler G, Wilson EO (1990) *The ants.* Belknap Press of Harvard University Press Cambridge, MA

4 Deneubourg JL, Goss S, Pasteels JM, Fresneau D, Lachaud JP (1987) Self-organization mechanisms in ant societies (II): learning in foraging and division of labor. *In*: JM Pasteels, JL Deneubourg (eds): *From individual to collective behavior in social insects.* Birkhäuser, Basel, 177–196

5 Wilson EO, Hölldobler B (1988) Dense heterarchies and mass communication as the basis of organization in ant colonies. *Trends Ecol Evol* 3: 65–68

6 Franks NR, Tofts C (1994) Foraging for work: how tasks allocate workers. *Anim Behav* 48: 470–472

7 Gordon DM (1996) The organization of work in social insect colonies. *Nature* 380: 121–124

8 Bonabeau E, Theraulaz G, Deneubourg JL, Aron S, Camazine S (1997) Self-organization in social insects. *Trends Ecol Evol* 12: 188–193

9 Robinson GE (1992) Regulation of division of labour in insect societies. *Annu Rev Entomol* 37: 637–665

10 Bonabeau E, Theraulaz G, Deneubourg JL (1996) Quantitative study of the fixed threshold model for the regulation of division of labour in insect societies. *Proc R Soc Lond B* 1565–1569

11 Bourke FG, Franks NR (1995) *Social evolution in ants.* Princeton University Press. Princeton

12 Wilson EO (1985) Between-caste aversion as a basis for division of labour in the ant *Pheidole pubiventris* (Hymenoptera: Formicidae). *Behav Ecol Sociobiol* 17: 35–37

13 Sorensen AA, Vinson SB (1981) Quantitative food distribution studies within labor colonies of the imported fire ant, *Solenopsis invicta* Buren (1). *Insect Soc* 28(2): 129–160

14 Cassill DL, Tschinkel WR (1998) Task selection by workers in the fire ant, *Solenopsis invicta. Behav Ecol Sociobiology*; *in press*

15 Cassill DL, Tschinkel WR (1998) Diet regulation by workers and larvae of the fire ant, *Solenopsis invicta. J Insect Behav*; *in press*

16 Franks NR, Sendova-Franks AB (1992) Brood sorting by ants: distributing the

workload over the work-surface. *Behav Ecol Sociobiol* 30: 109–123

17 Deneubourg JL, Goss S, Franks N, Sendova-Franks A, Detrain C, Chretien L (1991) The dynamics of collective sorting: robot-like ants and ant-like robots. *In*: JA Meyer, EO Wilson (eds): *Simulations of animal behavior: from animals to animals.* Harvard University Press, Cambridge, 356–365

18 Scherba G (1959) Moisture regulation in mound nests of the ant, *Formica ulkei* Emery. *Amer Midland Naturalist* 61: 499–508

19 Porter SD, Tschinkel WR (1993) Fire ant thermal preferences: behavioral control of growth and metabolism. *Behav Ecol Sociobiol* 32: 321–329

20 Cassill DL, Stuy A, Buck RG (1998) Emergent properties of food distribution among fire ant larvae. *J Theor Biol* 195: 371–381

21 Wilson EO (1976) Behavioral discretization and the number of castes in an ant species *Behav Ecol Sociobiol* 1: 141–154

22 Cassill DL, Tschinkel WR (1995) Allocation of liquid food to larvae via trophallaxis in colonies of the fire ant, *Solenopsis invicta.* Anim Behav 50: 801–813

23 Gordon DM, Paul REH, Thorpe K (1993) What is the function of encounter patterns in ant colonies? *Anim Behav* 45: 1083–1100

24 Sendova-Franks AB, Franks NR (1995) Division of labour in a crisis: task allocation during colony emigration in the ant *Leptothorax unifasciatus* (Latr.). *Behav Ecol Sociobiol* 36: 269–282

25 Cassill DL, Tschinkel WR (1996) A duration constant for worker-larva trophallaxis in ants. *Insect Soc* 43: 149–166

26 Howard DF, Tschinkel WR (1980) The effect of colony size and starvation on food flow in the fire ant *Solenopsis invicta* (Hymenoptera: Formicidae). *Behav Ecol Sociobiol* 7: 293–300

27 Porter SD, Tschinkel WR (1985) Fire ants polymorphism: the ergonomics of brood production. *Behav Ecol Sociobiol* 16: 323–336

28 Sorensen AA, Kamas F, Vinson SB (1983) The influence of oral secretions from larvae on levels of proteinases in colony members of Solensopsis invicta Buren (Hymenoptera: Formicidae). *J Insect Physiol* 29(2): 163–168

29 Howard DF, Tschinkel WR (1981) Food preference in colonies of the fire ant *Solenopsis invicta* (1). *Ins Soc* 28(2): 217–222

30 Abbott A (1978) Nutrient dynamics of ants. In: MV Brian (ed): *Production ecology of ants and termites.* Harvard University Press, Cambridge, pp 233–244

31 Sorensen AA, Busch TM (1985) Control of food influx by temporal subcastes in the fire ant, *Solenopsis invicta.* Behav Ecol Sociobiol 17: 191–198

32 Brian MV, Abbott A (1977) The control of food flow in a society of the ant *Myrmica rubra* L. *Anim Behav* 25: 1047–1055

33 Sudd JH (1967) *An Introduction to the Behavior of Ants.* St. Martin's Press, New York

34 Szlep-Fessel R (1970) The regulatory mechanism in mass foraging and the recruitment of soldiers in *Pheidole.* Insect Soc 17: 233–244

35 Glunn FJ, Howard DF, Tschinkel WR (1981) Food preference in colonies of the fire ant *Solenopsis invicta.* Insect Soc 28: 217–222

36 Nonacs P, Dill LM (1990) Mortality risk versus food quality trade-offs in a common currency: Ant patch preferences. *Ecology* 71: 1886–1892

37 Seeley TD, Camazine S, Sneyd J (1991) Collective decision-making in honey bees: how colonies choose among nectar sources.

Behav Ecol Sociobiol 4: 277–290

38 Farina WM (1996) Food-exchange by foragers in the hive—a means of communication among honey bees? *Behav Ecol Sociobiol* 38: 59–64

39 Pasteels JM, Deneubourg JL, Goss S (1987) Self-organization mechanisms in ant societies (I): Trail recruitment to newly discovered food sources. *In*: JM Pasteels, JL Deneubourg (eds): *From individual to collective behavior in social insects.* Birkhäuser, Basel, 177–196

40 Deneubourg JL, Goss S, Franks N, Pasteels JM (1989) The blind leading the blind: Modeling chemically mediated army ant raid patterns. *J Insect Behav* 2: 719–725

41 Deneubourg JL, Aron S, Goss S, Pasteels JM (1990) The self-organizing exploratory pattern of the Argentine ant. *J Insect Behav* 159–168

42 Detrain C, Pasteels JM (1991) Caste differences in behavioral thresholds as a basis for polyethism during food recruitment in the ant, *Pheidole pallidula* (Hymenoptera: Myrmicinae). *J Insect Behav* 4: 157–176

43 Beckers R, Deneubourg JL, Goss S, Pasteels JM (1990) Collective decision making through food recruitment. *Insect Soc* 37: 258–267

44 Deneubourg JL, Goss S (1989) Collective patterns and decision-making. *Ethol Ecol Evol* 1: 295–311

Models of information flow in ant foraging: the benefits of both attractive and repulsive signals

Tim R. Stickland, Nicholas F. Britton and Nigel R. Franks

Summary

Models of ant foraging and recruitment behaviour have demonstrated how simple algorithms can produce complex foraging patterns. In the absence of hierarchical control, self-organisation enables a colony to establish a foraging strategy, through the interaction of ants using simple signals that indicate success in foraging. In this paper we investigate a model of ant foraging, and demonstrate how the effectiveness of the foraging strategy of a colony can be increased by reducing the persistence of foraging signals. In addition we investigate the theoretical possibility that ants might employ "negative" signals to indicate their failure in foraging. Such behaviour has neither been recorded nor refuted in real ant species and remains an intriguing possibility, since we can demonstrate how this can be overwhelmingly the most effective strategy.

Introduction

The organisation of social insect colonies presents researchers with a provocative problem: the colonies display a range of complex behaviours which cannot be accounted for by any apparent central control or simple hierarchical structure. The concept of self-organisation has come to aid our understanding of the behaviour of insect societies, originally in observations and models of ant foraging behaviour [1–12]. The complex patterns and collective problem-solving abilities observed in foraging systems can emerge from simple rules of interaction among individuals. Following its introduction in this context, models including self-organisation have found much wider applicability [13–23].

A series of models employing abstract algorithms [9–11, 23] demonstrate how simple rules of individual behaviour can lead to complex colony behaviour,

C. Detrain et al. (eds) Information Processing in Social Insects

including collective problem solving that enables the colony to establish a "foraging strategy". By this we mean that a colony can select a food source according to a preferred balance of distance against quality, and with a trade-off of accuracy against the speed of decision making. These algorithms can be applied to a simple two-way choice between alternative foraging sites, or to a series of such choices that form a binary tree structure.

The relation between the behaviour of individuals and the foraging strategy that emerges can be highly dependent on colony size: an ant colony that grows in size (as real colonies can be expected to) will show a changing foraging strategy if the behaviour of the individuals is constant. It is possible that differently sized ant colonies of the same species, whilst showing identical behaviour on the part of the individual foragers, may exploit food resources differently (see especially [10]). In addition, the emergent phenomenon we call the "search limit" occurs under certain conditions, where foragers may effectively ignore a food source that is too far from the nest [10]. This is a significant constraint: contrary to basic expectations of optimal foraging, the quality of a distant food source (no matter how high) may have no effect on the probability of the colony selecting it.

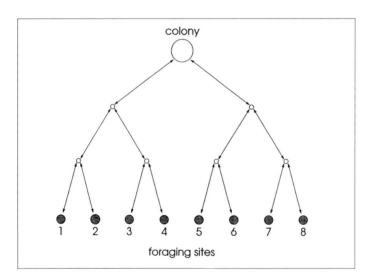

Figure 1 Binary tree model of path choice
At each branch point (small circle) each ant chooses between two paths. In this tree with three levels of branching, there are 8 possible foraging sites. The simulations employ a tree with seven levels and 128 foraging sites.

A binary tree structure (see Fig. 1) can be seen in many species of ants, most obviously those which forage in real (botanical) trees, where foragers are physically constrained to the structure [24]. Ground-foraging species such as seed harvesters [25, p. 285; 26] and certain army ants [6] create their own binary tree structures; though foragers are (presumably) able to move freely in space, their movements are mainly restricted to branching networks of trails. Edelstein-Keshet et al. [12] demonstrate how recruitment behaviour can coordinate the movements of individuals, and lead to the formation of trails. The important observation here is that trails do not (necessarily) restrict the movements of ants, but that ants restrict their movements to the trails.

When an ant colony forages on a branching structure such as a binary tree, there are many possible routes which the foragers can take; the purpose of the models we present here (and previously in [11]), is to investigate how the colony can collectively select the best single foraging route.

Recruitment behaviour

A key property of ant foragers is their ability to recruit nestmates. Foragers can influence the choices made by their nestmates, where a returning forager will encourage others to take the route from which it just returned. This property of individuals produces positive feedback, and enables the colony collectively to establish a single trail.

There is a great diversity of recruitment mechanisms (see [25, pp. 265–279] and [27, pp. 244–249] for reviews); at one extreme, a forager may lead a single ant by direct body contact ("tandem running"), and at the other extreme ants may be recruited en masse by the use of pheromone trails. The degree of influence a returning forager has on its nestmates is a major factor in determining emergent foraging patterns [5, 9–11, 23].

In certain species the recruitment behaviour of individuals is known to be "graded" according to the quality and type of the food source they visit [28–32]. Graded recruitment can enable the colony collectively to choose between food sources according to quality [5, 10] or the probability of finding food [23, 11], without the need for any comparative judgements by individuals. The reported observations and the models share the characteristic that more desirable food sources produce greater recruitment activity, so that more ants are encouraged to visit the better food source.

Recruitment behaviour is not restricted to ants. The "waggle dance" of honeybees is a very well documented example [33, pp. 80–106], and pheromone trails have been recorded in stingless bees and many species of termites [34, pp. 247–261]. In principle, self-organising patterns need not arise only from mutualistic behaviour, and certain properties may be shared with any system in which the behaviour of animals is influenced by their observations of others. Self-organising patterns, arising from individuals attracting each other to food sources, have even been suggested for humans "foraging" at restaurants [35, 36].

The persistence of recruitment signals

Earlier models of ant foraging [4, 5, 7, 8–11] assume that when a forager exerts an influence on its nestmates, that influence remains (though the effect may be reinforced or reversed by subsequent recruitment behaviour (but see [6]). In reality, a recruitment signal such as a pheromone trail will weaken if the pheromone evaporates. The existing models assume that such processes are slow enough to have no effect over the time taken to establish a foraging trail. This is a useful simplifying assumption, and has been justified (particularly by [2] and [4]) since the models do produce foraging behaviour of the type seen in real species.

In this paper we consider the consequences of recruitment signals that have a limited persistence. Superficially it might appear that the removal of a signal must be undesirable (an exception being in a variable environment where information might become "out of date", but these models consider a constant environment). However, in a complex system such as a foraging ant colony, we cannot assume that emergent effects will follow that which superficially appears probable. We concur with Traniello and Robson [27, p. 263] that in complex systems of this type it is difficult (and potentially misleading) to make intuitive predictions on the effect of changes in individual behaviour on the emergent group-level behaviour.

From this point we will refer to the process of a foraging signal becoming progressively weaker as "evaporation", envisaging recruitment by a pheromone trail, though in principle these models can be applied to any recruitment signal that becomes weaker over time.

The algorithm

The algorithm works on a binary tree structure (see Fig. 1). As with the previous model described in Stickland et al. [11], there is a bias value at each branch point in the tree, which is initially set at 50%:50%. Ants leave the nest and choose between the two paths probabilistically according to the bias, so the first ants to leave make purely random choices. At the end of each branch there is another two-way choice, until the ant reaches a foraging site at the end of the tree. At the foraging site there is a probability of the ant finding food; if it finds food, it is defined to be a successful forager.

After visiting a foraging site, the ant returns towards the nest; successful foragers update the bias value at each branch point as they pass, so that the next outward-bound ant to reach that point is more likely to take the same route. The amount of influence each ant has on the bias value is set by parameter I; for example, if $I = 10\%$, then the first successful forager to return will update the bias value from 50% to 60% (in favour of the route it took) at each branch point it passes. Ants arriving back at the nest immediately leave again, choosing a route probabilistically according to the current bias values, with no individual memory or fidelity to previous choices.

The rate at which ants initially leave the nest is set by parameter R; if R = 10 and the first ant leaves at t = 0, then following ants leave at t = 10, t = 20 and so forth. Time is accounted in periods; from t to t + 1 each foraging ant will move from one branch point to the next, that is, one level in the binary tree structure.

In the previous model, each bias value could only be changed by returning foragers. In the absence of ants, the value would remain the same. In this model each bias value follows an exponential decay towards its starting value of 50%. For a bias value b at each time iteration

$$b_{t+1} = b_t - \alpha(b_t - 50)$$

where α represents the proportion of bias lost by pheromone evaporation from time t to t + 1.

In all cases the binary tree has seven levels, and therefore a total of 128 foraging sites. As in the previous model, one of these foraging sites is defined to be "good", with a 0.9 probability that any forager visiting it will find food. At all of the other 127 sites there are lower and equal probabilities; for any given simulation these are set at either 0.6, 0.3 or 0.1.

Simulations are run in sets in which the values of I, R, α and the probabilities of finding food are all fixed. Within each set a series of simulations are carried out with variable numbers of foraging ants: 1, 21, 41, 61, 81, and 101 ants. For each value, 100 replicates are run. As noted above, the ants leave the nest one at a time, at internals defined by the value of R, until all the foragers have left the nest.

As the simulation runs, positive feedback tends to take each bias value to fixation. A bias value that reaches 100% or 0% is fixed at that value, so that ants will always take the fixed route (this simplification provides a definite end point to the simulation). When a route is fixed all the way from the base of the tree to one tip, the ants are all following a single foraging trail and the simulation is halted. If that trail leads to the good foraging site, the simulation is recorded as a success. The number of successes out of the 100 replicates, and the number of time steps each takes to run to completion, is recorded.

Behaviour of the algorithm

Despite evaporation reducing the recruitment effect of foragers, it can increase the probability of an ant colony fixing a trail to the best foraging site under certain circumstances. Results from simulations are given in Table 1; to give one example, a success rate of 81% with evaporating trails compares with only 35% in the absence of evaporation (I = 10%, R = 1 and α = 0.1, a colony of 101 ants, with a foraging probability of 0.9 at the good site and 0.1 at the others).

This increase in success rate is associated with a slower rate of decision making, that is, with an average time of 131 as opposed to 70 without evaporation in the example above. This is not surprising, since it will require more foraging trips to fix a trail when the recruitment signals left by ants are being continuously reduced by evaporation.

In the previous model [11] we also found that slower decision making, achieved by using low values for I, could increase the success rate. In comparison to the example quoted above, to gain a similar increase in success rate (83% from 35%) requires a reduction to I = 1%. This increases the mean decision mak-

Table 1 Summary of simulation results for the algorithm including evaporation

I	α	Success rate	Time
10%	0.10	81%	131
5%	0.10	100%	359
1%	0.10	*	*
10%	0.05	64%	99
5%	0.05	86%	155
1%	0.05	*	*
10%	0.01	42%	75
5%	0.01	61%	104
1%	0.01	94%	344
10%	0	35%	70
5%	0	51%	98
1%	0	83%	237

The success rate is the proportion of replicates in which the colony fixed a single foraging trail to the one good foraging site, where the probability of finding food is 0.9 at the one good site and 0.1 at the other 127 sites. I is the influence of ants on the path choice bias values, and α is the evaporation of this recruitment signal. The missing values (I = 1% with high values of α) indicate that trail formation was very slow, and exceeded the maximum processing time allowed for the simulation. For reference, values are also given for the previous model ([11]) without evaporation, denoted by α = 0. These results are for colonies of 101 ants, where R = 1.

ing time to 237. This is a much greater time penalty than is required to gain this level of success using evaporating trails.

Evaporation does not appear to provide any benefit where the food is not scarce, that is, where the probability of finding food is greater than 0.1 at most foraging sites. In addition, when the value of I is low, evaporation provides less of a gain (this is a relative statement: success rates always tend to be high when I is low). A higher value of R does not significantly affect the success rate, but slows down the decision making; in fact the time penalty for high values of R appears to be greater in the presence of evaporation.

In general, these results show that negative feedback can improve the efficiency of such systems. This role of negative feedback is likely to be extremely widespread. For example, Deneubourg et al. [37] showed through modelling that, in principle, selection of the most rewarding foraging zone could require an optimal rate of learning. Here, and in the previous model [11], we demonstrate that there is an optimum level of influence, that is, recruitment rate—equivalent to a rate at which there is "learning" by the colony.

Why can evaporation be beneficial?

Foraging ants employ recruitment signals as a deliberate act in response to a stimulus; we might reasonably expect that if natural selection has favoured this behaviour, then it must be of benefit to the colony. Because this behaviour is adaptive, it superficially seems (given a constant environment such as our model) that there should be no benefit in the constant erosion of the effects of this behaviour. Clearly this is not the case.

With previous models [9–11], it is clear that when ants have a high influence on nestmates (high value I), the success in choosing the best foraging site can be reduced. We have attributed this to the stochastic nature of the process: the collective decision making works because most of the influence of foragers is towards the most desirable foraging route, and when reinforced by positive feedback, the colony can direct all of its foraging effort along the best route.

It is of course quite possible that the first few ants to return to the colony may have taken the "wrong" route, and establish an initial path choice bias towards that route. Often the colony's collective preference for the "right" path will overcome an initial bias of this sort, but if the bias is very strong, it is also possible that positive feedback will dominate the decision-making process, so that the colony will choose the wrong path. The likelihood of this depends on the chance of a high initial bias being established in the wrong direction: when ants each have a greater influence (I is large), a high enough bias can be established by only a few ants, and so a wrong decision is more likely to be initiated.

The effect of evaporation is to reduce positive feedback. When pheromone evaporates, other foragers must reinforce it if the bias towards that foraging route is to endure, or get stronger. Consequently an initial bias in the wrong direction has less chance of becoming fixed purely through powerful positive feedback. The benefit is expressed where there would be the greatest risk of an incorrect path choice: when ants have a high influence (large I).

Naturally, when positive feedback is reduced, there is a time penalty, because it takes more ant foraging trips for all bias values to reach fixation. In situations where evaporation does not give much benefit (e.g. when food is not scarce), however, no such time penalty is seen. Since the penalty is only seen where a benefit is gained, this strategy cannot "backfire", which means the use of evaporating trails is a safe strategy.

Negative recruitment signals

Clearly a process such as trail pheromone evaporation (or a weakening of any other recruitment signals through time) can be beneficial, by limiting the extent to which positive feedback can drive the choice of foraging route to fixation. We modelled this process because it is known to occur in nature, but it is not necessarily the only process which can have this effect.

In principle, foragers could employ a negative recruitment signal, with the opposite effect to the "normal" recruitment signals we are familiar with. This can be envisaged as a pheromone which nestmates try to avoid, though of course many species employ recruitment signals that are not pheromone trails, and in principle any form of signal might be employed. The significant property is that unsuccessful foragers would reduce the probability of nestmates following the same route.

We constructed a model incorporating negative recruitment signals to test the properties of such as system. This model is run exactly as the evaporation model described above, except for the following differences:

- there is no "evaporation", so bias values can only be changed by the recruitment signals of ants;
- successful foragers change bias values exactly as in the previous model, by an amount set by parameter I_s;
- foragers that fail to find food will reduce the bias value towards the path they are returning from, by an amount set by parameter I_f;
- the value of I_s and I_f need not be equal, so normal recruitment signals can be stronger, or weaker, than negative signals.

Behaviour of the algorithm

Compared with the previous model described in Stickland et al. [11] and the model including evaporation described above, negative recruitment signals give a success rate that is massively higher. Even when $I_f = I_s/10$ (i.e. negative recruitment signals are only a tenth the strength of "normal" signals) the minimum success rate achieved was 92% compared with an 83% maximum without inverse recruitment. This outweighs the effects of I which was previously the predominant parameter.

When food is less scarce (0.9 versus 0.6 probability), weak negative recruitment signals give very little advantage (e.g. when $I_s = 5\%$ and $I_f = 0.5\%$, the success rate is only 4%). When the negative signals are strong, they can still be very effective (e.g. when $I_s = 5\%$ and $I_f = 10\%$, the success rate is 95%). A summary of simulation results is given in Table 2.

Negative recruitment signals slow down the decision-making process. This should be expected since bias values will be reduced by unsuccessful foragers, and therefore more foraging trips by successful foragers are required before bias values reach fixation. As in the evaporation model, the delay is less than that which would be suffered if an equivalent success rate was achieved by using a lower value of I. Interestingly, the delay is only observed when negative recruitment signals provide an increase in foraging success: again this is a "safe" strategy, where a penalty can only be incurred in circumstances which give rise to the benefits.

Why are negative signals so effective?

The effects of negative signals are similar to those of evaporation. Once again the essential property is that a strong bias in the "wrong" direction is unlikely to be established. If some foragers randomly choose a path to a poor foraging site and do manage to find food, then they may establish an initial strong bias in that direction; however, most foragers that subsequently take that route are likely to fail to find food, and so their negative influence will destroy the bias. Rather than positive feedback that occurs when ants have only positive influence, unsuccessful foragers now produce negative feedback.

Table 2 Summary of simulation results for the algorithm including negative recruitment signals

Recruitment influence		Success rate	Time
Foraging probability 0.9 vs 0.1			
$I_s = 10\%$	$I_f = 20\%$	93%	138
$I_s = 5\%$	$I_f = 10\%$	100%	170
$I_s = 1\%$	$I_f = 2\%$	100%	275
$I_s = 10\%$	$I_f = 10\%$	100%	133
$I_s = 5\%$	$I_f = 5\%$	100%	140
$I_s = 1\%$	$I_f = 1\%$	100%	279
$I_s = 10\%$	$I_f = 1\%$	92%	126
$I_s = 5\%$	$I_f = 0.5\%$	98%	136
$I_s = 1\%$	$I_f = 0.1\%$	100%	261
$I_s = 10\%$	$I_f = 0\%$	35%	70
$I_s = 5\%$	$I_f = 0\%$	51%	98
$I_s = 1\%$	$I_f = 0\%$	83%	237
Foraging probability 0.9 vs 0.6			
$I_s = 10\%$	$I_f = 20\%$	46%	162
$I_s = 5\%$	$I_f = 10\%$	95%	211
$I_s = 1\%$	$I_f = 2\%$	100%	403
$I_s = 10\%$	$I_f = 10\%$	19%	165
$I_s = 5\%$	$I_f = 5\%$	28%	275
$I_s = 1\%$	$I_f = 1\%$	23%	319
$I_s = 10\%$	$I_f = 1\%$	3%	49
$I_s = 5\%$	$I_f = 0.5\%$	4%	74
$I_s = 1\%$	$I_f = 0.1\%$	10%	180
$I_s = 10\%$	$I_f = 0\%$	2%	61
$I_s = 5\%$	$I_f = 0\%$	3%	73
$I_s = 1\%$	$I_f = 0\%$	7%	159

The success rate is the proportion of replicates in which the colony fixed a single foraging trail to the one good foraging site, where the probability of finding food is 0.9 at the one good site and 0.1 or 0.6 (as indicated) at the other 127 sites. For reference, values are also given for a previous model ([11]) without negative recruitment signals, denoted by $I_f = 0$. Each value was calculated from 1200 simulations: 100 replicates for each parameter set, with data amalgamated for $R = 1$, $R = 10$, and all numbers of foraging ants (1...101 in increments of 20), since these parameters are known to have only very weak or insignificant effects. The time taken to fix a trail is the average from 100 replicates for 101 ants, $R = 1$.

Intuitively, we might expect negative signals to be beneficial, since a forager that can leave two types of signal is capable of communicating more information to her nestmates. In fact, negative signals have rather more effect than simply making the path choices of nestmates more accurate: when these signals are used, negative feedback predominates and tends to prevent any trails forming. The bias values at all the branch points in the binary tree remain close to their starting value of 50%, and the ants swarm randomly throughout the structure. To an observer, it would simply appear that no recruitment signals are in use at all.

The absence of trail formation, exhibited as apparent "random" swarming of foragers, continues whilst the ants visit only poor foraging sites (these are 127 of the 128 possible foraging sites, so in the early stages virtually all foraging trips end up at poor sites). Positive feedback is only likely to have an effect when a forager discovers the single good foraging site: at this branch point one of the two routes leads to probable foraging success, and therefore produces positive feedback. When that branch point is fixed, the next branch point "down" the tree will similarly be subject to positive feedback, and will be fixed quickly. The net effect is for a single trail to be established at the good foraging site and then "grow" back towards the ant nest at the base of the tree.

In other words, the effect of negative signals is to postpone the establishment of any trails whilst the ants explore, and then to "switch on" trail formation when they discover a good food source. Though the individual foragers show a constant response to stimuli, their collective use of positive and negative feedback can cause a switch in the emergent behaviour seen in the colony.

Discussion

Clearly, the evaporation of trail pheromone and the use of negative foraging signals greatly increase the success rate of our models of ant foraging. Whereas it seems certain that pheromone evaporation must occur at some level in nature, to our knowledge negative recruitment signals have never been described in a real ant species.

This raises the question, why do ants not use negative signals? Of course the model is an abstraction, and it may be that negative signals are not a practical adaptation for real ants in nature. Equally, though our model suggests negative signals are more effective than trail evaporation, it could be that in nature the reverse is true. For instance, the models take no account of the cost of recruitment behaviour: whereas negative signals must require the investment of extra time and effort (and possibly pheromone) by foragers, the evaporation of trails is "free", and can benefit the ants without any additional investment.

The open-minded investigator should of course consider another possibility: though it has frequently been demonstrated that ants employ positive recruitment signals, this cannot be taken as proof that they never employ negative signals. It remains a possibility that certain species of ants do use negative signals.

Is the model biologically reasonable?

We contend that there is no reason why ants should not be as capable of negative recruitment signals as the well-documented "positive" signals. Negative signals have in fact been described in other hymenopterans in other contexts. Hassell [38, p. 200] states ,"Price (1970) [39] has clearly demonstrated that some female parasitoids leave scent… to prevent areas being re-searched by either the same female, or by others of the same, or even different species." Giurfa and Nunez [40] describe the use of negative signals by foraging honeybees, preventing repeat visits to flowers that had already been exploited.

Ants possess a wide behavioural repertoire, and it seems unreasonable to suggest that they cannot employ an additional signal. Numerous signals have been described: recruitment can occur from direct interaction between foragers, or via a pheromone trail; in many species a pheromone trail is used in conjunction with direct interactions, where a recruiting ant must "invite" other foragers to follow the trail. Hölldobler and Wilson [25, pp. 265–279] review recruitment behaviour, illustrating a great diversity; for instance the production of trail pheromone has been observed in 10 different anatomical structures, and the recruitment systems of Formicinae may have evolved independently of the Ponerines and Myrmicinae.

Traniello and Robson [27, especially pp. 244–249] also review the diversity of foraging signals employed by social insects. In particular, they document the difficulty in assaying pheromones; signals may consist of several pheromones, mediating separate behavioural responses involved in foraging (for instance, recruitment and orientation). The exact role of a signal can be sensitive to the context in which it is employed, depending on part on tactile or acoustic senses, and possibly ambient light conditions. Traniello and Robson conclude that "it is not realistic to refer to the secretion of a single exocrine gland or one pheromone component as being the chemical responsible for trail communication.... More than one gland may contribute to the chemical structure of a trail, and a given gland may synthesise and secrete more than one chemical product." We cannot assume that we are familiar with all of the chemical substances used in foraging signals, and neither is it a safe assumption that we fully understand the deployment of those substances with which we are familiar.

There is certainly no practical reason why an ant should not discourage others from following a trail by communication involving direct contact or pheromones. Where such a variety of positive as well as negative signals have been observed within an Order, it seems unlikely that a simple signal that produces a great benefit should not be utilised. If indeed negative signals have been rejected by natural selection, there is no reason that it should be because ants are simply incapable of using such a strategy.

The effect of trail evaporation

The advantage of negative signals over evaporation is that the reduction in bias values is "targeted" at routes where foragers usually fail to find food. This means that any random drift towards a poor foraging site is likely to be rapidly reversed; but trail evaporation will reverse all changes in bias values. In foraging systems where ants swarm in large numbers and there is a constant traffic along all foraging routes, this may not be a problem, since another forager will find (and probably reinforce) any trail leading to a good foraging site before it evaporates completely. If ants forage in smaller numbers such a trail could be lost before another forager comes along.

There is a potential advantage if evaporating trails can fulfil the function of inverse recruitment, in that the effects have no cost to the colony. Any form of active negative recruitment must have a cost in the form of energy or pheromones used. However, this matter is not clear-cut, since evaporating trails are likely to require reinforcement, which must have a cost. Further investigation of costs and benefits in this area should provide some valuable insights.

Do ants in fact use negative recruitment signals?

There is a wealth of recorded observations of foraging recruitment; clearly ants do engage in this behaviour. Negative signals have not to our knowledge been recorded, but this does not necessarily mean they do not exist. The effect of recruitment behaviour is to produce a foraging trail that is used by all foragers, so the positive effects of recruitment are clear. By contrast, it is somewhat coun-terintuitive to suggest that ants should discourage nestmates from taking the same route; a priori there is no reason to believe that this behaviour should be necessary. Further, since negative signals can prevent the formation of any trails, it is likely no observable phenomenon will occur during any phase in which these signals are in use.

Our model also suggests that it can be highly effective when at a lower level than positive recruitment. Even when negative influence is one-tenth the level of positive influence, the efficiency of the foraging system can be several times greater. The experimental detection of recruitment behaviour is complicated [27, pp. 245–246]; where signals may be unexpected and are likely to be at a low level, there must be a significant chance that they will be overlooked in observations of real ant species.

Conclusions

Where foraging recruitment signals have a limited persistence, because of the evaporation trail pheromone or gradual decline of other recruitment behaviour, we might reasonably expect this to have a detrimental effect on ant foraging. Clearly when signals disappear, either the information is lost to the colony or they must be reinforced at additional cost. However, these foraging algorithms demonstrate that a limited persistence can be beneficial.

In addition the theoretical possibility of negative recruitment signals allows even greater control of the foraging systems. It seems intuitively likely that the existence of additional signals should be adaptive, but the extent of the potential benefits are not intuitively clear. The models we present demonstrate this, as well as displaying a set of emergent properties that predict the collective behaviour that is likely to be observed: foragers search randomly until a good food source is discovered, at which point trail formation can be suddenly "turned on". This is a collective property: individuals apply the same rules according to local conditions throughout, and without any change in individual behaviour, the colony adopts a trail following strategy only at the appropriate moment.

The model suggests strongly that some form of negative recruitment, possibly only at a low level, should be observed. Clearly this is an abstract system and is intended to model only the key properties of real ant foraging systems (though as such it suggests that negative recruitment may be a very important property). Experimental evidence will, of course, be the final arbiter. In the absence of decisive evidence we contend that the apparently prevalent assumption, that foraging recruitment uses exclusively positive signals, should be brought into question.

Whether foraging signals decline passively by evaporation of pheromone, or are actively reduced by the ants, it remains clear that the rate of decline is important. The decline in signal strength is a control process which has a great impact on the properties of this type of system.

Acknowledgements

We especially thank Chris Tofts for discussions of these problems, and suggesting the probable similarity in the function of trail evaporation and negative recruitment signals. Jean-Louis Deneubourg and Guy Blanchard provided a number of valuable insights for which we are very grateful. NRF acknowledges the support of NATO Collaborative Grant no. 880344.

Note added in proof. Under certain circumstances the models in this chapter may be analysed using techniques of stochastic processes rather than simulation. The role of the parameters may then be seen more clearly [41].

References

1 Deneubourg JL, Pasteels JM, Verhaeghe JC (1983) Probabilistic behaviour in ants: a strategy of errors? *J Theor Biol* 105: 259–271

2 Deneubourg JL, Aron S, Goss S, Pasteels JM (1990) The self-organizing exploratory pattern of the Argentine Ant. *J Insect Behav* 3: 159–168

3 Pasteels JM, Deneubourg JL, Goss S (1987) Self-organisation mechanisms in ant societies (I): trail recruitment to newly discovered food sources. *In*: JM Pasteels, JL Deneubourg (eds): *From individual to collective behaviour in social insects.* Birkhäuser, Basel, 155–175

4 Goss S, Aron S, Deneubourg JL, Pasteels JM (1989) Self-organized shortcuts in the Argentine Ant. *Naturwissenschaften* 76: 579–581

5 Beckers R, Deneubourg JL, Goss S, Pasteels JM (1990) Collective decision making through food recruitment. *Insect Soc* 37: 258–267

6 Franks NR, Gomez N, Goss S, Deneubourg JL (1991) The blind leading the blind in army ant raid patterns—testing a model of self-organization (Hymenoptera, Formicidae). *J Insect Behav* 4: 583–607

7 Beckers R, Deneubourg JL, Goss S (1992a) Trails and u-turns in the selection of a path by the ant *Lasius niger. J Theor Biol* 159: 397–415

8 Beckers R, Deneubourg JL, Goss S (1992b) Trail laying behaviour during food recruitment in the ant *Lasius niger* (L). *Insect Soc* 39: 59–72

9 Stickland TR, Tofts C, Franks NR (1992) A path choice algorithm for ants. *Naturwissenschaften* 79: 567–572

10 Stickland TR, Tofts C, Franks NR (1993) Algorithms for ant foraging. *Naturwissenschaften* 80: 427–430

11 Stickland TR, Britton NF, Franks NR (1995) Complex trails and simple algorithms in ant foraging. *Proc R Soc Lond B* 260: 53–58

12 Edelstein-Keshet L, Watmough J, Ermentrout GB (1995) Trail following in ants: individual properties determine population behaviour. *Behav Ecol Sociobiol* 36: 119–133

13 Camazine S, Sneyd J (1991) A model of collective nectar source selection by honey bees: self-organisation through simple rules. *J Theor Biol* 149: 547–571

14 Seeley TD, Camazine S, Sneyd J (1991) Collective decision making in honeybees —how colonies choose among nectar sources. *Behav Ecol Sociobiol* 28: 277–290

15 Franks NR, Wilby A, Silverman W, Tofts C (1992) Self-organising nest construction in ants: sophisticated building by blind bulldozing. *Anim Behav* 44: 357–375

16 Franks NR, Sendova-Franks AB (1992) Brood sorting by ants: distributing the workload over the work-surface. *Behav Ecol Sociobiol* 30: 109–123

17 Hatcher MJ, Tofts C, Franks NR (1992) Mutual exclusion as a mechanism for information exchange within ant nests. *Naturwissenschaften* 79: 32–34

18 Tofts C, Franks NR (1992) Doing the right thing—ants, honeybees and naked mole rats. *Trends Ecol Evol* 7: 346–349

19 Bartholdi JJ, Seeley TD, Tovey CA, Vande Vate JH (1993) The pattern and effectiveness of forager allocation among flower patches by honey bee colonies. *J Theor Biol* 160: 23–40

20 Sendova-Franks AB, Franks NR (1993) Task allocation in ant colonies within variable environments (a study of temporal polyethism: experimental). *Bull Math Biol* 55: 75–96

21 Sendova-Franks AB, Franks NR (1994) Social resilience in individual worker ants, and its role in division of labour. *Proc R Soc Lond B*. 256: 305–309

22 Tofts C (1993a) Algorithms for task allocation in ants (a study of temporal polyethism: theory). *Bull Math Biol* 55: 891–918

23 Tofts C (1993b) The efficiency of ant path finding algorithms. *Proc 1st European Conference on Computer Simulation in Biology, Ecology and Medicine*

24 Veena T, Ganeshaiah KN (1991) Nonrandom search pattern of ants foraging on honeydew of aphids on cashew inflorescences. *Anim Behav* 41: 7–15

25 Hölldobler B, Wilson EO (1990) *The ants*. Belknap Press of Harvard University Press, Cambridge, MA

26 Lopez F, Serrano JM, Acosta FJ (1994) Parallels between the foraging strategies of ants and plants. *Trends Ecol Evol* 9: 150–153

27 Traniello JFA, Robson SK (1995) Trail and territorial communication in social insects. *In*: RT Carde, WJ Bell (eds): *Chemical ecology of insects* 2. Chapman and Hall, 241–286

28 Hangartner W (1969) Structure and variability of the individual odor trail in *Solenopsis geminata* Fabr. (*Hymenoptera, Formicidae*). *Z Vergl Physiol* 62: 111–120

29 Hölldobler B (1976) Recruitment behaviour, home range orientation, and territoriality in harvester ants, *Pogonomyrmex. Behav Ecol Sociobiol* 1: 3–44

30 Crawford DL, Rissing SW (1983) Regulation of recruitment by individual scouts in *Formica oreas* Wheeler (*Hymenoptera, Formicidae*). *Insect Soc* 30: 177–183

31 Breed MD, Fewell JH, Moore AJ, Williams KR (1987) Graded recruitment in a Ponerine ant. *Behav Ecol Sociobiol* 20: 407–411

32 Baroni-Urbani C, Nielsen MG (1990) Energetics and foraging behaviour of the European seed harvesting ant *Messor capitatus*. II. Do ants optimise their harvesting? *Physiol Entomol* 15: 449–461

33 Seeley TD (1985) *Honeybee Ecology: A Study of Adaptation in Social Life*. Princeton University Press, Princeton

34 Wilson EO (1971) *The insect societies*. Belknap Press of Harvard University Press, Cambridge, MA

35 Becker GS (1991) A note on restaurant pricing and other examples of social influences on price. J. Political Econ. 99: 1109–1116

36 Kirman A (1993) Ants, rationality and recruitment. *Quart J Econ* 108: 137–156

37 Deneubourg J-L, Goss S, Pasteels JM, Fresneau D, Lachaud JP (1987) Self-organisation mechanisms in ant societies (II): Learning in foraging and division of labour. *In*: JM Pasteels, JL Deneubourg (eds): *From individual to collective behaviour in social insects*. Birkhäuser, Basel, 177–196

38 Hassell MP (1978) *The Dynamics of Arthropod Predator-Prey Systems*. Princeton University Press, Princeton

39 Price PW (1970) Trail odours: recognition by insects parasitic on cocoons. *Science* 170: 546–547

40 Giurfa M, Nunez JA (1992) Honeybees mark with scent and reject recently visited flowers. *Oecologia* 89: 113–117

41 Britton, Stickland TR, Franks NR (1998) Analysis of ant foraging algorithms. *J Biol Sys* 6: 315–336

Information flow in the social domain: how individuals decide what to do next

Nigel R. Franks

The contributions in this section, in common with the entire theme of this book, address the question, how are social insect colonies organized? This is a major question because it is part of one of the most enthralling scientific quests—to understand how evolution by natural selection has built complexity and sophistication. This is a universal quest because organisms are the most complex and sophisticated organizations of all. The words, organization and organism, have an obvious common root. This surely indicates that the deep fascination of organisms (of all kinds, e.g. individualistic organisms or superorganisms) is associated with their organization. Indeed, the *Concise Oxford Dictionary* [1] presents as a key clause in its definition of an organism the following—an "organized body with connected interdependent parts". The coherent behaviour of organizations and organisms is a property of how their interdependent parts interact, how the parts communicate, and more specifically, how reliable information (from both signals and cues) flows among the parts of the whole. Thus the topic here is a fundamental one.

The goal of the contributions to this section (and indeed all the others in this book) is to understand something of the social physiology of insect societies. It may seem odd that many of us are so fascinated with a certain kind of physiology when physiology in biology as a whole is arguably out of fashion—uncomfortably squeezed by evolutionary biology at one end and molecular biology at the other. One manifestation of this pressure is the insistence by some that those who examine mechanisms are merely concerned with proximate "how" questions rather than the really important, evolutionarily ultimate, "why" questions. The folly of such reasoning can be highlighted by asking the question, what is the ultimate reason for the ecological success of social insects? Is it kin selection or is it a property of the diversity and efficiency of their patterns of organization? Kin selection explains wonderfully why organisms have evolved to work together as superorganisms [2]. Studies of social physiology, and especially of information

C. Detrain et al. (eds) Information Processing in Social Insects
© 1999, Birkhäuser Verlag Basel/Switzerland

flow and collective decision making, have the potential to explain both how social insects work together and why working together leads to enhanced efficiencies and novel ways of life and hence ultimately to the ecological success of social insects [3, 4]. Thus social physiologists are asking ultimate questions too.

One aspect of the ecological success of social insects is their biodiversity. Only a tiny fraction of social insect species has been studied in any detail. Given that each species yields ever more secrets the more it is studied, should we hope to discover deep insights through meticulous empirical studies of single societies or through broad-brushstroke theoretical studies? I believe that once again this is a false dichotomy—we need both approaches. The benefits of both are well illustrated by the contributions to this section.

Taken together, the contributions to this section point to at least one major insight. It is this: In small colonies the tendency is for workers to be independent decisionmakers, whereas in large colonies decision making by individual workers seems to be based much more on disseminated and carefully sampled information. Individuals in large social insect colonies, in certain situations, make a considerable investment of time and effort in sampling information before they decide what to do next. Typically they either sample many separate sources of information, or they tap into socially pooled information that effectively has already been collated. Both methods provide individuals, in large colonies, with relatively timely and accurate information. Workers do an appreciable amount of information sampling themselves when, for example, they directly or indirectly interact with many other nestmates (see Gordon, Cassil and Tschinkel, Stickland et al., all in this volume). Workers tap into information in the social domain when, for example, they asses queuing delays as they wait to offload resources to nestmates (see Jeanne, this volume; Anderson and Ratnieks, this volume).

Thus, certain decision-making processes in large social insect colonies involve individuals changing or tuning their behaviour in response to the accumulation of many small and separate influences rather than to all-or-nothing signals. In small colonies, by contrast, it appears that information flow among individuals, for example, from queuing delays, might be potentially so misleading, purely for stochastic reasons, that workers tend to select certain tasks independently. These insights may be generic given the breadth and diversity of examples, in this section, from which they are drawn. These are theoretical studies of queuing systems in honeybees and other social insects (Anderson and Ratnieks, this volume) models that explore rather abstract foraging algorithms (Stickland et al., this volume) and highly detailed empirical case studies, that is, building behaviour in wasps (Jeanne, this volume), larval feeding in fire ants (Cassil and Tschinkel, this volume), and task allocation in harvester ants (Gordon, this volume).

The contributions to this section are all concerned, in diverse ways, with group size and information flow inside the colony. One of the first questions this issue raises is whether social insect colonies have tightly determined species-specific sizes. Some do. For example, a number of social insect colonies reproduce by fission, often in a fair approximation to binary fission (e.g. honey bees and army ants)—in these species the size of the biggest colonies may only be two times the size of the smallest. Even in species in which small colonies bud off from much larger ones, colony sizes are likely to vary over less than one order of magnitude in their lifetimes. (See [5] for a discussion of the useful distinction between colony propagation through budding or fission.) By contrast, the workforces of colonies that are founded independently by solitary queens may grow from a population of one or two (reared by the queen) to hundreds of thousands (reared by one another). In such potentially large, independently founded, colonies are there really characteristic mature colony sizes, or are colony sizes highly indeterministic? Certain plants and fungi (see for example [6]) are characterized by indeterminate growth (they can respond with great plasticity to their local ecological conditions). Is indeterminate growth a feature of many, if not the majority, of social insect colonies? I believe such considerations are likely to be fundamental, because patterns of information flow are likely to change markedly as a colony grows unless colonies have evolved countermeasures (see [7]). Nevertheless, as a rule of thumb we can begin to think in terms of large and small colonies (say with a maximum of more than 1000 workers or with a maximum of about 100 workers or less, respectively).

Even such a crude classification can prompt some intriguing questions.

(1) Are there fundamental differences in social physiology—in terms of their systems of communication and information flow—between colonies that are always large (i.e. those that propagate by fission) compared with those that begin small (i.e. those that start independently) and become several orders of magnitude larger?

(2) In colonies that begin small and become large, do the ways in which individual workers behave, and interact with one another, change qualitatively as their society grows? That is, do the rules of thumb employed by individuals

change as a colony grows from having a few workers to many hundreds, thousands, or hundreds of thousands? Note that if the behaviour of individual workers is the same in colonies that have massively different population sizes, the strategic collective decision making of such colonies is likely to be quite different (see [7] and discussion later in this essay).

(3) Is it possible, in certain species in which colonies grow from very small to very large, that nest architecture is modular both in terms of infrastructure and social structure? For example, nest chambers (and the populations they contain) may be roughly the same size in small and large colonies, with variation only in the numbers of such chambers, so that dense interactions occur among groups, of workers and brood, of a similar size in large and small colonies.

(4) Do patterns of information flow change predictably with changes in the scale of societies? There are scaling laws in, for example, vertebrate physiology [8]. Are there scaling laws in social physiology? For example, is there a relatively smooth trend for decision making to be based at the individual level in small colonies and for it to be based at the collective level in large colonies? As a general trend, neuronal complexity may therefore be less in workers from large colonies than in workers from small colonies (see for example [9]).

Some of these questions are new, and answers may be a long way off; but for others we are beginning to see some progress. The contributions to this section, for example, include some mathematical models and some pioneering quantitative empirical studies that are beginning to get to grips with general issues of information flow in insect societies. Two books that have guided us to this important threshold are E.O. Wilson's *The Insect Societies* [10] and the older sibling of this current volume *From Individual to Collective Behaviour in Social Insects* [11].

Two of the chapters in this section consider task partitioning and queuing delays (Jeanne, this volume; Anderson and Ratnieks, this volume). Task partitioning occurs when "two or more individuals contribute sequentially to a particular task or piece of work, such as the collection, retrieval, and storage of one load of forage" (Anderson and Ratnieks, this volume; see also [12, 13]). Queuing delays may occur at the handover points in such partitioned tasks.

Robert Jeanne's chapter provides a lucid account of his pioneering studies of neotropical wasps. He makes a useful distinction between two groups of social vespids (within the subfamily Polistinae) (i) those with independent founders. in which reproductive females found colonies alone or in small groups (i.e. independent of workers) and (ii) swarm founders, in which colonies are initiated by a swarm of workers and queens. The former have mature colonies of up to 100 individuals, whereas the swarm founders have mature colonies which range from 100 to 1 000 000. Such colonies seem to have both very different ecologies and patterns of organization. First, as Jeanne points out "although swarm founders are less specious than the independent founders… individuals of swarm-founding species outnumber… independent founders by 100 or 1000 to 1 at some neotropical sites." Second, the larger colonies exhibit task partitioning, whereas the smaller ones do not (or do so to a much smaller extent).

Jeanne and his colleagues have looked at productivity, as a function of colony size, in *Polybia occidentalis* which propagates through reproductive swarms. They examined not only naturally occurring swarms but also absconding swarms experimentally provoked by nest damage. In this way they were able to study development of newly founded colonies of between 24 and 1500 adults. One really important finding from this study is that large colonies are more productive per capita than small ones both in terms of the production of nest cells and brood.

Jeanne then provides experimental data to discriminate among four alternative hypotheses for this increase in productivity, especially with regard to foraging for pulp, (for nest building), or for prey. Workers in bigger colonies do not appear to carry bigger loads or to be more successful per trip. Instead, they seem to make more trips per day, not because they work at a higher tempo, but because they experience shorter queuing delays as they wait to unload to receivers on the nest. This suggests that such perpetually rather large (swarm-founded) colonies experience ergonomic economies of scale as they grow bigger.

By contrast, in the wasp genera in which colonies are independently founded, start small, and never get very large (such as in *Polistes* and *Mischocyttarus*), there is little task partitioning, probably because queuing delays would be likely to be rather long and to vary stochastically so that they are not a good source of information for task reallocation (Jeanne, this volume; Anderson and Ratnieks, this volume).

Anderson and Ratnieks (this volume) expand on one of Jeanne's themes of task partitioning and queuing by providing both a review of task partitioning and of the role of caches (i.e. storage dumps that permit task partitioning without queuing) across a number of social insects from different groups. They explore in detail the reliability and the costs of the use of information from queuing delays for the reallocation of workers to changing task demands.

Anderson and Ratnieks's simulation modelling lends support for the principle that queuing delays should rapidly diminish with increasing colony size. Their study also suggests that long queuing delays can be a reliable source of information that individuals should be able to use to determine if they are in the group (either donors or receivers) that is too large for the other partnering group to cope with quickly. However, Anderson and Ratnieks point out that short queuing delays are less reliable as a source of information for an individual to determine if it is in the minority group. In social insects, queuing is likely to be "serve in random order" rather than "first come, first served", so an individual might be in the larger group and still only experience a short delay. Recognition of such asymmetries leads to the prediction that, all else being equal, it is the individuals that experience long queuing delays that should be proactive in task reallocation or recruitment decisions.

One of Anderson and Ratnieks's other findings is that even more reliable information about colony conditions can be obtained by individuals if they queue repeatedly to make contact with different partners at a transfer area. What is somewhat surprising in such a scheme is not that more reliable information can be gleaned from such behaviour (after all, each individual's sample size will be bigger), but that it can come at a remarkably low cost in terms of increased queuing delays. The explanation for the relatively small increase in queuing delays is that changing partners takes relatively little time. The advantage of multiple transfer behaviour is that members of the minority group become available more frequently to members of the waiting majority group if partners change more often. This reduces the variance around average queuing times and increases each individual's sample size of queuing times. Therefore, the reliability of queuing-time information increases greatly. This may well explain why nectar-foraging honeybees exhibit multiple transfer behaviour [14].

Individual collation of information and its use in task allocation is also a major theme in Gordon's chapter. She provides an account of her studies, both empirical and theoretical, of patterns of encounter and interaction among workers. Her work confronts the issue of how workers may avoid being swamped by too many encounters with nestmates as their densities increase. Such investigations should intrigue every human worker who feels their productivity and that of the organization for whom they work is being diminished by the constant input of untimely and often unimportant information. How do social insects avoid the social gridlock of too much communication?

Gordon also summarizes her very recent work on how interaction patterns may affect decisions to change tasks in individual harvester ants. The study recorded the tasks of focal ants and the tasks of all the ants each focal ant met over a roughly 20 min period. The site of all of this activity, and intense study, was the midden pile—the colony's refuse heap. As Gordon explains "There was a statistically significant association between the number and type of contact a focal ant experienced and its subsequent tasks" (Gordon, this volume p. 51). If an ant that was not performing midden work encountered high numbers of midden workers, it was more likely to join them in this task. But it would appear that it was not just the last encounter, but multiple encounters over a longer interval, that determined whether an individual changed its behaviour. Thus individual harvester ant workers, determining what to do next, may base their decisions on multiple sampling of what others are doing.

Multiple information sampling therefore seems to be a feature of harvester ants (Gordon, this volume) and honey bees (Anderson and Ratnieks, this volume) and may also occur in wasps (if they collate information from sequential queuing delays following separate bouts of foraging) (Jeanne, this volume).

The picture that begins to emerge is of social insect workers, in large colonies, basing their decisions either on directly interrogating more than one source of information or on tapping into sources of information such as queuing delays that represent the outcome of processes that effectively collate many sources of information. Such processes that increase real or virtual sample sizes should help to minimize sampling errors and lead to finely tuned decisions based on relatively timely and accurate information.

An extraordinary method of social fine tuning also involving multiple transfers is described by Cassil and Tschinkel. They discuss their amazing discovery that in the fire ant (*Solenopsis invicta*) workers feed their larvae absolutely minute discrete meals. Fire ants seem to have taken the notion of multiple transfers almost to its absolute limit. The volume of liquid food ingested by each larva, irrespective of its size, in one bout of feeding by a worker is 1.5 nanolitres! Replete donor workers have a vastly greater volume of food solution in their crops, but they partition it out, literally piecemeal, among large numbers of individual larvae. Individual larvae are fed by large numbers of workers, and it takes hundreds of feedings of such micromeals to satiate just one larva. In sum, Cassil and Tschinkel (this volume, p. 69) show "that [the] workers' feeding response is a simple binary decision—feed or do not feed this larva" and that "the hundreds of tiny morsels fed to larvae by workers allows larvae to fine-tune the nutritional mix of their diet by soliciting at different rates from different nutrients carried, unmixed in worker crops."

One of the interesting aspects of the *S. invicta* pattern of food flow is that it is predominantly a chain-of-demand system. This appears to be in contrast at least to certain aspects of honeybee organization which might be described as chain-of-supply systems. One reason for this apparent difference is that fire ants and many other ants can only store food inside themselves (or their larvae), so there are intrinsic limits to what can usefully be brought into the nest and not wasted, whereas honey bees can use the combs in the hive to store huge quantities of honey and pollen.

In common with all the other papers in this section, Stickland et al. (this volume) consider the influence of positive and negative feedback in the regulation of social processes. They do so, however, by using a highly simplified and abstract model of ant foraging. One of the achievements of their model is to show the theoretical advantages of the deployment not just of the well-known attractive foraging signals of ants but of yet to be discovered repulsive foraging signals. This work shows how models can point to novel areas of investigation.

Stickland et al. (this volume) also show that foraging effectiveness can also be tuned by varying the persistence of foraging information (e.g. the volatility of trail pheromones). They further suggest that, under certain circumstances, foraging by ants could be much more efficient if foragers convey to one another their lack of success associated with different foraging pathways. As Stickland et al. discuss, such usefully repulsive signals may await discovery in ants. However, such repulsive foraging signals have been found in other eusocial species (see, e.g. [15]). Though their speculation about ants deploying repulsive signals might be controversial, their findings that many small influences can lead to fine-tuned responses is exactly in accord with the discoveries reported by the other contributors to this section.

The contribution by Stickland et al. (this volume, taken together with earlier studies by Stickland et al. [7, 16, 17]), shows that interesting trade-offs can exist in the design of the foraging-recruitment systems of social insects. In many different species, social insect foragers have methods of informing their nestmates about the location, and sometimes the quality, of the food supplies they have discovered. Such foragers can influence other would-be foragers. But how strong an influence should one forager have upon another? How long should such influences persist? Should foragers just deploy positive influences in response to their advantageous discoveries or might they also beneficially deploy negative influences if they have met with failure? The abstract modelling of Stickland et al. suggests that colonies might evolve patterns of individual forager behaviour that can lead a colony to exploit a food source according to a preferred balance of distance against quality. In addition, there can be a trade-off of accuracy against the speed of decision-making in these foraging systems. In other words, a colony's foraging effort might tend to become locked onto one food site as a consequence of positive feedback from accumulating attractive signals from individual foragers that have visited such a site. There may be an advantage in quickly locking onto any suitable site (perhaps to prevent other competing colonies from monopolizing it). Here colonies might make snap decisions based on strong positive feedback from individual foragers (each has a high influence on others). Alternatively, colonies might select the best among all potential foraging sites by achieving a gradual "consensus" among many foragers each of which only has a small influence on its nestmates (through, for example, the application of pheromone foraging trails). Here colonies might make well-informed decisions, but could do so only relatively slowly.

This modelling also strongly suggests that as ant colonies grow, their foraging strategy in terms of such trade-offs might change even though the behaviour of their individual foragers remains the same because the number of foragers and their rate of departure from the nest will change as a function of colony size. Here, the foraging strategy will change because the cumulative collective signals resulting from positive feedback among foragers, influencing one another, can grow more quickly when more foragers are deployed or they are deployed at a greater rate. Hence, Stickland et al. [7] make the novel prediction that the effective foraging strategies of colonies may change as they grow unless individuals have different patterns of behaviour in colonies of different size.

There is a growing interest, in general within behavioural ecology, in the so-called ecology of information use [18, 19]. Major issues are how individuals collate information, for example, on successive time intervals between prey-capture events as they deplete patches, and hence how they make good decisions about giving-up times. How do individuals integrate and average information over time and how do they discard or devalue outdated information? Do animals use memory windows—deleting the influence of older events as newer data are added (see, for example, [20] and for an up-to-date review [19])?

As I hope this essay has made clear, such issues of time management and data updating are central to our understanding of information use in many social insects. Indeed, I predict that even though social insects are currently woefully, and inexplicably, neglected in reviews of the ecology of information use (e.g. [19]) they will become recognized as one of the very best "bench tests" for the empirical evaluation of theoretical models in this field.

A further fascinating issue in the ecology of information use is the extent to which selfish individuals use so-called public information (see [21]) to assess the quality of foraging sites. Starlings, for example, may not only use their own sampling data (e.g. on average periods between prey-capture events) to assess foraging patch quality, but they may also increase their sample sizes of inter-prey-capture intervals by monitoring the feeding rates of neighbouring starlings. In selfish foragers, however, such public information may be corrupted by some individuals tending to watch others rather than putting in equal sampling effort themselves [19]. Once again, for studies of the use of (uncorrupted) public information—or what I think is better termed "information in the social domain"—social insect studies should lead the field. When social insect nestmates are engaging in activities in which full cooperation and honest signalling is expected, such as in foraging, their adaptive use of social domain information and especially pooled information should be unrivalled by any other animals (with the possible exception of ourselves).

In sum, all the contributions in this section (in very different ways and for a wide variety of social insects) demonstrate either theoretically or empirically the effectiveness of information pooling or multiple information sampling in large social insect colonies. There is likely to be, in general, a fundamental trade-off between quick decision making and well-informed decision making, and many social insects seem to have achieved a compromise by putting a reasonable effort into information sampling. Note, however, that in a capricious and rapidly changing world perfectly well informed decision making will be impossible, because decisions involving many agents will always be based on old news. For this reason, hesitation may only promote further hesitation, whereas reasonably well informed decisions may lead to better ones.

The information age clearly began not with ourselves but with social insects. The study of how they collate and filter information to make collective decisions is still in its infancy. The contributions to this sections nevertheless strongly suggest that deep insights are possible through the study of information flow in social insects.

Acknowledgements

I wish to thank Sarah Backen, Iain Couzin, Melissa Cox, Elizabeth Langridge, Eamonn Mallon, Michael Mogie, Ana Sendova-Franks, and Andrew Spencer for their highly constructive criticism of an earlier draft of this essay. NRF's recent research has been generously supported by the Leverhulme Trust and a NATO Collaborative Research Grant (880344). I also wish to thank the Fondation Les Treilles and Jacques Pasteels, Jean-Louis Deneubourg, and Claire Detrain for their support of an excellent workshop.

References

1 *The concise oxford dictionary of current english* (1982) Oxford University Press, Oxford

2 Bourke AFG (1997) Sociality and kin selection in insects. *In*: JR Krebs, NB Davies (eds): *Behavioural ecology: an evolutionary approach.* Blackwells, Oxford, 203–227

3 Seeley TD (1995) *The wisdom of the hive: the social physiology of honey bee colonies.* Harvard University Press, Cambridge, MA

4 Bourke AFG, Franks NR (1995) *Social evolution in ants. monographs in behavioral ecology.* Princeton University Press, Princeton

5 Franks NR, Hölldobler B (1987) Sexual competition during colony reproduction in army ants. *Biol J Linn Soc* 30: 229–243

6 Rayner ADM, Franks NR (1987) Evolutionary and ecological parallels between ants and fungi. *Trends Ecol Evol* 2(5): 127–132

7 Stickland T, Tofts C, Franks NR (1993) Algorithms for ant foraging. *Naturwissenschaften* 80: 427–430

8 Schmidt-Nielsen K (1984) *Scaling: why is animal size so important?* Cambridge University Press, Cambridge

9 Franks NR (1989) Army ants: a collective intelligence. *Amer Sci* 77(2): 138–145

10 Wilson EO (1971) *The insect societies.* Harvard University Press, Cambridge, MA

11 Pasteels JM, Deneubourg J-L (1987) *From individual to collective behaviour in social insects.* Birkhäuser, Basel

12 Jeanne RL (1986) The evolution of the organization of work in social insects. *Monitore Zoologico Italiano (N.S.)* 20: 119–133

13 Jeanne RL (1991) Polyethism. *In*: KG Ross, RW Matthews (eds): *The biology of*

social wasps. Cornell University Press, Ithaca, 389–425

14 Kirchner WH, Lindauer M (1994) The causes of the tremble dance of the honeybee, *Apis mellifera. Behav Ecol Sociobiol* 35: 303–308

15 Goulson D, Hawson SA, Stout JC (1998) Foraging bumblebees avoid flowers already visited by conspecifics or by other bumblebee species. *Anim Behav* 55: 199–206

16 Stickland T, Tofts C, Franks NR (1992) A path choice algorithm for ants. *Naturwissenschaften* 79: 567–572

17 Stickland T, Britton NF, Franks NR (1995) Complex trails and simple algorithms in ant foraging. *Proceedings of the Royal Society. London* (B) 260: 53–58

18 Real LA (1994) *Behavioural mechanisms in evolutionary ecology*. Chicago University Press, Chicago

19 Giraldeau L-A (1997) The ecology of information use. *In*: JR Krebs, NB Davies (eds): *Behavioural ecology: an evolutionary approach*. Blackwells, Oxford, 42–68

20 Cowie RJ, Krebs JR (1979) Optimal foraging in patchy environments. *In*: RM Anderson, RD Turner, LR Taylor (eds): *British ecological symposium on population dynamics*. Blackwells, Oxford 183–205

21 Valone TJ (1989) Group foraging, public information and patch estimation. *Oikos* 56: 357–363

Part 2 Role and control of behavioral thresholds

Response thresholds and division of labor in insect colonies

Samuel N. Beshers, Gene E. Robinson and Jay E. Mittenthal

Summary

Division of labor in social insects can be explained at least in part by variation among workers in response thresholds for task-specific stimuli. In this
chapter we review the origins of the response threshold concept in physiology and ethology, and show how it can be used to explain division of labor.
Temporal polyethism and physical caste polyethism, as well as hormonal,
genetic, and learning effects on task performance, may be interpreted as the
results of changes in response thresholds. We then provide a more speculative
discussion of the implications of the response threshold concept for understanding the behavioral programs of workers and the organization of colonies.
We argue that the response threshold concept is a logical starting point for
developing explanations for the proximate basis of division of labor, and that
further development of this concept may be a necessary and critical step in the
study of social insect behavior. In addition, more experimental work is needed to demonstrate variation in response thresholds among workers, and to
show that this variation causes division of labor.

Introduction

Social insects are remarkable for the stereotyped but dynamic behavior of their
colonies. The emergent behavior patterns of colonies are believed to result from
the combined effects of independent actions by the workers that comprise a
colony [1], apparently with little or no centralized control. Workers are colony
members that engage in little, if any, personal reproduction and instead perform
all of the activities required for colony development, growth, and reproduction.

C. Detrain et al. (eds) Information Processing in Social Insects

Division of labor, in which workers specialize on different sets of tasks, is a prominent feature of colony level behavior. Task preferences of workers are typically associated with worker age, or, in the polymorphic ants and in termites, with worker size and morphology. Division of labor is characterized by both specialization and behavioral flexibility; workers devote most of their efforts to one or a restricted set of tasks, but they may switch tasks in response to changes in social context or colony needs [2–4].

The concept of a response threshold has been used to formulate behavioral rules that can account for specialization and flexibility in division of labor [5–12]. The basic idea is that when a stimulus exceeds a worker's threshold it elicits the response of performing the task associated with that stimulus. Variation in response thresholds among workers represents a simple underlying mechanism that can begin to explain patterns of variation in task performance among workers.

In this chapter we first review the use of response thresholds for explaining variation in task performance among social insect workers. We then discuss ways of extending the use of the threshold concept by exploring some of its consequences for understanding the behavioral organization of colonies. In particular, we address how workers vary in their responsiveness, how variation in responsiveness contributes to colony efficiency, and the mechanisms by which colonies can regulate their behavior by modulating worker response thresholds.

The response threshold concept and its application to division of labor in social insects

Roots of the response threshold concept in ethology

The idea of a "stimulus-response" event characterizes the interaction between an organism and its environment, and is a basic principle in physiology and psychology. Much of the history of comparative psychology and motivational psychology concerns theories of behavior as elaborations of stimulus and response, such as conditioning of responses [13] or building complex behavior from chains of stimulus-response events. The idea of a threshold, or minimum stimulus to which an animal will respond, arises from the description of stimuli and their effects.

The response threshold concept played a fundamental role in models developed by the pioneering ethologists to explain instinctive behavior [14, 15], in particular how it is that an animal performs a behavior at the appropriate time and place. The "innate releasing mechanism" was a hypothetical mechanism in the central nervous system that "released" a behavior when it received a suprathreshold stimulus, the "sign stimulus". Appropriate behavior for each set of circumstances encountered by an animal was ensured by responses to the correct stimuli.

A more elaborate model of instinctive behavior based on a hierarchical arrangement of innate releasing mechanisms was proposed by Tinbergen [15]. Each branch of the hierarchy represented an innate releasing mechanism and an associated response threshold that would lead the animal into a different behavioral state. By descending the levels of the hierarchy, an animal would approach specific behavioral "goals" such as eating or reproduction. An animal in a given state would be most responsive to stimuli associated with behaviors in that state, and less responsive to stimuli for other behaviors.

An example of the workings of the response threshold concept in the rules that guide behavior is seen in an analysis of feeding behavior in the sea slug *Pleurobranchaea californica* by R. Gillette and colleagues (Fig. 1) (unpublished observations). The animal responds to a chemosensory stimulus, taurine, with either feeding behavior or avoidance. Individual animals show consistent responses to the same stimulus level; when the feeding threshold is reached, avoidance is suppressed during active biting. The response threshold for feeding behavior is low when the animal is hungry and high when it is satiated. However, the response threshold for avoidance remains constant at a relatively low level; thus, weak stimuli elicit feeding in hungry animals and avoidance in satiated ones. This illustrates one way in which multiple thresholds may interact to produce behavioral rules.

Social insect biologists began to invoke the response threshold concept when they began to consider what behavioral rules guide the task choices of workers, and how these rules can account for division of labor and behavioral flexibility. The response threshold concept has helped to provide answers to these questions [1, 5–9], as well as testable hypotheses for experimental work [16, 17].

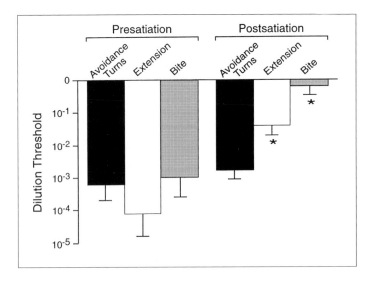

Figure 1 The response threshold to a feeding stimulus (taurine) in the sea slug Pleurobranchaea californica
A hungry animal has a low response threshold and will show feeding behaviors (extension and bite) at a taurine concentration of 10^{-3}–10^{-4}. After feeding, the response threshold is much higher. The response threshold for avoidance behavior is unchanged by hunger or satiation, but avoidance is suppressed by feeding behavior. At a concentration of about 10^{-3}, a hungry animal feeds but a satiated animal avoids food.

Response thresholds and variation in social insect behavior

Response threshold models of division of labor are based on the premise that the determinants of worker task performance are (i) the rules governing the behavior of a worker, which together comprise the worker's "behavioral program" [e.g. 18], and (ii) the labor needs of a colony. This means that the tasks performed by a worker at any point in time are not a fixed behavioral characteristic but are strongly influenced by the worker's environment [8, 9].

According to these models, the behavioral program of a worker includes a response threshold for every task in the repertoire. Each worker follows the rule of performing a task when it encounters a stimulus for that task that exceeds its response threshold. It is assumed that the "default state" of a worker—when it is not stimulated—is to remain quiescent and not to attempt to perform any task. The magnitudes of available stimuli, which are related to colony labor needs, affect individual responses. Although it may be possible for a worker to perform the entire repertoire, the tasks it actually performs depend on its thresholds in relation to the level of need for each task and to the thresholds of other workers in the colony, as will be discussed later.

Where is the response threshold located? We think of it as residing somewhere in the worker's central nervous system. There is evidence to support this view: honey bee workers become more responsive to alarm pheromone as they age, but this effect is not mediated by changes in antennal chemoreceptors [16, 19]. A threshold could exist either in a discrete location or as a distributed property of neuron ensembles. In principle, peripheral sensory systems or other physiological systems could affect response thresholds. Fixed differences in perceptual ability could contribute to response thresholds; some workers might not perform a task because they could not detect the stimulus. But then the workers would not show behavioral flexibility, which is a prominent feature of division of labor in social insects, and, as discussed below, is one of the primary reasons for invoking response thresholds to explain division of labor.

Actual measurements of response thresholds related to task performance are few. Responsiveness of honey bee workers to alarm pheromone increases with age [16]. Honey bee foragers perform waggle dances after collecting nectar above a threshold concentration of sugar, and thresholds vary dramatically among individual foragers. The foragers' thresholds are also modulated by the amount of nectar being brought into a hive, with thresholds being higher when more nectar is collected [20]. Recently, a correlation has been found between the threshold concentration of sucrose solution for extending the proboscis in honey bees and forager preferences for water, nectar, or pollen [21], but it is not yet clear how the sucrose stimulus is related to the tasks of foraging for water or pollen.

Variation in response thresholds can explain, at least in part, most of the important features of individual worker behavior in the context of division of labor: variation in task specialization, activity level, elitism, idiosyncrasy, and behavioral flexibility [10, 11]. Workers with low thresholds (relative to those of other workers) are most likely to perform tasks and may have high levels of *activity* for task performance; workers with uniformly high thresholds will be *inactive* in performing tasks. A worker with low thresholds for a small set of tasks shows *specialization*. *Elitism* is an unusually high frequency of task performance by a worker, either as a specialist or a generalist [2, 22, 23], and can be explained by very low thresholds for the tasks performed. *Idiosyncrasy* is variation in task performance that is not explained by age or size polyethism [22], and may be due to individual deviation from the response threshold norms for a particular age or size.

Behavioral *flexibility* at the level of the individual worker can be explained by higher stimulus levels that stimulate a worker to do new tasks. Higher stimulus levels can result from an increase in the need for a particular task, or from a loss of workers that were performing that task. Stimuli are assumed to increase in magnitude if tasks are not done [9, 10]. This type of flexibility, in which worker thresholds do not change, is believed to occur in response to short-term changes in the colony environment [8, 9]. Longer-term changes, such as those that result from drastic shifts in colony age demography, may induce long-lasting shifts in a worker's thresholds and task preferences. This can happen by modulation of a worker's trajectory of behavioral development [8, 10, 16].

Sources of variation in response thresholds

Intrinsic factors that are hypothesized to influence variation in response thresholds include worker age, size, genotype, early experience, and modulation of thresholds through positive or negative reinforcement following the performance of a task. Extrinsic factors such as interactions with other workers or changes in the nest environment may also cause variation in thresholds.

Age effects

Temporal polyethism can be explained by changes in response thresholds as workers mature [5, 7–9]. Adjustments in the patterns of age-related change in worker responsiveness may lead to the evolution of different systems of temporal polyethism [5].

Extensive evidence supports a close relationship between temporal polyethism and worker physiology in honey bees, as reflected by the juvenile hormone (JH) system (reviewed in [3, 24]; see Huang and Robinson, this volume). Young bees performing nursing tasks have low JH titers and older workers doing tasks such as foraging have high JH levels, but JH titers are correlated with a worker's behavioral state rather than its age [25]. Removal of the corpora allata, which synthesize JH, delays the transition to foraging, and this delay is eliminated by treatment with the JH analog methoprene [26]. These results support the interpretation of temporal polyethism as a process of behavioral development marked by transitions between at least two distinct behavioral and physiological states [27], and suggest that JH influences the rate of behavioral development of honey bee workers [24]. It is reasonable to think that changes in thresholds accompany these physiological changes. Consistent with this assumption are the demonstrations in honey bees of age-related changes in responsiveness to alarm pheromone and to mechanostimuli that elicit stinging behavior [16, 28, 29], and the finding that methoprene treatment causes premature sensitivity to alarm pheromone [16].

There are many known cases of behavioral flexibility in temporal polyethism (reviewed in [3, 9, 30]). Flexibility may take the form of "filling in" for missing workers, doing new tasks to meet a short-term increase in needs, or changes in the trajectory of behavioral development. Behavioral development can be delayed, accelerated, or reversed ([8, 9]; reviewed in [3, 31]). Flexibility and apparent behavioral development could in principle be explained by the combined effects of changing task needs and *fixed* thresholds (e.g. [32, 33]), or by changes in response thresholds induced by changes in task needs [8].

Size effects

Physical castes in ants often show striking differences in behavior, yet they may also show surprising behavioral flexibility [6, 34, 35]. In laboratory nests, major workers in the ant genus *Pheidole* usually have far smaller task repertoires than minor workers, and never care for brood. Removal of a large fraction of the minor workers from a colony causes the majors to begin performing brood care and other tasks needed to maintain the colony. A consistent pattern was found by Wilson in three *Pheidole* species: when the ratio of major to minor workers in a colony exceeded 1:1, major workers became active and cared for brood, and at higher ratios their frequency of task performance was greatly increased [34]. Simulations of a response threshold model with two castes yielded a pattern of response by major workers to changes in the major/minor ratio that was qualitatively similar to Wilson's empirical results [36].

The organization of prey retrieval in *Pheidole pallidula* appears to be based in part on differences in response thresholds to recruitment displays between minor and major workers. Only minor workers are recruited to patches of small prey that can be carried individually, but more vigorous displays are used to recruit major workers to large prey that must be dissected before being retrieved [12]. Major workers similarly appear to have higher thresholds than minor workers for responding to recruitment for colony defense [37].

Genetic effects

Robinson and Page [10, 11, 38] invoked response thresholds to explain how genetic variation among workers influences division of labor. In honey bees, differences among genotypic subfamilies in a colony have been found for numerous tasks including guarding, colony defense, corpse removal, and foraging (reviewed in [11]). For example, in colonies containing three different subfamilies, the subfamily composition of guards was significantly different from that of control workers in 9 out of 10 trials, involving five unrelated colonies [38]. Such differences can be explained by subfamily differences in response thresholds; the workers that do a task most often should belong to the subfamily with the lowest thresholds for that particular task.

This hypothesis predicts that when the need for a task is increased, more workers will respond. The new workers may belong to different subfamilies than those that were already doing the task, because the increased stimulus is likely to overlap the threshold distributions of other subfamilies. The result is that genotypic diversity among the workers doing the task is increased [10]. Support for this prediction was obtained by Fewell and Page [39] in a study of pollen foraging in honey bees. Genotypic diversity among pollen foragers increased when either the need for pollen was increased or the quality of available pollen was improved. However, Robinson and Page [17] found a lack of plasticity for corpse removal (undertaking) in honey bees. When they removed undertakers from colonies containing three subfamilies of workers, the undertakers were replaced to only a limited extent, and the colony rates of corpse removal were significantly lowered. The observed lack of plasticity suggests that either most workers have high thresholds for corpse removal, or potential undertakers were occupied by higher-priority tasks [17].

Differences in response thresholds may explain genotypic differences in defensive behavior in honey bees, including the stronger defensive responses of colonies of African honey bees, as compared with European bees [40].

Early experience effects

Many experiments have shown that experience soon after eclosion can affect a worker's later behavior and its integration into the colony [22, 41]. For example, a worker denied a certain experience, such as contact with larvae, may be less likely than other workers to care for brood later in life, or may even be unable to care for brood. Workers may have "sensitive periods" for acquiring certain experiences or abilities, which may influence the development of response thresholds. Such effects may actually facilitate the development of colony division of labor, by shunting workers away from those tasks with which they have no experience. The lack of experience would be caused by the presence of sufficient workers to do tasks during a worker's ontogeny.

Self-reinforcement effects

Deneubourg et al. [42] simulated foraging behavior in ants using the assumption that successful foraging increased the probability of a worker foraging again. This assumption represents a kind of internal positive reinforcement for successfully performing a task. The model of Deneubourg et al. showed that this mechanism could give rise to self-organized division of labor with respect to being a forager or not, and could also account for spatial fidelity of foragers. Self-reinforcement effects acting on response thresholds can result in the differentiation of the workers in a colony into specialists and generalists [43].

Similar models were used to explore the basis of elitism in bumblebees [23] and the genesis of division of labor and variation in response thresholds in the wasp *Polistes dominulus* [44].

Implications of postulating response thresholds

Postulating thresholds as an explanation for division of labor has important consequences for our understanding of the behavioral programs of workers, the behavior patterns of workers, and the behavioral organization and adaptive design of colonies. In this section we offer a speculative discussion of some of these consequences, to illustrate the potential of thresholds for illuminating our understanding and to suggest some possible directions for further research. We begin with the assumption that thresholds are fixed, and then later discuss how thresholds can be modulated.

Behavioral organization of individuals: hierarchies and priorities

How are response thresholds arranged in the behavioral program of a worker? Are they somehow ordered or clustered, or is each threshold independent of the others? We will consider this question in the context of temporal castes.

One possibility is that responsiveness to sets of tasks is regulated by a hierarchy of control mechanisms in the behavioral program of a worker. In honey bees, for example, there is a suite of behavioral and physiological changes associated with age-related behavioral transitions, particularly the transition from hive work to foraging (reviewed in [3]). Workers in one of these behavioral states may perform any of a range of tasks characteristic of the behavioral state, but may vary considerably in their selection of tasks and their level of task activity. This suggests that division of labor is regulated both at the level of clusters of tasks corresponding to different behavioral states and at the level of stimulus and response for different tasks within a behavioral state. Regulation at these two levels may correspond to the two types of behavioral flexibility: short-term behavioral flexibility without changes in thresholds would result from regulation at the level of stimulus and response, whereas long-term flexibility with changes in thresholds would result from regulation at the level of the behavioral state. With such a system, changing task needs could be accomodated by flexibility within a temporal caste, without compromising the long-term allocation of workers among temporal castes.

The hierarchy may even be extended one level further. "Performing a task" may itself be defined by a worker's internal state, rather than by its actions from one moment to the next. In one model of task allocation [45], a worker can be in a state corresponding to one of four tasks, and interactions among workers determine which task state a worker "chooses" to enter. Specific stimuli for performing tasks, and thresholds for responding to them, do not appear in the model. One way of interpreting such a model is that a worker could be considered to be "performing a task" for as long it was in the appropriate task state, whether or not it was performing a specific act associated with that task. A worker in the "foraging" state would be "foraging" whether or not it was retrieving a food item. This suggests that division of labor could be regulated by the higher-level events that cause a worker to enter a particular task state, rather than the lower-level stimuli that affect its behavior while it is in that state.

Robinson and Page [17] used the idea of hierarchy in a different sense to explain an observed lack of colony plasticity for rare tasks such as corpse removal. Although many workers might have the ability to remove a corpse from the hive, they might never do so if they were continuously occupied with tasks for which they had lower thresholds. This raises the question of how the thresholds for each worker are ordered: to which stimuli will a worker be most responsive? To distinguish this question from that of hierarchical organization, we call the ordering of a worker's thresholds its "priorities". The variation in task priorities among workers in a colony will have a strong effect on colony responses to task needs.

Behavioral organization of colonies and the colony threshold distribution

Another important issue is the question of how thresholds vary among individuals within a colony. If thresholds explain behavioral variation in division of labor, then what distributions of thresholds are present in colonies, and how do they give rise to the observed patterns of behavior? We believe that it is important to consider the thresholds for all tasks, for every worker in a colony (Fig. 2). This colony threshold distribution (CTD) should affect colony-level patterns of behavioral specialization, task allocation, responsiveness, and flexibility.

Little is known in general about patterns of specialization. For most tasks, there are specialists that perform the task frequently and nonspecialists that perform the task much less often [e.g. 46]. Specialist workers are those that perform one or only a few tasks, and are hypothesized to have low thresholds for the tasks they perform. The CTD should affect whether workers specialize on more than one task at a time, and what combinations of tasks are likely to be found in the same worker's repertoire.

Since workers will usually perform tasks for which they have low thresholds, the CTD also is related to the allocation of workers to different tasks. Variation in worker activity should also be determined largely by thresholds; the propor-

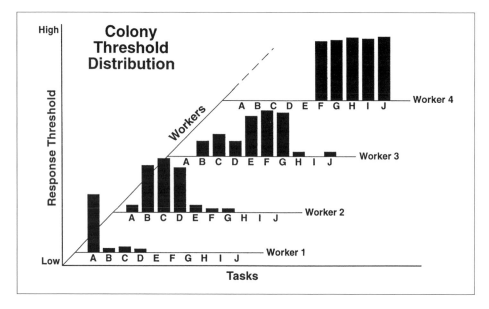

Figure 2 The colony threshold distribution (CTD) affects how the colony responds to varying levels of task-related stimuli

Each worker has a response threshold for every task, and the CTD includes the profile of thresholds for every worker. Threshold profiles for four workers are shown. Low thresholds correspond to high responsiveness to stimuli, and high thresholds to low responsiveness.

tion of active workers to inactive "reserves" will be determined by how many workers have generally high thresholds and so perform any tasks only when needs are greater than normal.

Workers with low thresholds should respond quickly to task stimuli, so the CTD will affect the latency of the colony response. The number of workers that respond to a task is high when the mean of the threshold distribution for that task is low [47, 48].

Colony flexibility, or *resiliency* [49], is a change in the allocation of workers to tasks in response to changing needs [10]. Colony flexibility is high if many workers have low thresholds for most tasks. Flexibility has been shown to be high for some tasks [39, 50] and low for others [17]; it may be that flexibility is generally lower for "rare" tasks such as corpse removal [11].

The relationship between the CTD and colony behavior is complicated by the effects of living socially. If one worker has a lower threshold for a task than another worker, it not only is likely to respond sooner to the task, but it reduces the stimuli for the task so that they may never reach the second worker's threshold. Thus workers with low thresholds may show *behavioral dominance* over workers with higher thresholds; by exerting negative feedback on task stimuli, they decrease the likelihood of task performance by other workers [10]. By their labor, workers reduce stimuli, but this effect is opposed by increases in stimuli as new task needs arise. Simulations suggest that an equilibrium will be reached at which the number of workers performing a task just matches the stimulus level for the task, and this equilibrium is affected by the mean and variance of the CTD [47, 48]. Thus the stimulus environment experienced by workers in a colony is to a large extent determined by their own actions.

In consequence, a worker's behavior may not directly reflect its response thresholds. The tasks performed by a worker depend in a complex way on the CTD and levels of needs for tasks. This in turn means that the CTD cannot be directly inferred from patterns of worker behavior. Progress in understanding how the CTD and colony behavior are related will therefore depend on direct measurements of response thresholds in individual workers.

The CTD is the basis for predicting colony behavior, but it may not be sufficient. Workers with extreme values of thresholds will behave consistently over a wide range of stimulus levels, and so their behavior is predictable. The behavior of workers with intermediate thresholds will be more strongly affected by the stimulus environment and thus harder to predict from the CTD alone; knowledge of task needs and of the dynamics of colony responses may also be required. The significance of the CTD may be that it constrains colony responses but does not completely determine them.

Response thresholds and colony efficiency

Division of labor in social insects is generally believed to increase ergonomic efficiency of individual workers and of colonies [2], but the concept of efficiency has not been well developed. Efficiency can be judged on the basis of specific performance measures, and for most systems these performance measures include economy, flexibility, reliability, and speed [51].

We suggest that variation in thresholds among workers affects performance at the level of the colony. At the individual level, a threshold affects whether or not a task is performed, but not the skill or speed of the performance. At the colony level, however, the threshold distribution does affect performance through its influence on the number of workers that respond to a task.

Because theory and experiment on this subject are lacking, we offer some tentative generalizations about the relationship between the CTD and measures of colony performance. Economy, flexibility, and speed should be directly affected by the CTD, whereas the implications of the colony threshold distribution for reliability are less clear.

Economy

Economy consists of maximizing the results per unit of available resources. One way a colony can achieve economy is to match closely the number of workers allocated to a task with the need for that task. This can be achieved with a mixed distribution of thresholds: workers with low thresholds to respond to slight needs, and workers with a range of higher thresholds to be recruited when the task need is greater [10].

Flexibility

Flexibility is the ability of a colony to reallocate its workers in response to changing needs. Flexibility depends on workers being both responsive to a task and available. Workers with low thresholds for all tasks would be responsive but also likely to be already performing tasks. This suggests that low thresholds may *permit* flexibility, but task switching depends on other factors such as interactions between workers. As argued in the previous section, the CTD constrains flexibility but may not be able to account for its occurrence, or may not explain particular dynamic patterns of flexibility. High response thresholds may be enough to prevent flexible responses, but low thresholds may not be enough to cause flexible responses.

Social insect colonies face trade-offs between specialization and flexibility. Complete flexibility (all workers can respond to all tasks) can be achieved with a uniformly low distribution of thresholds. However, too much flexibility can lead to lack of colony coherence and levels of response that are unrelated to needs. An alternative is "constrained plasticity" [3, 11], in which a colony reallocates workers as needs change, but only within limits imposed by the threshold distribution. Constrained plasticity represents a compromise between the advantages of flexibility and the benefits of division of labor.

Speed

Workers with low thresholds will respond more quickly to task-related stimuli, so having more workers with low thresholds will cause a faster colony response. Fast responses are generally desirable, but may not always be (see for example the Section on *Summation of inputs*).

Reliability

Colony reliability depends on the competence of individual workers and on the number of workers that respond to a task need [2]. Worker competence is not affected by thresholds. If worker competence is high, then increasing the number of workers with low thresholds should improve colony reliability.

Modulation of thresholds: insights from neurobiology

As discussed earlier, modulation of response thresholds can explain several properties of worker behavior, including temporal polyethism, specialization, and elitism. Modulation of response thresholds in response to changing colony needs may be essential to colony fitness. Without some form of regulation, the CTD would be disrupted by several factors, including fixed variation (e.g. genetic effects) among workers, the constant turnover of workers, changes in thresholds during ontogeny, and variation in experience among workers.

Analogies with neurons may lead to insights into the mechanisms and significance of modulation of response thresholds in social insects. Similarities between neurons and social insect workers have been noted by a number of authors [45, 52–54]. Neurons have thresholds that determine when they will respond to inputs with action potentials, and modulation of the responsiveness of individual neurons by neuron-neuron interactions is important in the functioning of neural systems [55]. Here we review examples of interactions that may modulate response thresholds in social insects. We consider them in the context of what is known about interactions between neurons.

Modulation of response thresholds through inter-individual interactions

Neurons may excite or inhibit other neurons. An excitatory signal increases the likelihood that the neuron receiving the signal will fire; an inhibitory signal decreases this likelihood.

Excitation

Many examples of excitatory interactions among workers are known. Recruitment displays for foraging are known in ants, bees, wasps, and termites [1], and for nest emigration in ants and honey bees [20, 30]. The honey bee tremble dance is performed by nectar foragers and serves to recruit additional receiver bees [56]. The honey bee "shaking" signal, also known as the "vibration signal" or "dorso-ventral abdominal vibration" (DVAV) is given in several contexts and appears to cause a nonspecific increase in the activity level of the recipient [57–62].

In at least one case excitatory interactions between queens and workers may have modulatory effects. Queens of the paper wasp *Polistes fuscatus* preferentially direct aggressive interactions at inactive workers, causing them to become active and perform tasks. This effect could be explained by modulation of worker thresholds, but could also be explained by workers encountering more task-related stimuli while moving around on the nest [63, 64]. It has also been reported that the presence of the queen can increase worker activity in some species [65, 66].

Excitatory signals vary in their specificity. Some, such as the honey bee waggle dance [67] or the odor trails of ants [30, 68] may transmit highly specific information. Others are examples of *modulatory communication* [52, 30], signals that "alter the probability of reactions to other stimuli by influencing the motivational state of the receiver" [30]. These may be general arousal signals, which derive specific meaning from the context in which they are performed and from the identities of the recipients. Some may have multiple functions; the honey bee tremble dance, for example, recruits additional nectar receiver bees but causes bees that are performing waggle dances to stop dancing and leave the hive to forage [56].

What these signals seem to have in common is that they cause a short-term adjustment of colony labor allocation by "pulling" additional workers to perform specific tasks [56, 59]. They may act as additional stimuli, to exceed the thresholds of workers that would not otherwise perform these tasks, or they may act to lower response thresholds to task-related stimuli.

Inhibition

The CTD may be created and maintained in part by social inhibition. Such inhibition has been demonstrated in honey bees; foragers slow down the pace of behavioral development in younger bees, delaying the age at onset of foraging ([31, 69]; see Huang and Robinson, this volume). These results suggest that workers inhibit one another's behavioral development, by a chemical or behavioral signal. The effect of this mutual inhibition is to create a balanced allocation of workers to tasks that is related to, but not directly dependent on, task needs. The allocation pattern is dynamically stable; a colony comprised of workers of almost any age distribution achieves this balanced allocation within a few days, and quickly restores it after perturbation by removal or addition of workers. This inhibition may act at the level of behavioral states to modulate response thresholds, resulting in a CTD that causes appropriate numbers of workers to be allocated to hive work and foraging.

Similar mechanisms may be found in other social insects. Using simulations, Theraulaz et al. [44] showed that inhibition through dominance interactions resulted in a stable CTD and division of labor resembling that in the wasp *P. dominulus*.

Major workers in species of the ant *Pheidole* are inhibited from participating in brood care in the presence of minor workers [34], due to avoidance of minor workers by majors [6] or to behavioral dominance [36]. This appears to be a short-term inhibition of task performance, rather than a modulation of the behavioral states of major workers, because major workers frequently approach the brood [6], and begin tending brood within 1 h after the removal of minor workers [34].

Summation of inputs

The output of a neuron is determined by the integration of excitatory and inhibitory inputs from other neurons. Similarly, a worker may integrate modulatory signals from many other workers. The effectiveness of recruitment displays for foraging, for example, may depend on the number of workers displaying. Workers with high thresholds for foraging could be aroused by simultaneous signals from several recruiters or by prolonged exposure to the displays of a series of recruiters.

Such integration of many signals has been proposed as a mechanism for the regulation of foraging in honey bees [61]. The shaking signal is used by foragers to arouse other foragers to follow dances or to leave the hive. The advantage of this mechanism may be that it acts to filter out "noise"—signals that convey inaccurate information. A few foragers could be successful even when little food is available near the hive, and their shaking signals could misrepresent the state of the environment to the colony. If many foragers are successful, however, food is likely to be plentiful. Since foragers have high thresholds, and respond to the shaking signal only after being shaken many times [61], they are most likely to respond when food actually is available.

Modulation of thresholds by self-reinforcement

Action potentials from a neuron may excite other neurons, which connect back and send excitatory signals to the first neuron. This positive feedback increases the likelihood that a neuron will generate further action potentials, causing an amplified or prolonged response in a neural system [70]. Positive feedback that lowers response thresholds in workers may occur by self-reinforcement ([23, 42, 43] see Section on *Early experience effects*). This mechanism could generate a stable, self-organized division of labor; however, empirical support for self-reinforcement is not yet available.

Patterns of interaction: pathways of information flow in colonies

Single, local events may combine to produce global, integrative effects at the level of the colony. The honey bee waggle dance is an excitatory signal; the combined effect of many waggle dances is a positive feedback effect that leads to rapid recruitment of foragers and efficient harvesting of a food source [20, 67]. In contrast, the decision of a forager to not waggle-dance is effectively a negative signal; the combined effect of negative and positive signals from dancing (or not dancing) allows a colony to maintain an appropriate level of foraging [20].

A major difference between colonies of social insects and nervous systems is that neurons have anatomically fixed locations, whereas workers are mobile and not tightly linked [20]. In theory interactions among all individuals in a colony are possible [53]. However, this is not likely. Not all encounters are equally probable, as workers tend to be found in certain areas within the nest and perform tasks there [71, 72]. The movement patterns of workers may differ as well as their locations. Some modulatory signals can be broadcast through a colony and affect nearly all of the workers, as appears to sometimes happen with the shaking signal in honey bees [73]. But in other cases modulatory signals could be directed at specific target workers, and so contribute to specific pathways of information flow.

Conclusions

Division of labor is a complex phenomenon that involves patterns and mechanisms at two levels of biological organization: the individual organism and the colony. One goal in developing models and hypotheses for understanding division of labor should be to find unifying explanations for phenomena at each of these two levels. Models based on response thresholds meet this requirement; they explain diverse patterns of behavior of individual workers and colonies as the products of a simple underlying mechanism.

The response threshold concept forms the core of a hypothesis about the behavioral programs of individual workers. For division of labor, the idea of the behavioral program of the individual worker is a key to solving "the central problem of insect sociology", the understanding of the mass behavior of colonies in terms of individual behavioral rules [1]. One virtue of response thresholds is that they have been used in models of the behavioral organization of nonsocial animals. Although the impetus for explaining division of labor by response thresholds comes primarily from demonstrations of colony level plasticity, a useful model of the behavioral program of an animal should explain both patterns of social behavior and patterns of task performance by individual workers. Another benefit is that consideration of response thresholds invites the integration of genetics, neurobiology, endocrinology, and behavior in the study of the mechanisms and evolution of social behavior. It is becoming increasingly clear that this synthesis is necessary to properly understand a behavioral system, especially one as complex as division of labor.

The response threshold concept leads to the concept of the colony threshold distribution, or CTD. Although for convenience we have discussed the CTD as though it were fixed, it is a dynamic pattern that varies according to unknown rules. The CTD at any point in time is a "snapshot" of a colony that summarizes the behavioral states of individual workers and the overall state of the colony. As the caste distribution function (CDF) summarizes the ergonomic potential of a colony [2], the CTD summarizes its behavioral potential.

The response threshold concept offers a phenomenological description of division of labor, but the hypothesis that division of labor in social insects is caused by differences in response thresholds among workers must be tested by actual measurements of thresholds, taken during relevant behavioral contexts. Obtaining such measurements may prove to be challenging. Quantitative information about thresholds, and how they vary among workers, is needed. Current models of division of labor based on response thresholds invoke both fixed [e.g. 10, 36, 47] and "fluid" response thresholds [e.g. 8, 42]. The need to reconcile these two conceptions reinforces the need for assays to measure thresholds directly, and for theory to assess how much fluctuation in individual thresholds is consistent with division of labor.

Response thresholds for task-specific stimuli are not the only possible mechanism for division of labor; models of division of labor caused by differences in the local environments experienced by workers [32], or by the patterns of interactions among workers [45], have been proposed. None of these models are mutually exclusive, and all may eventually contribute to our understanding of division of labor. All are based on plausible mechanisms, and it will be an empirical question which mechanisms are necessary parts of an explanation.

The response threshold concept is based on the idea of the animal as a stimulus-response machine. This is one of the simplest systems that can account for division of labor, and it is worth exploiting as a working hypothesis to guide research. We recognize that there are models based on very different conceptions of behavior, such as those that represent an animal as a hierarchy of control systems [74]; as general models of behavior these may have greater explanatory power than strict stimulus-response models. Even within the restricted context of division of labor, more complex models may be required, in which response thresholds are only part of the mechanism.

For example, the response threshold controls when an animal will start to do a task, but not when it stops. We have assumed that a worker stops doing a task because by doing the task it reduces the stimulus. However, a worker following this rule would constantly "switch on and off" for the task as the stimulus level varied, and for some tasks this would be very inefficient. This problem could be solved if a worker were to perform a task for a fixed length of time, or to follow some stochastic rule for the duration of task performance [36, 43]. Another way would be to have a second threshold for discontinuing performance of the task [75].

A response threshold may also be the simplest version of a cognitive template, or other internal representation, against which an animal compares its perceptions in "deciding" how to behave. A complex stimulus may require a pattern recognition mechanism. For example, a corpse may be recognized by many perceived features that identify it as an animal and as dead, or it may be recognized by a single chemical that, on exceeding a threshold concentration, releases the appropriate behavior pattern of carrying the corpse out of the nest [76, 77].

Despite these and other potential complications, we feel that the best approach is to begin with a simple model with basic assumptions that can be tested and validated, and then to add to the model only when necessary. Division of labor is complex; without starting from a simple model and scrutinizing every assumption and feature, we risk having models whose behavior we do not understand [78].

We have shown how postulating response thresholds leads to further insights into the behavioral programming of workers, the significance of division of labor for colony efficiency, and the mechanisms by which a colony integrates its workers into a single coherent entity. Response thresholds are characteristic of biological systems in general. Studies of response thresholds and how they are related to the behavior, physiology, and functional organization of colonies may lead to increased understanding of social insects and other biological organizations.

Acknowledgments

We thank G. Bloch, E.A. Capaldi, D.M. Gordon, M.M. Elekonich, Z.-Y. Huang, A. Lenoir, R.E. Page, Jr., S.S. Schneider, and D.J. Schulz for helpful discussion and comments, E. Bonabeau for manuscripts in press, and Rhanor Gillette for sharing his unpublished data. Supported in part by grants from USDA (to SNB) and NIH (to GER).

References

1 Wilson EO (1971) *The insect societies.* Belknap Press of Harvard University Press, Cambridge, MA

2 Oster GF, Wilson EO (1978) *Caste and ecology in the social insects.* Princeton University Press, Princeton

3 Robinson GE (1992) Regulation of division of labor in insect societies. *Annu Rev Entomol* 37: 637–665

4 Gordon DM (1989) Caste and change in social insects. *In*: PH Harvey, L Partridge (eds): *Oxford surveys in evolutionary biology,* vol 6. Oxford University Press, Oxford, 55–72

5 Wilson EO (1976) Behavioral discretization and the number of castes in an ant species. *Behav Ecol Sociobiol* 1: 63–81

6 Wilson EO (1985) Between-caste aversion as a basis for division of labor in the ant *Pheidole pubiventris* (Hymenoptera: Formicidae). *Behav Ecol Sociobiol* 17: 35–37

7 Wilson EO (1985) The sociogenesis of insect colonies. *Science* 28: 1489–1495

8 Robinson GE (1987) Regulation of honey bee age polyethism by juvenile hormone. *Behav Ecol Sociobiol* 20: 329–338

9 Calabi P (1988) Behavioral flexibility in Hymenoptera: a re-examination of the concept of caste. *In*: J Trager (ed): *Advances in myrmecology.* EJ Brill Press, Leiden, 237–258

10 Robinson GE, Page RE (1989) Genetic basis for division of labor in an insect society. *In*: MD Breed, RE Page (eds): *The genetics of social evolution.* Westview Press, Boulder, CO, 61–81

11 Page RE, Robinson GE (1991) The genetics of division of labour in honey bee colonies. *Adv Insect Physiol* 23: 117–169

12 Detrain C, Pasteels JM (1991) Caste differences in behavioral thresholds as a basis for polyethism during food recruitment in the ant *Pheidole pallidula* (Nyl.) (Hymenoptera: Myrmicinae). *J Insect Behav* 4: 157–176

13 Watson JB (1930) *Behaviorism.* University of Chicago Press, Chicago

14 Lorenz K (1950) The comparative method in studying innate behavior patterns. *Symp Soc Exp Biol* 4: 221–268

15 Tinbergen N (1952) *The study of instinct.* Oxford University Press, New York

16 Robinson GE (1987) Modulation of alarm pheromone perception in the honey bee: evidence for division of labor based on hormonally regulated response thresholds. *J Comp Physiol* 160: 613–619

17 Robinson GE, Page RE (1995) Genotypic constraints on plasticity for corpse removal in honey bee colonies. *Anim Behav* 49: 867–876

18 Downing HA, Jeanne RL (1988) Nest

construction by the paper wasp *Polistes*: a test of stigmergy theory. *Anim Behav* 36: 1729–1739

19 Allan SA, Slessor KN, Winston ML, King GGS (1987) The influence of age and task specialization on the production and perception of honey bee pheromones. *J Insect Phys* 33: 917–922

20 Seeley TD (1995) *The wisdom of the hive*. Harvard University Press, Cambridge, MA

21 Page RE, Erber J, Fondrk MK (1998) The effect of genotype on response thresholds to sucrose and foraging behavior of honey bees (*Apis mellifera* L.). *J Comp Physiol* 182: 489–500

22 Jaisson P, Fresneau D, Lachaud J-P (1988) Individual traits of social behavior in ants. *In*: RL Jeanne (ed): *Interindividual behavioral variability in social insects*. Westview Press, Boulder CO, 1–52

23 Plowright RC, Plowright CMS (1987) Elitism in social insects: a positive feedback model. *In*: RL Jeanne (ed): *Interindividual behavioral variability in social insects*. Westview Press, Boulder, CO, 419–431

24 Robinson GE, Vargo EL (1997) Juvenile hormone in adult eusocial Hymenoptera: gonadotropin and behavioral pacemaker. *Arch Insect Biochem Physiol* 35: 559–583

25 Robinson GE, Page RE, Strambi C, Strambi A (1989) Hormonal and genetic control of behavioral integration in honey bee colonies. *Science* 246: 109–112

26 Sullivan JP, Jassim O, Robinson GE, Fahrbach SE (1996) Foraging behavior and mushroom bodies in allatectomized honey bees. *Soc Neuro Abst* 22: 1144

27 Robinson GE, Huang Z-Y, Page RE (1994) Temporal polyethism in social insects is a developmental process. *Anim Behav* 48: 467–469

28 Collins AM (1980) Effect of age on the response to alarm pheromones by caged honeybees. *Ann Entomol Soc Amer* 73: 307–309

29 Kolmes SA, Fergusson-Kolmes L (1989) Stinging behavior and residual value of worker honey bees (*Apis mellifera*). *J N Y Entomol Soc* 97: 218–231

30 Hölldobler B, Wilson EO (1990) *The ants*. Belknap Press, Cambridge, MA

31 Huang Z-Y, Robinson GE (1996) Regulation of honey bee division of labor by colony age demography. *Behav Ecol Sociobiol 39:* 147–158

32 Tofts C, Franks NR (1992) Doing the right thing: ants, honeybees, and naked mole rats. *Trends Ecol Evol* 7: 346–349

33 Bonabeau E, Theraulaz G, Deneubourg J-L (1998) Fixed response thresholds and the regulation of division of labour in insect societies. *Bull Math Biol* 60: 753–807

34 Wilson EO (1984) The relation between caste ratios and division of labor in the ant genus *Pheidole* (Hymenoptera: Formicidae). *Behav Ecol Sociobiol* 16: 89–98

35 Wheeler DE, Nijhout HF (1984) Soldier determination in the ant *Pheidole bicarinata*: inhibition by adult soldiers. *J Insect Physiol* 30: 127–135

36 Bonabeau E, Theraulaz G, Deneubourg J-L (1996) Quantitative study of the fixed threshold model for the regulation of division of labor in insect societies. *Proc R Soc Lond B* 263: 1565–1569

37 Detrain C, Pasteels JM (1992) Caste polyethism and collective defense in the ant *Pheidole pallidula*: the outcome of quantitative differences in recruitment. *Behav Ecol Sociobiol* 29: 405–412

38 Robinson GE, Page RE (1988) Genetic determination of guarding and undertaking in honey bee colonies *Nature* 333: 356–358

39 Fewell JH, Page RE (1993) Genotypic variation in foraging responses to environmental stimuli by honey bees, *Apis mellifera*. *Experientia* 49: 1106–1112

40 Guzman-Novoa E, Page RE (1994) Genetic dominance and worker interactions affect

honeybee colony defense. *Behav Ecol* 5: 91–97

41 Lenoir A (1987) Factors determining polyethism in social insects. *In*: JM Pasteels, J-L Deneubourg (eds): *From individual to collective behavior in social insects*. Birkhäuser, Basel, 219–240

42 Deneubourg J-L, Goss S, Pasteels JM, Fresneau D, Lachaud J-L (1987) Self-organization mechanisms in ant societies (II): learning in foraging and division of labor. *In*: JM Pasteels, J-L Deneubourg (eds): *From individual to collective behavior in social insects*. Birkhäuser, Basel, 177–196

43 Theraulaz G, Bonabeau E, Deneubourg J-L (1998) Response threshold reinforcement and division of labour in insect societies. *Proc R Soc London B* 265: 327–332

44 Theraulaz G, Goss S, Gervet J, Deneubourg J-L (1991) Task differentiation in *Polistes* wasp colonies: a model for self-organizing groups of robots. *In*: J Meyer, SW Wilson (eds): *From animals to animats*. MIT Press, Cambridge, MA, 346–355

45 Gordon DM, Goodwin BC, Trainor LEH (1992) A parallel distributed model of the behaviour of ant colonies. *J Theor Biol* 156: 293–307

46 Trumbo S, Huang Z-Y, Robinson G (1997) Division of labor between undertaker specialists and other middle-age workers in honey bee colonies. *Behav Ecol Sociobiol* 41: 151–164

47 Page RE, Mitchell SD (1991) Self organization and adaptation in social insects. *In*: A Fione, M Forbes, L Wessels (eds): *PSA 1990*, vol 2. Philosophy of Science Association, East Lansing, MI, 289–298

48 Page RE, Mitchell SD (1998) Self organization and the evolution of division of labor. *Apidologie* 29: 101–120

49 Wilson EO (1983) Caste and division of labor in leaf-cutter ants (Hymenoptera: Formicidae: *Atta*). III. Ergonomic resiliency in foraging by *A. cephalotes*. *Behav Ecol Sociobiol* 14: 47–54

50 Calderone NW, Page RE (1992) Effects of interactions among genotypically diverse nestmates on task specialization by foraging honey bees (*Apis mellifera*). *Behav Ecol Sociobiol* 30: 219–226

51 Mittenthal J, Baskin A, Reinke R (1992) Patterns of structure and their evolution in the organization of organisms: modules, matching, and compaction. *In*: J Mittenthal, A Baskin (eds): *The principles of organization in organisms*. Addison Wesley, New York, 321–332

52 Markl H (1985) Manipulation, modulation, information, cognition: some of the riddles of communication. *In*: B Hölldobler, M Lindauer (eds): *Experimental behavioral ecology and sociobiology*. Sinauer Press, Sunderland MA, 163–194

53 Wilson EO, Hölldobler B (1988) Dense heterarchies and mass communication as the basis of organization in ant colonies. *Trends Ecol Evol* 3: 65–68

54 Seeley TD (1994) Honey bee foragers as sensory units of their colonies. *Behav Ecol Sociobiol* 34: 51–62

55 Nicholls JG, Martin AR, Wallace BG (1992) *From neuron to brain*. Sinauer Press, Sunderland, MA

56 Seeley TD, Kühnholz S, Weidenmüller A (1996) The honey bee's tremble dance stimulates additional bees to function as nectar receivers. *Behav Ecol Sociobiol* 39: 419–427

57 Schneider SS, Stamps JA, Gary NE (1986) The vibration dance of the honey bee. I. Communication regulating foraging on two time scales. *Anim Behav* 34: 377–385

58 Schneider SS (1987) The modulation of worker activity by the vibration dance of the honeybee, *Apis mellifera*. *Ethology* 74: 211–218

59 Painter-Kurt S, Schneider SS, Age and

behavior of honey bees that perform vibration dances on workers. *Ethology* 104: 457–473

60 Seeley TD, Weidenmüller A, Kühnholz S (1998) The shaking signal of the honey bee informs workers to prepare for greater activity. *Ethology* 104: 10–26

61 Nieh JC (1998) The honey bee shaking signal: function and design of a modulatory communication signal. *Behav Ecol Sociobiol* 42: 23–36

62 Schneider SS, Visscher PK, Camazine S (1998) Vibration signal behavior of waggle-dancers in swarms of the honey bee, *Apis mellifera* (Hymenoptera: Apidae). *Ethology* 104: 963–972

63 Reeve HK, Gamboa GJ (1983) Colony activity integration in primitvely eusocial wasps: the role of the queen (*Polistes fuscatus*, Hymenoptera: Vespidae). *Behav Ecol Sociobiol* 13: 63–74

64 Reeve HK, Gamboa GJ (1987) Queen regulation of worker foraging in paper wasps: a social feedback control system. *Behaviour* 102: 147–167

65 Breed MD, Gamboa GJ (1977) Behavioral control of workers by queens in primitively eusocial bees. *Science* 195: 694–696

66 Berton F, Lenoir A, Le Roux G, Le Roux A-M (1992) Effect of orphaning on the effectiveness of queen attraction and on worker behavioral repertoire in *Cataglyphis cursor* (Hymenoptera; Formicidae). *Sociobiology* 20(3): 301–313

67 von Frisch K (1967) *The dance language and orientation of honey bees*. Harvard University Press, Cambridge, MA

68 Traniello JFA, Robson SK (1995) Trail and territorial communication in social insects. *In*: RT Cardé, WJ Bell (eds): *Chemical ecology of insects*, vol 2. Chapman and Hall, London, 241–286

69 Huang Z-Y, Robinson GE (1992) Honey-bee colony integration: worker-worker interactions mediate hormonally regulated plasticity in division of labor. *Proc Natl Acad Sci USA* 89: 11 726–11 729

70 Delcomyn F (1997) *Foundations of neurobiology*. W H Freeman, New York

71 Seeley TD (1982) The adaptive significance of the age polyethism schedule in honeybee colonies. *Behav Ecol Sociobiol* 11: 287–293

72 Sendova-Franks AB, Franks NR (1995) Spatial relationships within nests of the ant *Leptothorax unifasciatus* (Latr.) and their implications for the division of labour. *Anim Behav* 50: 121–136

73 Schneider SS (1986) The vibration dance activity of successful foragers of the honeybee, *Apis mellifera* (Hymenoptera: Apidae). *J Kans Entomol Soc* 59: 699–705

74 Powers WT, Clark RK, McFarland RL (1960) A general feedback theory of human behavior: part I. *Percept Motor Skills* 11: 71–88

75 O'Donnell S (1998) Effects of experimental forager removals on division of labour in the primitively eusocial wasp *Polistes instabilis* (Hymenoptera: Vespidae). *Behaviour* 135: 173–193

76 Wilson EO, Durlach NI, Roth LM (1958) Chemical releasers of necrophoric behavior in ants. *Psyche* 65: 108–114

77 Gordon DM (1983) Dependence of necrophoric response to oleic acid on social context in the ant *Pogonomyrmex badius*. *J Chem Ecol* 9: 105–111

78 Robson SK, Beshers SN (1997) Division of labour and "foraging for work": simulating reality versus the reality of simulations. *Anim Behav* 53: 214–218

Role and variability of response thresholds in the regulation of division of labor in insect societies

Eric Bonabeau and Guy Theraulaz

Summary

In this chapter, the regulation of division of labor in social insect colonies is studied from the perspective of response thresholds. Response thresholds refer to likelihood of reacting to task-associated stimuli. Low-threshold individuals perform tasks at a lower level of stimulus than high-threshold individuals. Task performance reduces the level of task-associated stimuli. A model based on fixed response thresholds is described, its assumptions discussed, and some of its predictions are shown to be in agreement with empirical observations. This model can be modified to make thresholds variable, but experimental results are scarce to guide the design of a model with variable thresholds. It is suggested, however, that the response threshold framework can help us understand the regulation of division of labor in social insects from a unifying perspective.

Introduction

The organization of division of labor is a major factor contributing to the impressive ecological success of social insects [1, 2]. Social insects are all characterized by one fundamental type of division of labor, reproductive division of labor, a main ingredient in the definition of eusociality: only one or a few individuals account for most of the reproduction [3]. Beyond this primary form of division of labor between reproductive and worker castes, there most often exists a further division of labor among workers, who tend to perform specific tasks for some amount of time, rather than be generalists who perform various tasks all the time. We will be concerned with this aspect of division of labor in the present chapter. Workers are classically divided into age or morphological "subcastes"

C. Detrain et al. (eds) Information Processing in Social Insects
© 1999, Birkhäuser Verlag Basel/Switzerland

[2]. Age subcastes correspond to individuals of the same age that tend to perform identical tasks: this phenomenon is called temporal polyethism. In some species, workers can have different physical characteristics: workers that belong to different morphological castes tend to perform different tasks. But even within an age or morphological caste, there are differences between individuals in their frequency and sequence of task performance.

It has become clear in the last 2 decades that the emergence, maintenance, and regulation of division of labor result from the interplay of several factors: "how an individual colony member behaves is influenced by many factors, including its genetic makeup, trophic history, age, experience, social environment, and external environment" [4]. All these factors determine how information processing and decision making are performed at the individual and colonial levels, and how both levels are connected. The important question is indeed how individual behavior is integrated within, and partly determines, colony behavior, and how colony behavior in turn influences the behavior of individuals. In particular, one of the most striking aspects of division of labor is plasticity: the ratios of workers performing the different tasks that maintain the colony's viability and reproductive success can vary (that is, workers switch tasks) in response to internal perturbations or external challenges. This flexibility is implemented at the level of individual workers, which certainly do not possess any global representation of the colony's needs. How does information flow throughout the colony, and how do the various above-mentioned factors influence information pathways?

Robinson (1987) [5] proposed to consider division of labor in social insects from the viewpoint of response thresholds (see also [6]). When some stimulus exceeds the response threshold of an individual, that individual is likely to respond to the stimulus. Responding to a task-associated stimulus by performing the task may reduce the intensity of the stimulus. Therefore, individuals with high thresholds are unlikely to perform the task when other individuals, with lower thresholds, maintain the stimulus intensity below their thresholds; when, however, individuals with low thresholds fail to perform the task, because of predation, swarming or other perturbations, those individuals that have high thresholds may engage in task performance because stimulus intensity exceeds their thresholds. This shows, rather informally, that response thresholds could explain the short-term flexibility observed in many species of social insects. It was also suggested [5, 6] that response thresholds could reconcile such flexibility with the predictions of more "classical" ergonomic arguments [2], that had until then been interpreted as supporting a static view of division of labor, where castes, physical or temporal, would be associated with a rigid set of tasks, without flexibility. These two aspects, he proposed, correspond to constraints operating at two clearly separated time scales: viability and survival, on the short term, leading to flexibility, and long-term adaptations, leading to "optimal" caste or class ratios in "average" ecological conditions. Although within- and between-caste task switching and flexibility are observed (e.g. [5–9]), class ratios have certainly been selected for in the course of evolution on the basis of efficiency and energetic factors. For example, Wilson [10] observed flexibility in *Atta cephalotes* ants when one class of workers was removed; replacement workers performed similarly to "specialized" workers in terms of frequency and effectiveness of task performance, but were 30% less efficient with respect to oxygen consumption [6, 10]. It is therefore tempting to think that "optimal" caste ratios predicted on the basis of energetic factors [2] are relevant over evolutionary time scales, and that observed caste ratios have been shaped more or less accordingly, but, over short time scales, changing ecological conditions make flexibility an absolute necessity — flexibility also has been selected for. "Where colony survival is at issue, considerations of behavioral and/or metabolic efficiency should be secondary. By this reasoning, the caste distribution function only approximately reflects long-term selective pressures, because flexibility among classes allows short-term compensation for any 'slop' or inappropriateness in caste ratios" [6]. The same argument applies in support of West-Eberhard's [11] hypothesis of centrifugal task allocation (younger workers, because they can reproduce, tend to stay in the nest where they can lay eggs, while older workers are less likely to be able to reproduce and can therefore perform more dangerous tasks outside the nest): selection acting at the level of the individual exerts a long-term pressure leading to a more or less marked centrifugal pattern of temporal task allocation, and short-term flexibility, driven by immediate colony requirements, tends to blur this pattern.

Response thresholds can definitely reconcile both views, that is, both time scales. The present chapter is aimed at substantiating this statement, by making the notion of response threshold more accurate, reviewing some experimental evidence for response thresholds, presenting a formal model the results of which are in qualitative as well as quantitative agreement with experimental observations, and studying what factors may cause response threshold variability. Threshold variability, not only between, but also within individuals might be essential to our understanding of patterns of temporal polyethism and deviations from them, including accelerated or retarded development, or behavioral reversion. We believe that response thresholds can provide students of social insects with a unifying picture of how division of labor is regulated.

How are tasks allocated?

How do workers find or gather the information necessary to decide whether or not they switch tasks or engage in task performance?

There are undoubtedly many ways for workers to find information about the colony's needs and/or what task(s) they should perform. The nature of task-related stimuli may vary greatly from one task to another, and so can information sampling techniques. Various possibilities have been proposed, and some of them substantiated. Finding information may involve direct interactions among workers (trophallaxis, antennation, etc.) [12, 13], nest "patrolling" [14], or more or less random exposure to task-related stimuli.

For Gordon [12], direct interactions among workers are the essential ingredient of information gathering. According to her, the encounter rate (measured via the number of antennal contacts per ant) is actively regulated by ants. This regulation prevents excessive numbers of individuals from performing a task, especially at high density. Seeley [15, 16] and Jeanne [17] showed, in honey bees *Apis mellifera* and tropical eusocial wasps *Polybia occidentalis* respectively, that contact rates, or, more precisely, waiting times, could be used as a cue to determine whether a task should or not be performed: a forager that comes back to the hive or to the nest has to transfer its load to another individual, a storer bee or a builder wasp; if the forager has to wait too long to transfer its load (which indicates that foraging is no longer needed, or that more storers or builders are needed), it stops foraging and may switch tasks. It is interesting to notice that, in the case of honey bees [15, 16] waiting or search time is correlated with the spatial distribution of nectar unloadings: nectar foragers have to go deeper into the hive in order to find an available storer bee when the influx of foragers is high than when it is low. This means that nectar foragers may use spatial (location of unloading) as well as temporal (waiting time) information to make decisions. Between-caste aversion also relies on direct contact among workers that belong to different morphological subcastes: studying brood care in *Pheidole pubiventris*, Wilson [18] noticed that majors pay a greater attention to the brood when less minors are present, which is due, according to him, to the fact that majors actively avoid minors while in the vicinity of the immature stages. Another example where contacts may be important can be found in the context of food or defensive recruitment, whereby an individual stimulates another individual or a set of individuals into performing a given task. Here, information about the existence of a food source, or about the presence of danger, is directly transmitted among individuals. Recruitment may also take place through the deposition of chemical substances, pheromones, which provide cues for other workers to follow. Pheromones are used in many different contexts including defense, alarm, foraging, or exploration.

Often, direct or indirect contacts among workers have to do with the transmission of information: it remains to know how information is found in the first place. The nest itself is a primary source of information. For example, Seeley [16] suggests that foragers coming back to the hive may have to wait a long time to unload their nectar when storer bees are unable to find cells to store nectar: the number of cells that have been filled therefore serves as an implicit regulator of foraging and storing activities. The stimulatory field inside the nest is certainly extremely rich, due to the presence, among other things, of the queen(s), many workers, and the brood, which may all emit pheromones to express their needs. The queen can be a source of information: for example, the organization of some activities, such as worker foraging in wasps [19], has been interpreted as resulting from active regulation and control by the queen, the central organizer of a large amount of information which is redistributed to the workers through stimulations. Some experiments show that tasks are sometimes spatially organized [20–25]. When this is the case, it is clear that the probability of being exposed to stimuli (either directly or through worker interactions) associated with task B when performing task A depends on the spatial relationships between both tasks.

Patrolling outside the nest can allow workers to sense important environmental cues: for example, Passera et al. [26] showed that the presence of individuals from another colony in *P. pallidula* leads to an increased production of soldiers; the only available information in their experiment was in the form of the neighboring colony's odor, which was found by patrollers who brought it back to the nest (notice, however, that this response, which takes place on a time scale of several weeks, seems to be exceptional in polymorphic species [27]: caste ratios are most often "hard-wired", stress does not usually lead to the production of more soldiers, and peace does not lead to a reduction of the number of soldiers).

In summary, relevant information comes from the nest and the environment (social and external), and flows throughout the colony. In addition, task performance by an individual certainly tends to reduce the intensity of task-associated stimuli, and therefore modifies the stimulatory field of other colony members: information about the necessity of performing such or such task is certainly transmitted through a combination of direct and indirect channels. Indirect information transfer through modifications of the environment is a very important concept, as it can lead in principle to coordinated activities, flexibility, and colony-level responsiveness without direct contacts among individuals. Grassé [28] first understood the importance of this concept and coined the term "stigmergy" to describe it. Stigmergy is easily overlooked, as it does not explain the detailed mechanisms by which individuals coordinate their activities, but it certainly does provide a general mechanism that relates individual and colony-level behaviors: individual behavior modifies the environment, which in turn modifies the behavior of other individuals. Calabi [6] and Bonabeau et al. [29] implicitly used the notion of stigmergy to build their models.

How do workers actually decide to engage task performance, switch tasks, continue performing their current task, or become inactive, once they have the information?

Response thresholds: definition and experimental evidence

The first question we have to answer is: what is a response threshold? Let s be the intensity of a stimulus associated with a particular task. s can be a number of encounters, a chemical concentration, or any quantitative cue sensed by individuals, as discussed in section 2.1. A response threshold θ, expressed in units of stimulus intensity, is an internal variable that determines the tendency of an individual to respond to the stimulus s and perform the associated task. More precisely, θ is such that the probability of response is low for $s \ll \theta$ and high for $s \gg \theta$. It is important to emphasize that different response thresholds, in the sense we use them here, may either reflect actual differences in behavioral responses, or differences in the way task-related stimuli are perceived. More precisely, an individual may appear to be more responsive to a given stimulus than another individual: this may be due to a stronger motivation to respond, the stimulus being perceived at the same level of intensity, or to a difference in the way the stimulus is perceived, the motivations of both individuals being identical. The two possibilities are not distinguished in our exposition. One family of response functions $T_\theta(s)$ that can be parametrized with thresholds that satisfy the "threshold requirement" is given by

$$T_\theta(s) = \frac{s^n}{s^n + \theta^n} \tag{1}$$

where $n > 1$ determines the steepness of the threshold. Figure 1 shows several such response curves, with $n = 2$, for different values of θ. The meaning of θ is clear: for $s \ll \theta$, the probability of engaging task performance is close to 0, and for $s \gg \theta$, this probability is close to 1; at $s = \theta$, this probability is exactly $1/2$. Therefore, individuals with a lower value of θ are likely to respond to a lower level of stimulus. The notion of a threshold is often associated with a change in concavity in the response curve, as is the case, for example, for response curves given by equation (1) with $n > 1$, where the inflection point is given by

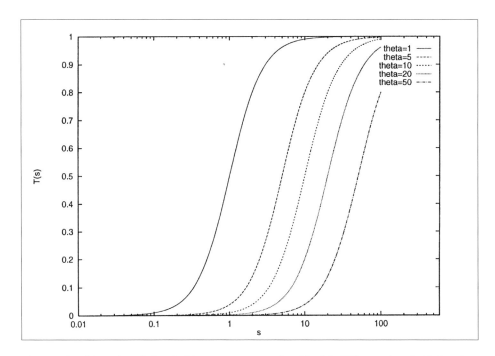

Figure 1 Semi-log plot of threshold response curves (n = 2) with different thresholds (θ = 1, 5, 10, 20, 50)

$s = \theta[(n-1)/n + 1)]^{1/n}$. But the definition of a threshold given above does not require a change in concavity. An important example is when the response function is exponential [30]:

$$T_\theta(s) = 1 - e^{-s/\theta} \qquad (2)$$

Figure 2 shows $T_\theta(s)$ given by equation (2) for different values of θ. We see, here again, that the probability of engaging task performance is small for $s \ll \theta$, and is close to 1 for $s \gg \theta$. Although there is no change in concavity in the curve, this response function produces behaviors which are comparable to those produced by response functions based on equation (1) [64]. Threshold models encompass exponential response functions: the essential ingredient is the existence of a characteristic value θ of stimulus intensity. Exponential response functions are potentially important because they may in principle be encountered quite frequently. For example, imagine a stimulus that consists of a series of encounters with, say, items to process. If, at each encounter with an item, an individual has a fixed probability ρ of processing the item, then the probability that

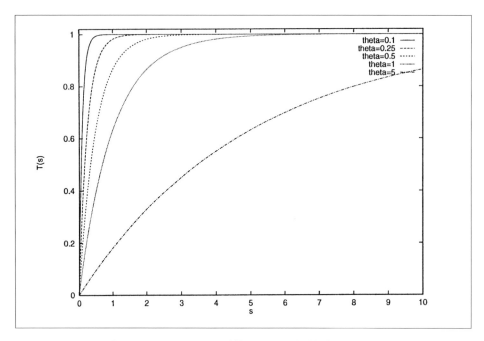

Figure 2 Exponential response curves with different thresholds (θ = 0.1, 0.25, 0.5, 1, 5).

the individual will not respond to the first N encountered items is given by $(1 - \rho)^N$. Therefore, the probability P(N) that there will be a response within the N encounters is given by $P(N) = 1 - (1 - \rho)^N = 1 - e^{N \ln(1 - \rho)}$, which is exactly equation (2) with $s = N$ and $\theta = -1/\ln(1 - \rho)$. For example, the organization of cemeteries in ants provides a good illustration of this process. The probability of dropping a dead body (or a dead item, that is, a thorax or an abdomen) has been studied experimentally by Chrétien [31] in the ant *Lasius niger*: the probability that a laden ant drops an item next to a N-cluster can be approximated by $P(N) = 1 - (1 - p)^N$, for N up to 30, where $p \approx 0.2$ (Fig. 3). Here, the intensity of the stimulus is the number of encountered dead bodies, and the associated response is dropping an item. Another situation in which exponential response functions may be observed is when there are waiting times involved: let us assume that tasks A and B are causally related in the sense that a worker performing task A has to wait for a worker performing task B to, say, unload nectar or pulp, or any kind of material. If a task A worker has a fixed probability p per unit waiting time of giving up task A performance, the probability that this worker will still be waiting after t time units is given by

$$P(t) = 1 - (1 - p)^t = 1 - e^{t \ln(1 - p)}.$$

Very few experiments have been aimed at showing the existence of response thresholds in social insects—although many such experiments have been with other animals (see Beshers and Robinson, this volume). Such experiments

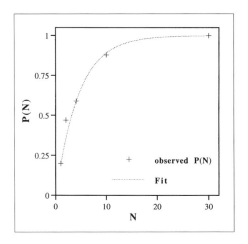

Figure 3 Probability P(N) of dropping a dead body next to an N-cluster as a function of N (after [3])

Fit $P(N) = 1 - (1 - p)^N$ with $p = 0.2$ is shown.

require controlling (or at least being able to vary and measure) the intensity of the stimuli workers are responsive to, a task that can be very difficult. Most experimental curves show the probability of response of an individual as a function of its size, or weight, and so on, at fixed stimulus intensity: although these curves can teach a lot, they cannot prove the existence or absence of response thresholds.

Robinson and colleagues [1, 5, 32, 33] showed the existence of hormonally regulated behavioral response thresholds to alarm pheromones in honey bees (*Apis mellifera*). Treatment of young worker honey bees with a juvenile hormone (JH) analog increases their sensitivity to alarm pheromones, which play a role in nest defense. JH is produced by the corpora allata glands, which are known to grow with age. Robinson [5] also showed that JH treatment of bees stimulates their production of alarm pheromone. He further noted that isolated JH-treated bees do not respond to presented alarm pheromones, but do respond when in group: although alternative explanations are possible, this may be due to the fact that increased production of alarm pheromones by treated bees allows a threshold to be reached when several such bees are put together.

If it takes a forager too long to unload her nectar to a storer bee, it gives up foraging with a probability that depends on its search time in the unloading area. It will then start a "tremble dance" [16] to recruit storer bees (the tremble dance also inhibits waggle dancing). If, on the other hand, its in-hive waiting or search time is very small, it starts recruiting other foragers with a waggle dance. If its in-hive waiting or search time lies within a given window, it is likely not to dance at all and return to the food source. If one plots the probability of either waggle or tremble dancing as a function of search time, a clear threshold function can be observed [16, Fig. 6.12, p. 169].

A series of experiments by Detrain and colleagues [34–36] indicate the existence of differential response thresholds in the ant *Pheidole pallidula* in at least two activities, foraging and nest defense. The intensity of behavioral stimuli (measured by trail concentration and the number of tactile invitations in the case of foraging, supplemented by the number of intruders in the case of defense) required to induce the effective recruitment of majors is greater than for minors for both tasks, indicating that majors have higher response thresholds. An interesting discussion of the adaptive significance of these findings is given by Detrain and Pasteels [35, 36]. They also hypothesize that JH titers [37] or the volume and number of cells of corpora allata could affect behavioral thresholds.

Finally, Schatz [38] presents evidence for response thresholds in the primitive ant species *Ectatomma ruidum*: when presented with an increasing number of prey, specialized "stinger ants" (or killer ants) start to become involved in the retrieval process (in addition to transporters, to which dead prey are transferred), the number of such ants being dependent on the number of prey in a characteristic sigmoidlike manner (Fig. 4). This suggests that within-caste specialization among hunters is indeed based on response thresholds.

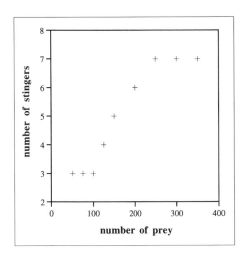

Figure 4 Number of stingers involved in prey retrieval as a function of the number of prey in the ant Ectatomma ruidum, *for a nest comprised of 130 workers (after [38]).*

Mathematical model

Let us assume that one task only need to be performed. This task is associated with a stimulus or demand, the level of which increases if it is not satisfied. Let us consider two types of workers, type 1 and type 2, or group 1 and group 2, or minors and majors. Let n_1 and n_2 be the respective numbers of workers of type 1 and type 2 in the colony, N the total number of workers in the colony ($n_1 + n_2 = N$), $f = n_1/N$ the fraction of type-1 workers in the colony, N_1 and N_2 the respective numbers of type-1 and type-2 workers engaged in task performance, and x_1 and x_2 the corresponding fractions ($x_i = N_i/n_i$). The average deterministic equations describing the dynamics of x_1 and x_2 are given by:

$$\partial_t x_1 = T_{\theta_1}(s)(1 - x_1) - p x_1 \qquad (3)$$

151

$$\partial_t x_2 = T_{\theta_2}(s)(1 - x_2) - p x_2 \qquad (4)$$

where θ_i is the response threshold of type-i workers, and s the integrated intensity of task-associated stimuli. The first term on the right-hand side of equations (3) and (4) describes how the $(1 - x_i)$ fraction of inactive type i workers responds to the stimulus intensity or demand s, with a threshold function $T_{\theta_i}(s)$ given by equation (1). Let us assume that individuals can assess the demand for a particular task when they are in contact with the associated stimulus, and that each insect encounters all stimuli with equal probability per time unit, and can respond to these stimuli. Let us define $z = \theta_1^2/\theta_2^2$. If $z > 1$, type-1 workers are less responsive than type-2 workers to task-associated stimuli. The second right-hand-side term in equations (3) and (4) expresses the fact an active individual gives up task performance and becomes inactive with probability p per unit time, that we take identical for both types of workers. $1/p$ is the average time spent by an individual in task performance before giving up this task. It is assumed that p is fixed, and independent of stimulus. Therefore, individuals involved in task performance spend $1/p$ time units working even if their work is no longer necessary. Individuals give up task performance after $1/p$, but may become engaged again immediately if the stimulus is still large. The dynamics of the stimulus intensity s is described by:

$$\partial_t s = \delta - \frac{\alpha}{N}(N_1 + N_2) \qquad (5)$$

that is, since $(N_1 + N_2)/N = f x_1 + (1 - f) x_2$

$$\partial_t s = \delta - \alpha f x_1 - \alpha(1 - f) x_2 \qquad (6)$$

where δ is the (fixed) increase in stimulus intensity per unit time, and α is a scale factor measuring the efficiency of task performance. Identical efficiencies in task performance are assumed for type-1 and type-2 workers. That efficiencies do not vary significantly is a plausible assumption provided the time scales of experiments are sufficiently short, whereas learning may take place over longer time scales. The amount of work performed by active individuals is scaled by N, as can be seen on equation (5), to reflect the idea that the demand is an increasing function of N, that we take to be linear here. In other words, colony requirements scale linearly with colony size. Under this assumption, our results should be independent of colony size. In the stationary state, where all δ_t's are equal to 0, the value of x_1 converges toward a stationary value x_1^s, which is found to be given by

$$x_1^s = \frac{\chi + \left(\chi^2 + 4f(p+1)(z-1)(\delta/\alpha)\right)^{1/2}}{2f(p+1)(z-1)} \tag{7}$$

where $\chi = (z-1)\left(f + (p+1)\dfrac{\delta}{\alpha}\right) - z$ as been defined for convenience. Results obtained with this model are presented in the next section.

Response thresholds: predictions

Although the notion of response threshold is appealing, and seems to be supported by experiments at least for some tasks, can a response threshold model make predictions? Bonabeau et al. [29] have shown that it can reproduce experimental results of Wilson [7], who artificially varied the ratio of majors to minors in several polymorphic ant species (*Pheidole*) and observed a dramatic increase in task performance by previously inactive majors as the ratio exceeded some value; the involvement of majors occurred within an hour of minors' removal. Assuming differential response thresholds for minors and majors, this phenomenon can be readily explained by the reasoning described in the introduction. Not only is this model in qualitative agreement with Wilson's observations [7], it can also reproduce his results quantitatively [29]. Figure 5 shows how the frequency of task performance varies as a function of the fraction f of majors in the colony, as given by equation (7), for particular values of the parameters, and a comparison with Wilson's results [7] (who measured, among other things, the number of acts of social behavior and self-grooming per major in *Pheidole guilelmimuelleri*). Figure 6 shows that the very same curve is obtained, with different values of the parameters, if $T_{\theta_i}(s)$ is given by the exponential response function in equation (2) rather than by equation (1).

Besides the nine species of *Pheidole* studied by Wilson [7], there are other examples of flexibility, whereby individuals perform tasks that do not belong to the normal repertoire of their physical or age caste, that the threshold model can certainly explain. Wilson [10, 39, 40] studied flexibility in the ants *Atta cephalotes* and *A. sexdens*, in which there is a continuum of size classes rather

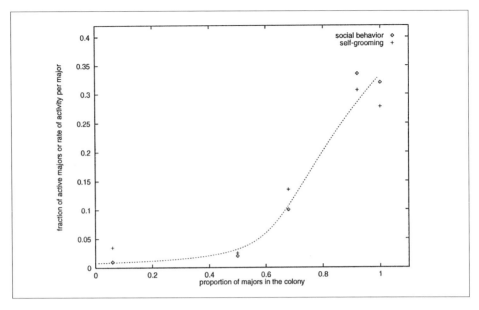

Figure 5 *Fraction of active majors as a function of the proportion f of majors in the colony, as given by equation (7) with z = 64, p = 0.2, δ = 1, and α = 3*
Comparison with results from [7] (scaled so that curves of model and experiments lie within the same range): number of acts of social behaviour and self-grooming per major within time of experiments in Pheidole guilelmimuelleri.

than simply two physical castes as is the case in *Pheidole*. He showed that the experimental removal of a size class stimulates individuals belonging to adjacent size classes into performing the tasks of the missing size class. Lenoir [41] found that young workers of the ant *Tapinoma erraticum* tend to be stimulated into foraging activities when the ratio of old to young workers goes below 1. Calabi [6, 42, 43] found that young workers of the ant *Pheidole dentata*, when raised in the absence of older minors, forage significantly earlier than when older minors are present; conversely, old minors in colonies without young minors perform brood care, a behavior that is not observed when young minors are present. Carlin and Hölldobler [44] found interesting interspecific differences, possibly related to differences in response thresholds, in mixed species colonies of *Camponotus* ants. *C. pennsylvanicus* performs brood care and works inside the nest when raised with *C. americanus* or *C. noveboracensis*, but works outside the nest when raised with *C. ferrugineus*. This suggests, within the context of the threshold model, that *C. pennsylvanicus* has a lower response threshold to brood care than

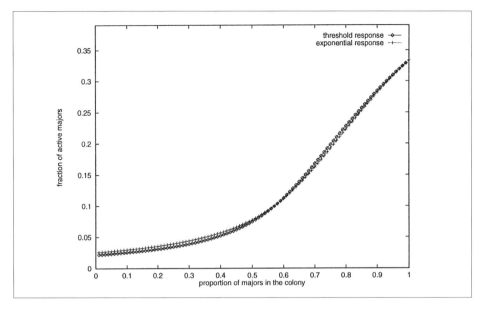

Figure 6

Comparison between the fraction of active majors as a function of the proportion of majors in the colony obtained with an exponential response function with $\theta_1 = 0.1$, $\theta_2 = 1$, $\alpha = 3$, $\delta = 1$, and $p = 0.2$ (curve obtained by numerical integration of equations (2–4, 6), and a threshold response function with $z = 64$, $\alpha = 3$, $\delta = 1$, $p = 0.2$ [curve obtained from equation (7)].

C. americanus or *C. noveboracensis*, but higher than *C. ferrugineus*. More studies would be welcome, however, since other factors could play a role in this unusual example of interspecific task allocation.

The complexity of threshold models directly results from the complexity of assumed information sampling techniques. A threshold model can be quite complex if one takes into account spatial, causal, or topological relationships among tasks: in effect, tasks are not uniformly distributed in space (e.g. brood care occurs within the nest, whereas foraging occurs outside the nest), so that performing a given task may enhance or prevent contacts with other task-associated stimuli (including nestmates), some tasks are causally related (e.g. foraging and storing), so that individuals performing a task may switch to the next and so on. In other words, sampling is unlikely to be homogeneous, and some workers may end up being specialized simply because they always encounter the same stimuli, and do not encounter the stimuli that would stimulate them to perform other tasks. There may as well be learning processes on top of that, so that, for example, workers tend to become more and more sensitive to stimuli associated with the tasks they are currently performing.

One essential question that remains to be answered is the following: How are response thresholds determined, how are threshold differences between individuals determined, and (how) do they vary over time? This question will be discussed in section 3.

Between-caste aversion

In order to explain the increased involvement of majors in minor-related tasks when the fraction of majors in the colony increases [7], Wilson [18] introduced the notion of between-caste aversion, a phenomenon that he apparently observed in *P. pubiventris*. The fact that majors pay a greater attention to the brood when fewer minors are present results, according to him, from active avoidance of minors by majors in the vicinity of the brood. He noticed that majors did not avoid minors in any other part of the nest (which, in passing, casts doubt on the generality of between-caste aversion as a basis for division of labor). Wilson's observations [18] can be explained by a threshold model. Majors could use contacts with minors in the vicinity of the brood as a stimulus to assess indirectly the degree of satisfaction of the brood. It is hard to determine whether majors avoid minors, or respond to chemical (or other) cues carried by minors. Indeed, Wilson [18] reports that majors showed the clearest responses after making direct antennal contacts with minors: this suggests that instead of identifying minors, they may be sensitive to cues carried by minors from the brood area. It is also perfectly possible that majors have higher response thresholds to brood stimuli, so that brood care by minors maintains larval demand below threshold. The response threshold approach explains Wilson's observations [18] qualitatively and quantitatively, and does not raise the same issues as between-caste aversion. In effect, between-caste aversion does not seem to be an efficient way of dividing labor among workers (notwithstanding its lack of generality): if larvae are satiated, fewer minors will take care of the brood, so that more majors will access the brood area and take over brood care, which is not necessary; if larvae are hungry, many minors are present, preventing majors from reaching the brood, while they could be useful. All these remarks make the threshold hypothesis more likely.

Foraging for work

In the foraging-for-work (FFW) model, introduced by Tofts and Franks [45–47], individuals actively seek work, and continue in their current role if they find work within that role; if they don't find work within their current role, or fail to find work too frequently, they randomly move to adjacent zones until they find work. The FFW model can be viewed as a special threshold model, in the sense that each individual responds to any stimulus greater than zero: all individuals are characterized by $\theta = 0$. The FFW model assumes that tasks are spatially ordered: the simplest possible ordering is a one-dimensional line, where spatial locations are represented as a function of their distance from the center of the nest. This ordering induces neigborhood relationships among tasks associated with spatial locations: for example, brood care, which necessarily takes place in the center of the nest, is far away from foraging, which takes place outside the nest, but foraging is close to nest defense. A globally centrifugal motion of individuals as they age (from the middle of the nest to outside the nest) [11] can be caused by an inflow of newly born individuals, who first actively seek work in the nest, pushing older individuals to seek work outside the center of the nest. By this mechanism, the FFW model can in principle produce temporal polyethism. This model is remarkable in that it does not require any kind of biological clock, which would physiologically transform individuals at the "right" time, to generate temporal polyethism. Sendova-Franks and Franks [48] argue that the FFW model is fully consistent with their observations of *Leptothorax* ants, which are characterized by weak temporal polyethism. Their argument is further supported by more recent experiments [23], in which they showed the existence of individual-specific spatial fidelity zones (SFZ) which appear to be independent of the individual's age: workers tend to perform tasks within their SFZ, irrespective of their age. In other examples, however, it appears that the FFW model alone is unable to explain observed patterns of temporal polyethism [49, 50].

How and why do thresholds vary between and within individuals?

If one assumes that response thresholds, as diverse in nature as they may be, regulate task allocation in social insects, one has to identify sources of variations of these thresholds, and how such variations influence patterns of task allocation. We see basically three, interrelated, main causes of threshold variability between and within individuals: genetics, aging, and learning. While there is a growing body of experimental evidence showing the effects of genotypic characteristics on response thresholds and task allocation, the two other factors are certainly plausible but remain to a large extent unexplored.

Genetics

It is not absurd to think that individuals with close genotypic characteristics (e.g. belonging to the same patriline in honey bees) may have similar response thresholds and are therefore predisposed to perform the same tasks. Conversely, individuals with different genotypes are more likely to have different response thresholds. There is now convincing evidence that there is a genetic component to division of labor in honey bees and ants [33, 49, 51–56]. For example, Robinson and Page [54, 55] have shown that honey bee workers belonging to different patrilines may have different response thresholds. Assume for simplicity that workers of patriline A engage in nest guarding as soon as there are fewer than 20 guards, whereas workers of patriline B start performing this task when there are fewer than 10 workers guarding the hive's entrance: workers of patriline B have a higher response threshold to perform this task. More generally, these authors have shown in a series of papers that response thresholds are partly determined by genes. Genetic variability is therefore a source of response threshold variability. According to Calderone and Page [49, 51, 52] genes may have effects on (i) the rate of behavioral ontogeny, and (ii) the probability of task performance independent of the rate of behavioral ontogeny. This suggests that, in addition to varying between individuals, response thresholds may vary within a given individual at a rate that depends on the individual's genotypic characteristics. In a genetically diverse colony, not only are there individuals with low and high response thresholds, some individuals may also have an "accelerated development" and other individuals may exhibit "retarded development" [51, 53]. However, it could be that individuals exhibiting "accelerated development" have lower thresholds with respect to tasks that are usually performed by older individuals, and are therefore attracted toward performing those tasks, whereas individuals exhibiting "retarded development" have higher thresholds with respect to stimuli associated with those tasks.

Aging and learning

There exists a huge body of literature on the association between age and task performance. Age polyethism, or age-based division of labor, appears to be common in social insects. But is age the (only) relevant factor in age polyethism? Because the answer to this question is far from obvious, some authors suggest to use the more cautious expression "temporal polyethism" [57]. Distinctions must be made between absolute age (or normal age), physiological age, which depends by definition on how an individual's physiological state has been affected by that individual's personal history (e.g. the fate of an individual (for instance, gyne, minor, or major) in a polymorphic species may be determined in part by genes and in part by the individual's diet at crucial points in time [58]), and relative age, which is defined relative to other colony members. Physiological and relative ages are both relative quantities, often more relevant than absolute age [6, 22, 41, 48, 59, 60], which means that stimuli provided by the environment and other colony members play an important role in shaping behavioral ontogeny, that is, in the context of response thresholds, in modifying response thresholds. In this context, aging would involve the systematic modification of thresholds with time, so that, for example, a young individual would have a naturally high response threshold with respect to nest defense, but that threshold would progressively become lower as the individual ages, until the probability of response to a defense-related stimulus becomes significant. Learning would involve, for example, reduction of a response threshold when the associated task is being performed (task "fixation" [20, 48]: ants become increasingly entrained on certain tasks they practice, and seek out such tasks in preference to others" [48]), and increase of the response threshold when the task is not being performed ("unlearning") [30, 61–63]. Both mechanisms, aging and learning, are likely to be involved in how task allocation tends to vary with age, and experiments will have to be devised to determine which mechanism is at play in different situations. In order to help design such experiments, Theraulaz et al. [63] have modified the fixed response threshold model to include the kind of threshold reinforcement mentioned above. Theraulaz et al. [63] made predictions as to what should be observed when specialists of a given task are removed from the colony and reintroduced after a varying amount of time: the colony does not recover the same state as prior to the perturbation and the difference between before and after the perturbation is more strongly marked as the time between separation and reintroduction increases. Empirical evidence is currently too weak to support the model's assumptions, but this prediction provides an unambiguous way of testing the reinforcement hypothesis [63].

Conclusion

This chapter is an attempt to unify studies of division of labor within the framework of the response threshold model. Our approach has been to make the idea of response threshold more operational, more accurate, and, we hope, somewhat more testable and amenable to experimental validation. The threshold model, or, should we say, threshold models, are a promising avenue of research for understanding task allocation in social insects, and greatly help formulating relevant questions about polyethism. We urge students of social insects to undertake experiments to show the existence of response thresholds, which will provide a stronger basis for future investigations.

Aknowledgments

E.B. is supported by the Interval Research Fellowship at the Santa Fe Institute. G.T. is supported in part by a grant from the Conseil Régional Midi-Pyrénées.

References

1 Robinson GE (1992) Regulation of division of labor in insect societies. *Annu Rev Entomol* 37: 637–665

2 Oster G, Wilson EO (1978) *Caste and ecology in the social insects.* Princeton University Press, Princeton

3 Keller L (ed) (1993) *Queen number and sociality in insects.* Oxford University Press, Oxford

4 Jeanne RL (1991) Polyethism. *In*: KG Ross, RW Matthews (eds): *The social biology of wasps.* Comstock, Cornell, Ithaca, NY, 389–425

5 Robinson GE (1987) Modulation of alarm pheromone perception in the honey bee: evidence for division of labour based on hormonally regulated response thresholds. *J Comp Physiol* 160: 613–619

6 Calabi P (1988) Behavioral flexibility in Hymenoptera: a re-examination of the concept of caste. *In*: JC Trager (ed): *Advances in myrmecology.* Brill Press,

Leiden, 237–258

7   Wilson EO (1984) The relation between caste ratios and division of labor in the ant genus *Pheidole* (Hymenoptera: Formicidae). *Behav Ecol Sociobiol* 16: 89–98

8   Lenoir A (1987) Factors determining polyethism in social insects. *Experientia Suppl* 54: 219–241

9   Gordon DM (1989) Dynamics of task-switching in harvester ants. *Anim Behav* 38: 194–204

10 Wilson EO (1983b) Caste and division of labor in leaf-cutter ants (Hymenoptera: Formicidae: *Atta*). IV. Colony ontogeny of *A. cephalotes*. *Behav Ecol Sociobiol* 14: 55–60

11 West-Eberhard M-J (1981) Intragroup selection and the evolution of insect societies. *In*: RD Alexander, DW Tinkle (eds): *Natural selection and social behaviour*. Chiron Press, New York, 3–17

12 Gordon DM (1996) The organization of work in social insect colonies. *Nature* 380: 121–124

13 Huang ZY, Robinson GE (1992) Honeybee colony integration: worker-worker interactions mediate hormonally regulated plasticity. *Proc Natl Acad Sci USA* 89: 11 726–11 729

14 Lindauer M (1952) Ein Beitrag zur Frage der Arbeitsteilung im Bienenstaat. *Z Vergl Physiol* 34: 299–345

15 Seeley TD (1989) Social foraging in honey bees: how nectar foragers assess their colony's nutritional status. *Behav Ecol Sociobiol* 24: 181–199

16 Seeley TD (1992) The tremble dance of the honey bee: message and meanings. *Behav Ecol Sociobiol* 31: 375–383

17 Jeanne RL (1996) Regulation of nest construction behaviour in *Polybia occidentalis*. *Anim Behav* 52: 473–488

18 Wilson EO (1985) Between-caste aversion as a basis for division of labor in the ant *Pheidole pubiventris* (Hymenoptera: Form-

icidae). *Behav Ecol Sociobiol* 17: 35–37

19 Reeve HK, Gamboa GJ (1987) Queen regulation of worker foraging in paper wasps: a social feedback control system (*Polistes fuscatus*, Hymenoptera: Vespidae). *Behaviour* 102: 147–167

20 Wilson EO (1976) Behavioral discretization and the number of castes in an ant species. *Behav Ecol Sociobiol* 1: 141–154

21 Seeley TD (1982) Adaptive significance of the age polyethism schedule in honeybee colonies. *Behav Ecol Sociobiol* 11: 287–293

22 Sendova-Franks AB, Franks NR (1994) Social resilience in individual worker ants and its role in division of labour. *Proc R Soc Lond B* 256: 305–309

23 Sendova-Franks AB, Franks NR (1995) Spatial relationships within the nests of the ant *Leptothorax unifasciatus* (Latr.) and their implications for the division of labour. *Anim Behav* 50: 121–136

24 Sendova-Franks AB, Franks NR (1995) Division of labour in a crisis: task allocation during colony emigration in the ant *Leptothorax unifasciatus* (Latr.). *Behav Ecol Sociobiol* 36: 269–282

25 Franks NR, Sendova-Franks AB (1992) Brood sorting by ants: distributing the workload over the work surface. *Behav Ecol Sociobiol* 30: 109–123

26 Passera L, Roncin E, Kaufmann B, Keller L (1996) Increased soldier production in ant colonies exposed to intraspecific competition. *Nature* 379: 630–631

27 Wilson EO, Hölldobler B (1988) Dense heterarchies and mass communications as the basis of organization in ant colonies. *Trends Ecol Evol* 3: 65–68

28 Grassé P-P (1959) La reconstruction du nid et les coordinations inter-individuelles chez *Bellicositermes Natalensis* et *Cubitermes* sp. La théorie de la stigmergie: essai d'interprétation du comportement des termites constructeurs". *Insect Soc* 6:

41–81

29 Bonabeau E, Theraulaz G, Deneubourg J-L (1996) Quantitative study of the fixed threshold model for the regulation of division of labour in insect societies. *Proc R Soc Lond B* 263: 1565–1569

30 Plowright RC, Plowright CMS (1988) Elitism in social insects: a positive feedback model. *In*: RL Jeanne (ed): *Interindividual behavioral variability in social insects*. Westview Press, Boulder, 419–431

31 Chrétien L (1996) Organisation spatiale du matériel provenant de l'excavation du nid chez *Messor barbarus* et des cadavres d'ouvrières chez *Lasius niger* (Hymenopterae: Formicidae). PhD Thesis, Université Libre de Bruxelles

32 Robinson GE (1987) Regulation of honey bee age polyethism by juvenile hormone. *Behav Ecol Sociobiol* 20: 329–338

33 Breed MD, Robinson GE, Page RE (1990) Division of labor during honey bee colony defense. *Behav Ecol Sociobiol* 27: 395–401

34 Detrain C, Pasteels JM, Deneubourg J-L (1988) Polyéthisme dans le tracé et le suivi de la piste chez *Pheidole pallidula* (Formicidae). *Actes Coll Insect Soc* 4: 87–94

35 Detrain C, Pasteels JM (1991) Caste differences in behavioral thresholds as a basis for polyethism during food recruitment in the ant *Pheidole pallidula* (Nyl.) (Hymenoptera: Myrmicinae). *J Insect Behav* 4: 157–176

36 Detrain C, Pasteels JM (1992) Caste polyethism and collective defense in the ant *Pheidole pallidula*: the outcome of quantitative differences in recruitment. *Behav Ecol Sociobiol* 29: 405–412

37 Wheeler DE, Nijhout HF (1981) Soldier determination in ants: new role for juvenile hormone. *Science* 213: 361–363

38 Schatz B (1997) Modalités de la recherche et de la récolte alimentaire chez la fourmi *Ectatomma ruidum* Roger: flexibilités

individuelle et collective. PhD dissertation, Université Paul Sabatier, Toulouse, France

39 Wilson EO (1980) Caste and division of labor in leaf-cutter ants (Hymenoptera: Formicidae: *Atta*). I. The overall pattern in *A. sexdens*. *Behav Ecol Sociobiol* 7: 143–156

40 Wilson EO (1983) Caste and division of labor in leaf-cutter ants (Hymenoptera: Formicidae: *Atta*). III. Ergonomic resiliency in foraging by *A. cephalotes*. *Behav Ecol Sociobiol* 14: 47–54

41 Lenoir A (1979) Le comportement alimentaire et la division du travail chez la fourmi *Lasius niger*. *Bull Biol France and Belgique* 113: 79–314

42 Calabi P (1986) Division of labor in the ant *Pheidole dentata*: the role of colony demography and behavioral flexibility. PhD dissertation, Boston University, Boston

43 Calabi P, Traniello JFA (1989) Behavioral flexibility in age castes of the ant *Pheidole dentata*. *J Insect Behav* 2: 663–677

44 Carlin NF, Hölldobler B (1983) Nestmate and kin recognition in interspecific mixed colonies of ants. *Science* 222: 1027–1029

45 Tofts C, Franks NR (1992) Doing the right thing: ants, honeybees and naked mole-rats. *Trends Ecol Evol* 7: 346–349

46 Tofts C (1993) Algorithms for task allocation in ants (a study of temporal polyethism: theory). *Bull Math Biol* 55: 891–918

47 Tofts C, Franks NR (1994) Foraging for work: how tasks allocate workers. *Anim Behav* 48: 470–472

48 Sendova-Franks AB, Franks NR (1993) Task allocation in ant colonies within variable environments (a study of temporal polyethism: experimental). *Bull Math Biol* 55: 75–96

49 Calderone NW, Page RE (1996) Temporal polyethism and behavioural canalization in the honey bee, *Apis mellifera*. *Anim Behav*

51: 631–643

50 Robinson GE, Page RE, Huang Z-Y (1994) Temporal polyethism in social insects is a developmental process. *Anim Behav* 48: 467–469

51 Calderone NW, Page RE (1988) Genotypic variability in age polyethism and task specialization in the honey bee *Apis mellifera* (Hymenoptera: Apidae). *Behav Ecol Sociobiol* 22: 17–25

52 Calderone NW, Page RE (1991) The evolutionary genetics of division of labor in colonies of the honey bee (*Apis mellifera*). *Amer Naturalist* 138: 69–92

53 Frumhoff PC, Baker J (1988) A genetic component to division of labour within honey bee colonies. *Nature* 333: 358–361

54 Page RE, Robinson GE (1991) The genetics of division of labour in honey bee colonies. Adv. *Insect Physiol* 23: 117–169

55 Robinson GE, Page RE (1988) Genetic determination of guarding and undertaking in honeybee colonies. *Nature* 333: 356–358

56 Stuart RJ, Page RE (1991) Genetic component to division of labor among workers of a leptothoracine ant. *Naturwissenchaften* 78: 375–377

57 Franks NR, Tofts CMN (1994) Foraging for work: how tasks allocate workers. *Anim Behav* 48: 470–472

58 Wheeler DE (1986) Developmental and physiological determinants of caste in social Hymenoptera: evolutionary implications. *Amer Naturalist* 128: 13–34

59 Jaisson P, Fresneau D, Lachaud J-P (1988) Individual traits of social behaviour in ants. *In*: RL Jeanne (ed): *Interindividual behavioral variability in social insects.* Westview Press, Boulder, CO, 1–51

60 Van der Blom J (1993) Individual differentiation in behaviour of honey bee workers (*Apis mellifera* L.). *Insect Soc* 40: 345–361

61 Deneubourg J-L, Goss S, Pasteels JM, Fresneau D, Lachaud JP (1987) Self-organization mechanisms in ant societies (II): learning in foraging and division of labour. *In*: JM Pasteels, J-L Denubourg (eds) *From Individual to Collective Behavior in Social Insects.* Birkhäuser, Basel, 54: 177–196

62 Theraulaz G, Goss S, Gervet J, Deneubourg J-L (1991) Task differentiation in *Polistes* wasp colonies: a model for self-organizing groups of robots. *In*: JA Meyer, SW Wilson (eds): *From animals to animats.* MIT Press, Cambridge, MA, 346–355

63 Theraulaz G, Bonabeau E, Deneubourg J-L (1998) Threshold reinforcement and the regulation of division of labour in social insects. *Proc R Soc Lond B* 265: 327–333

64 Bonabeau E, Theralauz G, Deneubourg JL (1998) Fixed response thresholds and the regulation of division of labour in insect societies. *Bull Math Biol* 60: 753–807

Social control of division of labor in honey bee colonies

Zhi-Yong Huang and Gene E. Robinson

Summary

Honey bee workers change jobs as they age. By rearing workers under different social conditions and manipulating the age structure of a colony, it was found that workers regulate their onset of foraging through a negative feedback mechanism, which we describe verbally in an "activator-inhibitor" model. The level of juvenile hormone (the hypothesized activator) in a worker influences the probability of it becoming a forager. Workers emerge with low levels of hormone, which are programmed to increase. Once the hormone reaches a critical level in a worker, the worker becomes a forager. The level of an inhibitor also becomes high in foragers, and unlike the activator, the inhibitor is transferred from bee to bee so that other individuals are inhibited from becoming foragers. Experimental results support the predictions of this model. Workers become foragers at earlier ages when colonies are manipulated to presumably decrease the levels of inhibitor, and they become foragers at older ages when the levels of inhibitor are increased. Foragers, with high juvenile hormone titers, show a drop in hormone titers when a manipulation is made that presumably increases inhibitor levels drastically. The activator-inhibitor model provides a heuristic tool for understanding the division of labor in honey bee colonies.

Introduction

Social insects have been extensively studied because of their economic importance, ecological success, and modes of social organization—such as division of labor—that are analogous in some ways to those found in human society. Division of labor among workers is related to differences in morphology, size, genotype or age [1–3, see Dreller and Page, this volume]. Many social insects are characterized by an age-related division of labor in which workers change jobs as they age. In honey bee colonies, workers perform tasks inside the nest, such as brood care ("nursing"), comb building, and food processing for the first 2–3 weeks of adult life, and carry out tasks outside the hive such as foraging and defense for the final 1–3 weeks of life.

Age-related division of labor in insect colonies shows great plasticity. In honey bee colonies, behavioral development from nest work to foraging can be accelerated, delayed, or even reversed in response to changing colony conditions [3]. For example, if a colony is experimentally composed of 1000–2000 newly emerged bees (a single-cohort colony), some workers become foragers at approximately 7 days of age [4–6], which is 2 weeks younger than typical foraging age. Conversely, if an experimental colony is composed entirely of foragers, some foragers "revert" to caring for the brood and queen instead of continuing to forage for food [5, 7]. Plasticity in aged-related division of labor also occurs under more natural conditions. A few workers were observed to initiate their foraging as young as 5–7 days of age [8, 9], perhaps because of a shortage of older workers in the colony. Delayed behavioral development occurs in colonies with a population of mostly older bees, such as in a colony established from a reproductive swarm. New workers do not emerge until 3 weeks after the colony is established, so just prior to their emergence the youngest workers nursing the brood are at least 21 days old [6].

A central question in insect sociobiology is how the activities of thousands of individual workers are integrated to forge a productive colony. The acquisition of information regarding colony needs is an important process in colony integration (Fig. 1). How do workers obtain information that results in changes in their behavioral development? Our approach to this problem has been to use endocrine and behavioral analyses and an experimental framework inspired by developmental biology. Looking to developmental biology is consistent with the venerable metaphor of the insect colony as a superorganism [1, 10]; in this case the control of division of labor is seen as a "superorganismic" trait [11]. This paper reviews our current understanding of how age-related division of labor is controlled in the honey bee colony. To facilitate these studies, we have focused on nursing and foraging, which occur at the beginning and end of behavioral development, respectively. One emphasis of this review is a further elaboration of a descriptive model that explains how social interactions can control the rate of

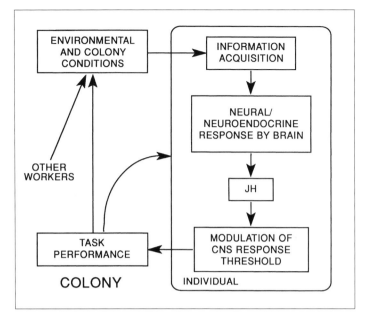

Figure 1 Hypothesized relationships between environmental and colony conditions and task performance by workers (modified from [3])

behavioral development. Other recent reviews deal more extensively with endocrine [12, 13]; genetic [14] and neural [12, 15] mechanisms.

Juvenile hormone and behavioral development in honey bees

Juvenile hormone (JH) is involved in the regulation of behavioral development in adult worker honey bees [3]. JH blood titers and rates of biosynthesis increase with age; they are low in bees that are nursing and performing other activities in the hive, and high in foragers [6, 16–19] and other bees whose tasks are outside the nest (guards, undertakers and soldiers, see [19]). Treating bees with JH, JH mimic, or JH analog induces precocious foraging [6, 20–25]. Removal of the corpora allata (CA), the glands that produce JH, delays the onset of foraging; this delay is eliminated by treating with JH analog [26]. These results suggest that high JH has a causal role in the timing of foraging behavior.

167

Plasticity in behavioral development also appears to be mediated by JH. Precocious foragers have high JH titers despite their young age, overaged nurses have low titers, and bees that revert from foraging to nursing show a drop in JH titer [6, 27, 28]. These results suggest that changes in colony conditions act on the endocrine system to cause changes in behavioral development [23]. Our working model is that colony conditions are perceived through one or more sensory modalities and cause a neural or neuroendocrine response by the brain, which results in changes in JH titers. These changes act directly or indirectly to modulate response thresholds for task-related stimuli, which influence what tasks a bee will perform (Fig. 1).

JH also is known to influence exocrine glands that are related to division of labor. For example, the hypopharyngeal glands play two entirely different roles depending on worker age. In nurses they are well developed and produce a proteinaceous secretion fed to larvae [29], whereas in foragers these glands are atrophied and produce α-glucosidase, an enzyme that converts nectar to honey [30]. JH appears to mediate the morphological and functional changes of the hypopharyngeal glands. Nurses, with low JH titers, have large glands that show high rates of protein synthesis [31, 32]. Treatment with JH or JH analog causes premature regression of the glands [20, 21, 33, 34], premature decrease in protein synthesis activity [31], and a premature increase in α-glucosidase activity [25]. Removal of the CA prevents gland regression, but the process is restored by applying JH [35]. Furthermore, when JH titers decrease in bees that revert from foraging to nursing, hypopharyngeal glands also regenerate to sizes similar to that of nurses [28]. JH also mediates alarm pheromone production. Methoprene, a JH analog, causes premature production of 2-heptanone and isoamyl acetate [22]. JH does not seem to regulate wax secretion by wax glands, because removal of the CA showed no effect [36].

Pathways of information acquisition in a honey bee colony: worker-worker versus worker-nest interaction

Cell development within a multicellular organism is mediated by interactions either with other cells or with the extracellular matrix. Considering workers as cells, they could either acquire information on colony needs indirectly while interacting with other adult workers, perhaps via trophallaxis or other forms of social contact, or while interacting directly with the nest and its contents while walking about inside the hive. Trophallaxis plays a role in information transfer in the context of social foraging in several social insects [37–39]. It is not known whether it also plays a role in division of labor, but this idea was suggested over 40 years ago [8]. Similarly, Lindauer [40] hypothesized that bees walking on the comb are actually "patrolling" for information on colony needs, but this too remains speculative.

Huang and Robinson [41] reared individual workers in social isolation in the laboratory. After 1 week, workers showed prematurely elevated rates of JH biosynthesis, similar to those of normally aged foragers (Fig. 2). Socially isolated bees also foraged significantly earlier than did colony-reared bees, when both groups were placed together in a third colony. Results from bees reared in small groups in the laboratory suggest that accelerated development in isolated bees was not due to an absence of nest environment, but rather an absence of a social

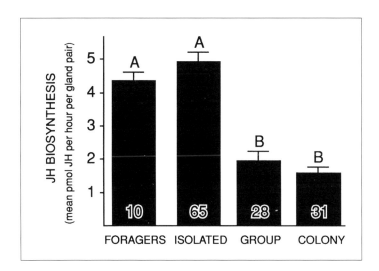

Figure 2 Effect of social and nest environment on juvenile hormone biosynthesis (mean ± SE) in worker bees
Newly emerged workers were reared either in isolation ("isolated"), in groups of four bees ("group"), or in a colony ("colony"). Juvenile hormone biosynthesis rates in vitro were measured after 7 days. Data for normal-aged foragers also presented for comparison. Numbers on each bar indicate sample sizes. Means with different letters on top indicate a significant difference by Tukey's test at the 5% level (modified from [3])

environment. Bees reared in small groups showed rates of JH biosynthesis typical of 7-day-old colony-reared bees, even though they lacked exposure to any nest stimuli.

To determine whether the social environment affects behavioral development in a colony setting, a "transplant" assay was developed [41]. In this assay, a group of foreign bees is transplanted into a single-cohort colony, and the rate of behavioral development in the resident bees is quantified. When foragers were used as the transplants, precocious development was inhibited in the resident bees, apparently due to the inhibition from the transplanted old bees (Fig. 3). This was a specific effect of transplanting foragers, because resident bees did develop precociously when young bees were transplanted. Because the transplanted older bees were foragers, there was a possibility that their nectar or pollen loads could have been used by the resident bees as cues for the presence of old bees. This possibility was eliminated by transplanting "soldiers", a group of bees that are behaviorally, genetically, and physiologically distinct from foragers [19, 42]. Soldiers inhibited behavioral development in a manner similar to foragers.

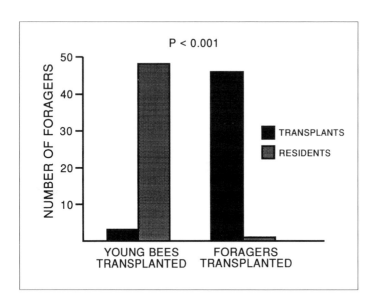

Figure 3. Effect of transplanted bees on the number of resident bees that forage precociously
(data from [41]).

Was the inhibitory effect of foragers mediated by their interaction with resident bees, or by changes in the nest environment caused by their foraging activity? To test these alternatives, we performed transplant experiments with the colony entrance closed, thereby eliminating any change of the nest due to foraging. Inhibition of behavioral development in resident bees was still observed. These results suggest that the behavioral development of young bees is affected directly by the presence of foragers.

Results of recent experiments confirm that changes in nest conditions are not playing a strong role in controlling behavioral development in honey bees. When bees in single-cohort colonies are starved, workers accelerate their development even faster than in regular (well-fed) single-cohort colonies, starting foraging as early as 4 days of age [43]. However, this starvation response apparently is not mediated by an assessment of food stores. A colony was set up in which bees were fed from a sugar feeder but no food stores were allowed to accumulate. The rate of behavioral development of these bees was similar to bees in a colony with ample food stores but significantly slower relative to bees in starved colonies [43]. These results demonstrate that the effect of starvation on behavioral development is not mediated via a direct interaction of workers with colony food stores. Rather, starvation-mediated changes in social interactions, personal nutritional status, or both, are suggested.

The activator-inhibitor model

A descriptive model was proposed to explain how worker-worker interactions influence honey bee behavioral development [41]. We hypothesized that there is an interplay between JH, which we designate as an intrinsic "activator" that promotes behavioral development, and an "inhibitor", an as-yet unidentified factor(s) transferred among workers that retards development. This factor may be a behavior, a chemical, or both. Activator levels are programmed to increase with age, but this increase is modulated by the amount of inhibitor a worker receives. There is evidence that the "default" trajectory of JH is a rapid increase during the first week of life, but social interactions apparently delay this increase. For example, 7-day-old workers deprived of social interactions show the same level of JH biosynthesis as that of normally aged foragers (Fig. 2). The model assumes that the production of the activator and the inhibitor are coupled, with older bees producing or transferring higher levels of inhibitor than younger bees. According to this model, genetic variation for production of, or sensitivity to, the activator and inhibitor can explain previously observed genetic variation in rates of behavioral development [6, 44, 45].

This model can explain all three forms of plasticity in behavioral development. Precocious development by some workers in a colony deficient in older bees is a consequence of young workers interacting relatively less frequently with older workers, receiving less inhibitor, and as a result showing an accelerated rate of activator increase. Delayed or reversed development in a colony with predominantly old bees is a result of workers receiving unusually high levels of inhibitor from older workers, delaying or reversing their increase of activator. Normal development of workers requires a "normal" amount of the inhibitor transferred among workers through social interactions. Therefore, the rate of behavioral development for honey bees is not a fixed, intrinsic, property of the individual worker; rather, it is a consequence of the effects of social interactions.

The activator-inhibitor model spans three levels of biological organization (Fig. 4). Within the colony, bees interact with one another. Via these interactions, older individuals (or more accurately, individuals with higher inhibitor levels) can inhibit the development of younger individuals (individuals with lower inhibitor levels). Within an individual bee, there are three hypothesized interacting subsystems. Sensory perception of social inhibition, by whatever modality it occurs, would be perceived by a peripheral sensory system, which transduces social inhibition into signals that influence the neuroendocrine system of the brain. The brain sends signals to an inhibitor production system that controls the level of inhibitor in the worker. Within the brain, results from other insect species suggest that neurosecretory cells in the pars intercerebralis receive these signals and produce substances that stimulate or inhibit production of JH by the CA [46]. JH is then hypothesized to act on a JH responsive system in the brain to influence responses to task-related stimuli [23]. At present this model specifies JH as the sole activator. However, JH is perhaps only one of several factors regulating behavioral development, since removal of the CA only delays, but does not eliminate, foraging behavior [26]. The activator-inhibitor model could easily be modified to accommodate multiple activators.

The activator-inhibitor model for honey bee behavioral development was inspired by models in developmental biology. Turing [47] pioneered the concept of modeling biological pattern formation using two interacting substances with

172

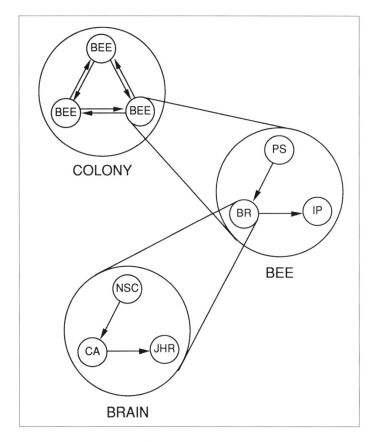

Figure 4. The workings of the hypothesized activator-inhibitor model span three levels of biological organization. Within a colony, workers interact with one another

Within an individual bee, a peripheral sensory (PS) system perceives the social inhibition and transduces it into signals that influence the neuroendocrine system of the brain (BR). The brain sends signals to an inhibitor production (IP) system that controls the level of inhibitor in the worker. Within the brain, neurosecretory cells (NSC) receive signals from PS and produce substances that stimulate or inhibit production of juvenile hormone by the corpora allata (CA). Juvenile hormone may then act on a juvenile hormone responsive (JHR) system in the brain to influence responses to task-related stimuli that affect task perform-ance.

different diffusion rates. This concept was later developed as an activator-inhibitor model for explaining head and foot generation in *Hydra* [48–50]. In the *Hydra* model, both the hypothesized activator and the inhibitor are diffusible, although the activator is usually thought to act at short range and is autocatalytic, whereas the inhibitor can antagonize the effect of the activator over a longer range. In our model, the hypothesized inhibitor is "diffusible", that is, transferred among workers, but the activator works within the individual and does not diffuse among workers. The activator and the inhibitor, therefore, do not work at the same levels of organization in our model, whereas in other activator-inhibitor models they do.

Empirical evidence for the activator-inhibitor model

In the following section we review results of recent experiments [28] designed to test predictions of the activator-inhibitor model. Rate of behavioral development was measured in "triple-cohort colonies" with precisely controlled proportions of young, middle-aged, and old bees. In one set of experiments, experimental colonies had foragers artificially removed to simulate predation, wheras control colonies had the same number of workers removed, but evenly across all three age cohorts. Behavioral development in the middle-aged cohort was accelerated in response to forager depletion. We suggest that this was due to the fact that young workers were exposed to fewer bees that had high inhibitor levels during social interactions. But the observed accelerated development could also have been due to changes in the nest environment, because forager depletion probably also caused a decrease in the amount of fresh nectar and pollen brought into the nest.

A second experiment was performed to distinguish between these two possibilities. Experimental colonies had their foragers almost completely confined inside the hive with a water sprinkler for a few days, whereas control colonies were undisturbed. If changes in the nest environment were the proximate cause of accelerated development, we expected that the experimental colonies should have accelerated foraging, because they lacked fresh nectar or pollen. However, we found that middle-aged bees in forager-confined colonies showed delayed, rather than accelerated, development. These results are reminiscent of the results of transplant experiments discussed above [41], but in this case the colony age structure is more typical (though overall populations were smaller then in more natural colonies). These results are consistent with the activator-inhibitor model: younger bees in the forager-confined colonies would be exposed to more inhibitor than usual because they are more likely to come in contact with old bees.

Another prediction of the activator-inhibitor model is that reversion will occur in a colony composed of all old bees, even if it is devoid of brood. It has been shown previously that in a colony made entirely of foragers, some bees show a decrease in JH titers and revert to nursing activities, but in these experiments the colony always contained open brood, and it was assumed that the stimulus from the brood induced some foragers to become nurses [7, 27]. The activator-inhibitor model predicts that because foragers in an old-bee colony are interacting only with other foragers, all of which are presumed to have (or to transfer) high amounts of inhibitor, some foragers should experience a drop in JH titers because they are exposed to unnaturally high amounts of inhibitor. Such hormonal reversion indeed does occur in old-bee colonies without brood within 24 h of colony establishment [28].

Calderone [51] suggested that the activator-inhibitor model predicts that bees reared in groups in the laboratory for the first several weeks of adulthood will show an earlier onset of foraging, and pointed out that results from one study [52] are not consistent with this prediction. However, the activator-inhibitor model does not make this prediction. The activator-inhibitor model predicts that workers reared in groups in the laboratory for the first several weeks of adulthood will show greater interindividual variance in age at first foraging relative to colony-reared bees. This is because the bees will differentiate strongly, despite their identical ages, as they do in a single-cohort colony. Some bees will develop precociously, whereas the majority will show normal development because they will be inhibited by the precocious bees. Results from a new analysis of the data presented in Calderone and Page [51] are consistent with this prediction (Tab. 1A). In three out of four trials, bees reared in groups in the laboratory for the first 1–2 weeks of adulthood showed a significantly higher variance in age at first foraging than did colony-reared bees. Calderone's [51] prediction probably was based on results showing that bees reared as isolated individuals in the laboratory show an earlier onset of foraging when introduced into a colony [41]. Huang and Robinson [41], however, also showed that bees reared in groups in the laboratory had low mean rates of JH biosynthesis (Fig. 2), suggesting that many of them would not have foraged precociously. A similar analysis on the data of Huang and Robinson [41] also showed that in three out of three tests, bees reared in groups showed a significantly higher variance in rates of JH biosynthesis (Tab. 1B). Thus, the differentiation in group bees lacking foragers is not only seen at the behavioral level, but also at the underlying hormonal level.

Table 1 *Effect of laboratory rearing on variance in rates of: behavioral development (A) and JH biosynthesis (B)*

(A) Rearing condition	Age at first foraging mean ± SE (n)		Variance	Test for equality of variance	
				F-statistic	P value
High-pollen strain					
175 bees/group	27.44 ± 0.81	(68)	43.96	2.11	0.001
colony-reared	25.56 ± 0.37	(153)	20.81		
125 bees/group	30.47 ± 0.61	(100)	36.84	1.55	0.015
colony-reared	30.38 ± 0.49	(100)	23.77		
Low-pollen strain					
175 bees/group	34.88 ± 0.64	(84)	34.00	1.41	0.033
colony-reared	28.01 ± 0.39	(160)	24.18		
125 bees/group	30.19 ± 0.63	(79)	30.96	1.35	0.076
colony-reared	33.10 ± 0.47	(105)	22.97		

(B) Rearing condition	Rates of JH biosynthesis mean ± SE (n)		Variance	F-statistic	P value
12 bees/group	1.91 ± 0.23	(48)	2.49	2.56	0.004
7 bees/group	2.10 ± 0.47	(14)	2.87	2.95	0.007
4 bees/group	1.97 ± 0.27	(29)	1.97	2.03	0.031
colony-reared	1.59 ± 0.18	(31)	0.97		

Towards the identification of a worker inhibitor

One major weakness in our understanding of the control of behavioral development in honey bees is that no worker inhibitor has yet been identified. However,
recent studies implicate substances produced by worker mandibular glands as a
primer pheromone transmitted via direct social contact [53]. Using the transplant
assay described above, it was shown that older bees with mandibular glands
removed are less able to inhibit precocious foraging in a single-cohort colony relative to intact bees. This reduction of inhibition appears to be due to removal of
the mandibular glands because sham-operated bees were as potent as intact bees.
That direct social contact is required for inhibition was shown in another set of
experiments. Newly emerged bees were reared for 7 days in a typical colony in
one of three ways: individually in cages with a double screen, individually in
cages with a single screen, or freely inside the hive (control bees). Bees in double-screen cages had no direct physical contact with other colony members; bees
in single-screen cages could antennate and exchange food with colony members,
and control bees had normal, unrestricted access to colony members. Bees reared
in double-screen cages showed significantly higher levels of JH than control
bees, and when placed in a colony, they were also more likely to become precocious foragers than control bees. Bees reared in single-screen cages, however,
were only partially inhibited, both hormonally and behaviorally. These results
suggest that direct physical contact is required for worker-worker inhibition, but
further experiments are necessary to prove this. Inhibitory chemicals produced in

---

Table 1 (Legend)
(A) Mean (±SE) age at first foraging for two strains of honey bee workers that were either
reared in the laboratory (group size as indicated) or in a test colony. Bees were reared for the
first 6 (175-bee group) or 12 (125-bee group) days of adulthood and then introduced into the
same test colonies or reared in the colonies upon emergence. Mean, standard error and sam-
ple size (in parentheses) are from Table 1 of Calderone and Page [52]. Variance was calculated
as [SE*√n – 1]². F-statistic was calculated by dividing the variance of laboratory-reared bees by
that of the corresponding colony-reared bees. (B) Mean (±SE) rates of JH biosynthesis in bees
reared either in laboratory (group size as indicated) or in a colony. Bees were reared for the
first 6 days of adulthood and tested on day 7. Mean, standard error and sample size (in paren-
theses) are from Figure 1 of Huang and Robinson [41]. F-statistic was calculated by dividing the
variance of each group by that of colony-reared bees. Other group sizes were tested for JH [41],
but only groups with JH not significantly different from colony-reared bees were tested for
equality of variance.

the mandibular glands of workers may be passed around during either antenna-tion or exchange of food, in a manner similar to the pheromone produced by the mandibular glands of the queen [54].

Negative feedback mechanisms and social control

According to the activator-inhibitor model, an important component in the control of division of labor in honey bee colonies is a negative feedback mechanism based on social interactions. This type of control mechanism is common among biological systems, from slime molds to mammals. In slime molds, inhibitors released by cells are implicated in regulating the proportion of spore and stalk cells [55]. In rodents, inhibition of reproduction is socially mediated, and primer pheromones have been implicated [56, 57]. In the only known eusocial mammal, the naked mole rat, reproduction in workers is apparently inhibited under normal colony conditions, because both males and females show reproductive development when they are removed from their colonies [58, 59].

Social inhibition also is quite common in insect colonies. Queens are thought to produce pheromones that result in inhibition of reproductive development in termites [60], or inhibition of queen rearing in honey bees [61, 62]. The number of drones in a honey bee colony also is regulated [63], and it has been suggested that this is based on an inhibitory pheromone emanating from adult drones [64, but see 65]. Evidence suggests that a pheromone produced by fire ant queens inhibits the fecundity of other queens in polygyne colonies [66]. In termites and ants, a mechanism similar to the activator-inhibitor model might be involved in the regulation of soldier production. Adult ant soldiers inhibit the production of new soldiers in a colony, in a process mediated by JH [67, 68] that apparently involves a pheromone.

Similarities between the activator-inhibitor model for honey bees and more proven social regulatory mechanisms in other systems do not, of course, provide evidence that such a mechanism also operates in honey bee colonies. However, if further experimental work continues to support the activator-inhibitor model, these similarities can be used to gain further insight into the mechanisms of social regulation in both honey bees and other systems.

Colony needs and the structure of the worker force

There are at least two ways for a colony of social insects to control the structure of its worker force. A colony might regulate the proportion of workers in different behavioral states based solely on the current needs of the colony [69, 70]. Alternatively, a colony might always maintain groups of workers in different behavioral states, even if there is no need for particular tasks to be performed at certain times. We suggest that the second alternative would result in a more stable work force, although it probably requires a relatively large colony population. This would be more efficient in the context of temporal polyethism. This is because complex physiological changes occur as workers move from one state of behavioral development to another. As discussed earlier, in honey bees there are several changes in exocrine gland activity that occur during behavioral development. Young workers also have lower levels of brain biogenic amines [71, 72] and differences in brain structure [73] compared with foragers. Maintaining a standing force of physiologically specialized workers for a particular task, even if the workers are at times underemployed, minimizes the time and energy required to develop physiological competence once such workers are needed. This argument assumes that these physiological differences do in some way relate to competence, but the evidence is limited.

Evidence from honey bees suggests that workers physiologically specialized to perform a particular task can be found in a colony, even in the absence of current need. In the reversion experiments conducted in broodless colonies described earlier [28], foragers that showed hormonal reversion had large hypopharyngeal glands, indistinguishable from those of normal nurse bees. Perhaps there was a need for some nurse bees to care for the queen, but the widespread reversion among workers suggests that the queen was not the main cause of this reversion. We also have observed that a colony in a temperate environment can quickly mobilize a foraging force in the winter, if external conditions suddenly become favorable [74]. These results are consistent with the idea that there is a flexible development program that underlies age-related division of labor [75]. These results are much more difficult to interpret using the foraging for work model [69, 70].

Social regulation of temporal polyethism via an activator-inhibitor mechanism is one way of maintaining a stable structure of the workforce that is not directly dependent on the current needs of the colony. The activator-inhibitor model also predicts appropriate changes in the colony's labor force in the face of changing environmental conditions. A loss of foragers leads to an acceleration of behavioral development in honey bee colonies [28], a response that is predicted both by the activator-inhibitor model and any model based on responses to the current needs of the colony. But when foragers are inactive, a delay in behavioral development in young bees is observed [28], and this response is consistent only with the activator-inhibitor model. A model based on the current needs of the colony would predict instead an acceleration of behavioral development. The response specified by the activator-inhibitor model probably makes better sense because the problem is not a shortage of foragers but their inability to forage due to external conditions. These observations suggest that it is necessary in the future to understand better the relationship between the structure of a colony's labor force and its response to changing conditions, and that such an understanding requires both physiological as well as ecological analyses.

The activator-inhibitor model and self-organization

Self-organization is a process in which a pattern or structure emerges spontaneously from the dynamic interactions among individual components in a system, without either central control or predetermined blueprint. Many processes in biology have been explained from a self-organization perspective, including some aspects of the organization of insect societies (division of labor in wasps [76]; social foraging in ants and bees [77, 78]; and pattern formation of honey bee combs [79]).

The activator-inhibitor model also reflects a process of self-organization. According to the model, a colony always produces foragers, regardless of initial conditions. It predicts, for example, that if there are no foragers (such as in a single-cohort colony) or insufficient foragers (due to predation or other factors), in the resulting social environment of decreased inhibition workers are more likely to become foragers. Conversely, in a colony composed entirely of old bees, the high level of inhibition would result in some bees reverting to preforaging status. Because of this self-organizational property, colonies with no foragers or with mostly foragers would not exist in a stable manner, but would redevelop quickly to achieve a division of labor.

From our description of the activator-inhibitor model, it would appear that it only predicts that a colony will have preforagers and foragers. Indeed, all experiments so far have dealt only with nurses and foragers, the two most differentiated groups of workers in the honey bee colony. However, preliminary simulations of the model (Z.-Y. Huang, Y. Oono and G. Robinson, unpublished results) result in broader behavioral heterogeneity.

Beyond the activator-inhibitor model

It is already becoming apparent that the control of age-related division of labor in honey bee colonies is more complex than suggested by the activator-inhibitor model. JH was originally thought of as an activator that triggers foraging behavior, but the allatectomy results described earlier suggest that JH controls behavioral development in concert with other, as yet unidentified, factors. This is consistent with results showing that in middle-aged bees low JH levels are associated with comb building and food storing, whereas high JH levels are associated with guarding and undertaking [19]. There must be other factor(s), besides JH, to regulate division of labor. The tremble dance, a behavior performed by nectar foragers when they find it difficult to unload their harvest, causes some nurse bees to shift into food storing, a task associated with middle age [80]. This suggests that behavioral development can be stimulated by behavioral interaction. Dorsoventral abdominal vibrations exert short-term modulatory effects on the likelihood of foraging [81], but it is not known whether behavioral development is influenced.

The hypothesized worker inhibitor is not yet identified, but at least one other inhibitor has already been isolated. Queen mandibular pheromone inhibits JH biosynthesis in caged bees and in small colonies [82], and also delays behavioral development in workers in field colonies [83]. Evidence for an inhibitory brood pheromone also has been obtained recently (Y. Le Conte and G.E. Robinson, unpublished results).

It is possible that environmental factors influence the production or efficacy of activators and inhibitors. For example, it is reported recently that starvation results in very high rates of JH biosynthesis in worker honey bees maintained in isolation in the laboratory [84]. This is consistent with the findings of accelerated behavioral development by starved bees in small colonies [43].

The regulation of age-related division of labor in colonies of honey bees and other social insects is beginning to be understood as a multifactorial process. The activator-inhibitor model provides a heuristic tool for exploring the intricate underlying mechanisms.

Acknowledgment

We thank S. L. Beshers for ideas on colony needs and the structure of the worker force, P. K. Visscher for statistical advice, and S. L. Beshers, J. L. Deneubourg, C. Detrain and D. J. Schulz for reviewing this manuscript. Research reported here was supported by grants from the National Institute of Health, the National Science Foundation, and the United States Department of Agriculture to GER.

References

1 Wilson EO (1971) *The insect societies.* Belknap Press, Cambridge, MA

2 Page RE, Robinson GE (1990) The genetics of division of labour in honey bee colonies. Adv. *Insect Physiol* 23: 117–169

3 Robinson GE (1992) Regulation of division of labor in insect societies. *Annu Rev Entomol* 37: 637–665

4 Nelson FC (1927) Adaptability of young bees under adverse conditions. *Amer Bee J* 67: 242–243

5 Milojévic BD (1940) A new interpretation of the social life of the honeybee. *Bee World* 21: 39–41

6 Robinson GE, Page RE, Strambi C, Strambi A (1989) Hormonal and genetic control of behavioral integration in honey bee colonies. *Science* 246: 109–112

7 Page RE Jr, Robinson GE, Britton DS, Fondrk MK (1992) Genotypic variability for rates of behavioral development in worker honeybees (*Apis mellifera* L.). *Behav Ecol Sociobiol* 3: 173–180

8 Ribbands CR (1952) Division of labour in the honeybee community. *Proc R Soc Lond B* 14: 32–42

9 Sakagami SF (1953) Arbeitsteilung der Arbeiterinnen in einem zwergvolk, bestehend aus gleichaltrigen Volksgenossen. *J Fac Sci Hokkaido Univ Ser* 11: 343–400

10 Seeley TD (1997) Honey bee colonies are group-level adaptive units. *Amer Naturalist* 150S: 22–41

11 Ratnieks FLW, Reeve HK (1992) Conflict in single-queen hymenoptera societies: the structure of conflict and processes that reduce conflict in advanced eusocial species. *J Theor Biol* 158: 33–65

12 Fahrbach SE, Robinson GE (1996) Juvenile hormone, behavioral maturation, and brain structure in the honey bee. *Dev Neurosci* 18: 102–114

13 Robinson RE, Vargo EI (1997) Juvenile hormone in adult eusocial Hymenoptera:

gonadotropin and behavioral pacemaker. Arch. *Insect Biochem Physiol* 35: 559–583

14 Robinson GE, Huang Z-Y (1998) Colony integration in honey bees: genetic, endocrine, and social control of division of labor. *Apidologie* 29: 159–170

15 Robinson GE, Fahrbach SE, Winston ML. Insect societies and the molecular biology of social behavior. *Bioessays* 19: 1099–1108

16 Rutz W, Gerig L, Wille H, Lüscher M (1976) The function of juvenile hormone in adult worker honeybees, *Apis mellifera. J Insect Physiol* 22: 1485–1490

17 Fluri P, Lüscher M, Wille H, Gerig L (1982) Changes in weight of the pharyngeal gland and haemolymph titres of juvenile hormone, protein and vitellogenin in worker honey bees. *J Insect Physiol* 28: 61–68

18 Robinson GE, Strambi A, Strambi C, Paulino-Simões ZL, Tozeto SO, Barbosa JMN (1987) Juvenile hormone titers in Africanized and European honey bees in Brazil. *Gen Comp Endocrinol* 66: 457–459

19 Huang Z-Y, Robinson GE, Borst DW (1994) Physiological correlates of division of labor among similarly aged honey bees. *J Comp Physiol* 174: 731–739

20 Jaycox ER (1976) Behavioral changes in worker honey bees (*Apis mellifera* L.) after injection with synthetic juvenile hormone (Hymenoptera: Apidae). *J Kans Entomol Soc* 49: 165–170

21 Jaycox ER, Skowronek W, Gwynn G (1974) Behavioral changes in worker honey bees (*Apis mellifera*) induced by injections of a juvenile hormone mimic. *Ann Entomol Soc Amer* 67: 529–534

22 Robinson GE (1985) Effects of a juvenile hormone analogue on honey bee foraging behaviour and alarm pheromone production. *J Insect Physiol* 31: 277–282

23 Robinson GE (1987) Regulation of honey bee age polyethism by juvenile hormone.

Behav Ecol Sociobiol 20: 329–338

24 Robinson GE, Ratnieks FLW (1987) Induction of premature honey bee (Hymenoptera: Apidae) flight by juvenile hormone analogs administered orally or topically. *J Econ Entomol* 80: 784–787

25 Sasagawa H, Sasaki M, Okada I (1989) Hormonal control of the division of labor in adult honeybees (*Apis mellifera* L.) I. Effect of methoprene on corpora allata and hypopharyngeal gland, and its α-glucosidase activity. *Appl Entomol Zool* 24: 66–77

26 Sullivan JP, Jassim O, Robinson GE, Fahrbach SE (1996) Foraging behavior and mushroom bodies in allatectomized honey bees. *Soc Neuro Abst* 22: 1144

27 Robinson GE, Page RE, Strambi C, Strambi A (1992) Colony integration in honey bees: mechanisms of behavioural reversion. *Ethology* 90: 336–350

28 Huang Z-Y, Robinson GE (1996) Regulation of honey bee division of labor by colony age demography. *Behav Ecol Sociobiol* 39: 147–158

29 Jung-Hoffman I (1966) Die Determination von Königin und Arbeiterin der Honigbiene (*Apis mellifera* L.). *Bienenforsch* 8: 296–322

30 Simpson J, Riedel IBM, Wilding N (1968) Invertase in the hypopharyngeal glands of the honey bee. *J Apic Res* 7: 29–36

31 Brouwers EVM (1983) Activation of the hypopharyngeal glands of honeybees in winter. *J Apicultural Res* 22: 137–141

32 Huang Z-Y, Otis GW (1989) Factors determining hypopharyngeal gland activity of worker honey bees (*Apis mellifera*). *Insect Soc* 36: 264–276

33 Beetsma J, Ten Housten A (1974) Effects of juvenile hormone analogues in the food of honeybee colonies. *Z Angew Entomol* 77: 292–300

34 Rutz W, Gerig L, Wille H, Lüscher M (1974) A bioassay for juvenile hormone (JH) effects of insect growth regulators

(IGR) on adult worker honeybees. *Mitt Schweiz Entomol Gesell* 47: 307–313

35 Imboden H, Lüscher M (1975) Allatektomie bei adulten Bienen-arbeiterinnen (*Apis mellifica*). *Rev Suisse Zool* 82: 694–699

36 Muller EJ, Hepburn HR (1992) Temporal and spatial patterns of wax secretion and related behavior in the division of labour of the honeybee (*Apis mellifera capensis*). *J Comp Physiol* 171: 111–115

37 Seeley TD (1992) The tremble dance of the honey bee: message and meanings. *Behav Ecol Sociobiol* 31: 375–383

38 Jeanne RL (1990) Polyethism. *In*: KG Ross, RW Matthews (eds): *The Social Biology of Wasps*. Cornell University Press, Ithaca, NY, 389–425

39 Sorensen AA, Busch TM, Vinson SB (1985) Control of food influx by temporal subcastes in the fire ant, *Solenopsis invicta*. *Behav Ecol Sociobiol* 17: 190–198

40 Lindauer M (1952) Ein Beitrag zur Frage der Arbeisteilung im Bienenstaat. *Zeitschr Vergl Physiol* 36: 299–345

41 Huang Z-Y, Robinson GE (1992) Honeybee colony integration: worker-worker interactions mediate hormonally regulated plasticity in division of labor. *Proc Natl Acad Sci USA* 89: 11 726–11 729

42 Breed MD, Robinson GE, Page RE (1990) Division of labor during honey bee colony defense. *Behav Ecol Sociobiol* 27: 395–401

43 Schulz DJ, Huang Z-Y, Robinson GE (1998) Effect of colony food shortage on the behavioral development of the honey bee, *Apis mellifera*. *Behav Ecol Sociobiol* 42: 295–303

44 Kolmes SA, Winston ML, Fergusson LA (1989) The division of labor among worker honey bees (Hymenoptera: Apidae): the effects of multiple patrilines. *J Kans Entomol Soc* 62: 80–95

45 Giray T, Robinson GE (1994) Effects of intracolony variability in behavioral development on plasticity of division of labor in honey bee colonies. *Behav Ecol Sociobiol* 35: 13–20

46 Tobe SS, Stay B (1985) Structure and regulation of the corpus allatum. *Adv Insect Physiol* 18: 305–432

47 Turing AM (1952) The chemical basis of morphogenesis. *Phil Trans R Soc B* 237: 5–72

48 Gierer A, Meinhardt H (1972) A theory of biological pattern formation. *Kybernetik* 12: 30–39

49 MacWilliams HK (1983) *Hydra* transplantation phenomena and the mechanism of hydra head regeneration. II. Properties of the head activation. *Dev Biol* 96: 239–257

50 Meinhardt H (1993) A model for pattern formation of hypostome, tentacles, and foot in Hydra: how to form structures close to each other, how to form them at a distance. *Dev Biol* 157: 321–333

51 Calderone NW (1997) Proximate mechanisms of age polyethism in the honey bee, *Apis mellifera* L. *Apidologie* 29: 127–158

52 Calderone NW, Page RE (1996) Temporal polyethism and behavioural canalization in the honey bee, *Apis mellifera*. *Anim Behav* 51: 631–643

53 Huang Z-Y, Plettner E, Robinson GE (1998) Effects of social environment and worker mandibular glands on endocrine-mediated behavioral development in honey bees. *J Insect Physiol*; *in press*

54 Winston ML, Slessor KN (1992) The essence of royalty: honey bee queen pheromone. *Sci Am* 80: 374–385

55 Bloom L, Kay R (1988) The search for morphogens in *Dictyostelium*. *Bioessays* 9: 187–190

56 Batzli GO, Getz LL, Hurley SS (1977) Suppression of growth and reproduction of microtine rodents by social factors. *J Mammal* 58: 583–590

57 Darney KJ Jr, Goldman JM, Vandenbergh

JG (1992) Neuroendocrine responses to social regulation of puberty in the female house mouse. *Neuroendocrinol* 55: 434–443

58 Faulkes CG, Abbott DH, Jarvis JUM (1990) Social suppression of ovarian cyclicity and wild colonies of naked mole-rats, *Heterocephalus glaber. J Reprod Fert* 88: 559–568

59 Faulkes CG, Abbott DH (1990) Social control of reproduction in breeding and non-breeding male naked mole-rats (*Heterocephalus glaber*). *J Reprod Fert* 93: 427–435

60 Lüscher M (1961) Social control of polymorphism in termites. *Symp R Entomol Soc Lond* 1: 57–67

61 Winston ML, Higo HA, Slessor KN (1990) Effect of various dosages of queen mandibular gland pheromone on the inhibition of queen rearing in the honey bee (Hymenoptera: Apidae). *Ann Entomol Soc Amer* 83: 234–238

62 Engels W, Adler A, Rosenkrnanz P, Lubke G, Francke W (1993) Dose-dependent inhibition of emergency queen rearing by synthetic 9-ODA in the honey bee, *Apis mellifera carnica. J Comp Physiol* 163: 363–366

63 Free JB, Williams IH (1975) Factors determining the rearing and rejection of drones by the honeybee colony. *Anim Behav* 23: 650–675

64 Omholt SW (1988) Drone production in honeybee colonies: controlled by a longlasting inhibitory pheromone from the drones. *J Theor Biol* 134: 309–318

65 Henderson CE (1994) Influence of the presence of adult drones on the further production of drones in honey bee (*Apis mellifera* L) colonies. *Apidologie* 25: 31–37

66 Vargo EL (1992) Mutual pheromonal inhibition among queens in polygyne colonies of the fire ant *Solenopsis invicta*.

Behav Ecol Sociobiol 31: 205–210

67 Passera L (1974) Différenciation des soldats chez la fourmi *Pheidole pallidula* Nyl. (Formicidae: Myrmicinae). *Insect Soc* 21: 71–86

68 Wheeler DE, Nijhout HF (1984) Soldier determination in *Pheidole bicarinata*: inhibition by adult soldiers. *J Insect Physiol* 30: 127–135

69 Tofts C, Franks NR (1992) Doing the right thing: ants, honeybees and naked mole-rats. *Trends Ecol Evol* 7: 346–349

70 Franks NR, Tofts C (1994) Foraging for work: how tasks allocate workers. *Anim Behav* 48: 470–472

71 Harris JW, Woodring J (1992) The effects of stress, age, season, and source colony on levels of octopamine, dopamine and serotonin in the honeybee (*Apis mellifera* L.) brain. *J Insect Physiol* 38: 29–35

72 Taylor DJ, Robinson GE, Logan BJ, Laverty R, Mercer AR (1992) Changes in brain amine levels associated with the morphological and behavioural development of the worker honeybee. *J Comp Physiol* 170: 715–721

73 Withers GS, Fahrbach SE, Robinson GE (1993) Selective neuroanatomical plasticity and division of labour in the honeybee. *Nature* 364: 238–240

74 Huang Z-Y, Robinson GE (1995) Seasonal changes in juvenile hormone titers and rates of biosynthesis in honey bees. *J Comp Physiol* 165: 18–28

75 Robinson GE, Page RE, Huang Z-Y (1994) Temporal polyethism in social insects is a developmental process. *Anim Behav* 48: 467–469

76 Theraulaz G, Goss S, Gervet J, Deneubourg JL (1990) Task differentiation in *Polistes* wasp colonies: a model for self-organizing groups of robots. *In*: JA Meyer, S Wilson (eds): *Simulation of animal behavior: from animals to animats.* MIT Press, Cambridge, MA, 346–355

77 Beckers R, Goss S, Deneubourg JL, Pasteels JM (1989) Colony size, communication and ant foraging strategy. *Psyche* 96: 239–256

78 Camazine S, Sneyd J (1990) A model of collective nectar source selection by honey bees: self-organization through simple rules. *J Theor Biol* 149: 547–571

79 Camazine S (1990) Self-organizing pattern formation on the combs of honey bee colonies. *Behav Ecol Sociobiol* 28: 61–76

80 Seeley TD, Kühnholz S, Wiedenmuller A (1996) The honey bee's tremble dance stimulates additional bees to function as nectar receivers. *Behav Ecol Sociobiol* 39: 419–427

81 Schneider SS, Stamps JA, Gary NE (1986) The vibration dance of the honey bee. II. The effects of foraging success on daily patterns of vibration activity. *Anim Behav* 34: 386–390

82 Kaatz H-H, Hildebrandt H, Engels W (1992) Primer effect of queen pheromone on juvenile hormone biosynthesis in adult worker honey bees. *J Comp Physiol* 162: 588–592

83 Pankiw T, Huang Z-Y, Winston ML, Robinson GE (1998) Queen mandibular gland pheromone influences worker honey bee (*Apis mellifera* L.) foraging ontogeny and juvenile hormone titers. *J Insect Physiol*; *in press*

84 Kaatz H, Eichmuller S, Kreissl S (1994) Stimulatory effect of octopamine on juvenile hormone biosynthesis in honey bees (*Apis mellifera*): physiological and immunocytochemical evidence. *J Insect Physiol* 40: 865–872

Genetic, developmental, and environmental determinants of honey bee foraging behavior

Claudia Dreller and Robert E. Page Jr.

Summary

Honeybees forage for pollen, nectar, water, and propolis. The decision of an individual to forage either for nectar or pollen, the foraging rate, load size, and recruitment behavior are influenced by its genotype, development, and the internal conditions of the colony. The collective activities of foragers build the intake rate of the colony, which influences the major stimulus factors determining pollen foraging behavior: the quantities of stored pollen and brood. The current knowledge of the functional pathways from genes through neurophysiological process to individual behavior that serve as the determinants for social behavior are presented.

Introduction

The evolution of division of labor in insect societies has perplexed evolutionary biologists since Darwin first considered it a serious "difficulty" for his fledgling theory of evolution by natural selection [1]. Workers in social insect colonies differ anatomically and behaviorally; they also interact in complex ways that lead to coordinated colony organization. The collective activities of workers result in apparently coordinated colony-level allocation in workers into specific tasks, and in the allocation of nest space. A good example is the social organization of workers resulting in the colony's allocation of foragers into pollen and nectar collecting. Whenever flowers are in bloom, a honeybee colony must deploy its foragers among different food patches. This entails acquiring information about the colony's need for nectar and pollen combined with information about available patches in the environment. Individual foragers have to decide whether they collect pollen or nectar on the basis of this information.

C. Detrain et al. (eds) Information Processing in Social Insects
© 1999, Birkhäuser Verlag Basel/Switzerland

Foraging decisions

Division of labor takes place between workers that remain in the nest and those that forage. Foragers collect propolis, a sticky resinous substance, water, pollen, and nectar. Little is known about propolis and water foraging decisions; however, many recent studies demonstrate that pollen and nectar foraging behavior is influenced by the genotype of the individual, its nest and foraging environment, and its stage of behavioral development.

Effects of genotype on foraging decisions

The genotype of a forager has a significant effect on whether she collects pollen or nectar. Several different kinds of studies have shown this. Robinson and Page [2], Calderone et al. [3], and Dreller et al. [4] showed that foragers derived from different mates of naturally mated queens differed in their likelihood of collecting pollen. Studies using artificially selected strains of bees provide the same results. These strains had been selected for several generations in Davis for their pollen hoarding behavior as described in Hellmich et al. [5] and were clearly different for colony-level performance as well as for individual behavior with respect to pollen collecting and storing [6]. Colonies from high pollen strains (high pollen bees) have larger stores of pollen compared with low pollen strains (low pollen bees). When bees derived from these strains selected for high and low pollen hoarding [5, 6] were cofostered (raised together in the same colony), the bees from high strains are more likely to forage for pollen than those from low strains [7–9], demonstrating that pollen foraging behavior is a heritable trait.

Genotype also affects the foraging rates, load sizes, and recruitment behavior of individual foragers. Observations on high and low strain foragers in an indoor flight cage, where only artificial nectar sources were provided, revealed that nectar foragers from the low pollen strain made more trips for nectar per day and collected larger loads per trip than did nectar foragers from the high pollen strain [10] (Fig. 1). On the other hand, high strain pollen foragers made more trips for pollen than did low strain pollen foragers. Recruitment behavior is also affected. The dances of high strain pollen foragers were more vigorous than dances of low strain pollen foragers [11] (Fig. 1). Therefore, the genotype of a worker not only influences whether she will be a pollen or nectar forager, but also her rate of nec-

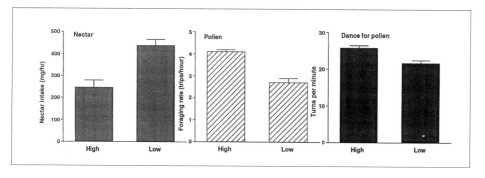

Figure 1 The left histogram shows the intake rates per hour of high and low strain foragers that were foraging for nectar (data modified from [10])

The center histogram demonstrates that high strain pollen foragers take more pollen trips per hour than do low strain pollen foragers (data modified from [10]). The right histogram shows that high strain pollen foragers perform more vigorous recruitment dances-measured by dance reversals per minute for pollen than do low strain pollen foragers. The bars indicate SE.

tar and pollen uptake and her recruiting potential, which then might change the overall foraging activity within a colony.

Honeybees are able to inform their nestmates about the location of a profitable food source. Foragers are, therefore, either recruited by dances or they search independently as scouts for new patches. Recent studies [12] have shown that scouts also search for pollen or nectar according to their genetically based preference. In one of these experiments, cofostered foragers from the high and low pollen strain were individually tagged and tested for their behavior as scouts in an unselected colony. There was a significant difference between strains with respect to the preference to scout for pollen or nectar. Foragers from the high pollen strain tended to come back from a scouting trip with pollen more often than did foragers from the low pollen strain (Fig. 2A), whereas scouts from the low pollen strain preferred to search for nectar sources. Also, the genotype of a worker affects her attendance at recruitment dances (C. Dreller, unpublished data). High and low strain bees were cofostered, individually tagged, and introduced into a colony maintained in an observation hive. Dances of untagged workers foraging at natural floral sources were observed. Each contact of a tagged bee with a dancer and the number of wagging runs the contact lasted were recorded. At the same time the number of pollen and nectar dances performed in the hive was estimated by scanning the dance floor every 10 min. This number represents

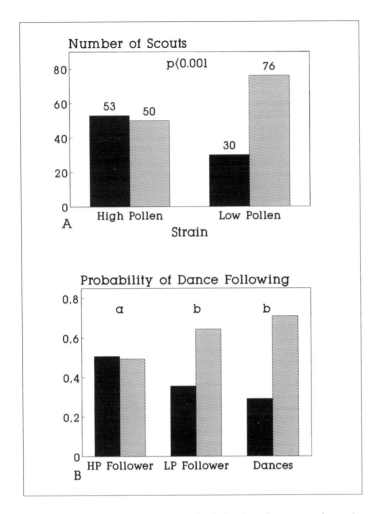

Figure 2 Effect of the genotype on the behavior of scouts and recruits
(A) The number of scouts of the high and the low pollen strain which search for pollen (black bars) or nectar (shaded bars). The number of individuals sampled is given on top of bars. The distribution of nectar and pollen scouts, compared by chi^2-test, is different between both strains.

(B) The percentage of bees of the high and low pollen strain which follow for more than one wagging run pollen (black bars) or nectar dances (shaded bars) and the probability of nectar and pollen dances performed in the hive during observation time. a and b indicate which distributions are different based on chi^2-test. High pollen (HP) strain bees follow more often pollen dances than expected from the random probability (p < 0.001), while low pollen (LP) strain bees follow nectar and pollen dances as expected from the random probability (p > 0.3).

the random probability for a follower to bump into a pollen or nectar dance. During the 20-h observation period 112 pollen dances and 273 nectar dances were observed (Fig. 2B). Compared with this random probability, high pollen strain bees preferred to follow pollen dances: 41 high pollen bees followed pollen dances and 40 high pollen bees followed nectar dances, which is significantly different from the random probability (chi^2-test, $p < 0.001$). Low strain bees, however, were observed more often following dances of nectar foragers: 21 bees were observed following dances for pollen and 38 bees followed nectar dances, which represents the random probability of bumping into nectar and pollen dances (chi^2-test, $p > 0.3$). Conclusively, information processing is also affected by the genotype of the foragers.

In summary, the genotype of a worker affects all aspects of her foraging behavior: resource choice, load size, foraging rate, recruitment, and scouting behavior.

Hunt et al. [13] (see also [14]) constructed a genomic linkage map of the honey bee and located two major genomic regions that affect quantities of stored pollen. These regions, designated *pln1* and *pln2*, contain major quantitative trait loci (QTL) that explained 59% of the total phenotypic variance for stored pollen that was observed in a backcross population derived from the high and low strains of Page and Fondrk [6]. Genetic markers linked to these regions were subsequently shown to correlate with pollen and nectar foraging behavior of individual workers [9, 13]. These results demonstrated that alleles of major genes differentially affect foraging decisions of individual workers and result in colony differences in quantities of stored pollen.

Effects of nest environment

In a colony, the numbers of foragers collecting pollen and nectar are influenced by the colony's "need" and current supply. Stimuli that provide information about need and supply can be transmitted by interaction between foragers and nestmates, or the foragers can make their own individual assessment. Pollen is the primary source of proteins fed to larvae. As a consequence, the primary factors which influence pollen foraging are the pollen reserves and the amount of open brood present in the colony.

Quantities of stored pollen are regulated by colonies through negative feedback. When excess pollen is added to a colony, pollen foraging activity decreases until the excess pollen has been depleted by the nurse bees and the quantity of stored pollen returns near to its previous level [15–18]. When pollen is removed from a colony, the number of pollen foragers, trip frequencies, and the sizes of their loads increase until the stored pollen is restored to the previous balance between foraging intake and nurse bee consumption [18–20]. Increasing quantities of brood, on the other hand, stimulate more pollen foraging activity [21–26]. Therefore, stored pollen acts like a negative stimulus for pollen foraging, whereas the presence of brood acts like a positive stimulus.

The regulation of nectar foraging has been studied in the greatest detail. It appears that foragers indirectly receive information about the colony's need for nectar from their nestmates [27–29]. Returning nectar foragers unload to other bees, and the time they need to do so ("search time") is correlated with the probability to recruit. If the nectar intake rate of a colony is high, the time to unload is high and the probability of dancing low. As a consequence, fewer bees are recruited to the resource. If the inflow of nectar and the quality of the food source are high, then the search time to unload is low and the probability of recruitment by dancing is high. This mechanism results in colonies increasing and decreasing nectar foraging effort without the foragers having direct information about the nectar supply within the colony.

Scott Camazine [30] proposed a similar mechanism for the assessment of pollen reserves. Pollen foraging activity of honeybee colonies appears to be regulated by two factors, the amount of brood which acts as a positive stimulus, and the quantity of stored pollen which acts as an inhibitory stimulus. Camazine [30] suggested that the two factors are integrated into a single inhibitory signal. Nurse bees are consumers of pollen and deplete the pollen reserves. They convert the proteins in the pollen into glandular secretions that are fed to the larvae. When there are plentiful pollen reserves but few young larvae to feed, nurse bees have an excess of glandular proteins that are available for feeding to foragers. According to Camazine's hypothesis, these glandular proteins then inhibit pollen foraging behavior. When pollen reserves are low relative to the number of larvae, there is little excess protein available to feed foragers and, therefore, no inhibition. We refer to this as the "indirect inhibitor" hypothesis because neither the brood nor the pollen directly provide cues that affect foraging behavior. An alternative hypothesis, the "direct multifactor" hypothesis, maintains that the quantities of stored pollen and quantities of young larvae themselves provide inhibitory and activating stimuli. This hypothesis is derived directly from empirical results demonstrating the effects of adding or removing brood and stored pollen from colonies [15–18, 20–26] and is supported by the work of Dreller and Tarpy (in press). They observed returning pollen foragers in observation hives. Two 4-frame observation hives were simultaneously established. Both observation hives contained the same amount of sealed brood (two frames), unsealed brood (one frame), pollen and honey (one frame), but the position of the frame with unsealed brood was different. One hive had the unsealed brood frame on the bottom (frame 1) and the other had the unsealed brood frame in the third position (frame 3). Experimental conditions were switched between colonies after 2 days of observation. Individually tagged, returning foragers with pollen loads spent more time on the frame with open brood, and they inspected more cells there than on other frames (Fig. 3A, B). Furthermore, most foragers unloaded the pollen on the frame that contained unsealed brood independent of its position in the hive (Fig. 3C). The searching and unloading behavior changed immediately according to experimental changes in the position of the frame. These observations suggest that pollen foragers have the opportunity to directly assess the colony's need and supply for pollen and do actively seek out the area of the brood nest where that information is available. Therefore, the exact mechanism(s) by which foragers assess the pollen supply are still unclear.

Recruitment is an important part of the stimulus environment with respect to the colony's need for nectar and pollen. Most foragers of a colony are recruited to resources by attending dances, rather than by discovering resources on their own [31]. The number of bees foraging at a particular food source is dependent on dance behavior. Dances for richer nectar sources last longer and, therefore, result in higher numbers of recruits [32, 33]. The dance also correlates with the value of food sources, providing potential qualitative information to new recruits. The "vigor" of dances increases with increasing quality of nectar [34–36] and pollen [11]. In addition, bees from different genetic backgrounds vary in their dance vigor and for the probability that they will perform recruitment dances, even when foraging for the same resources [11] (Fig. 1), demonstrating genetic variation among foragers for their "evaluation" of pollen resources.

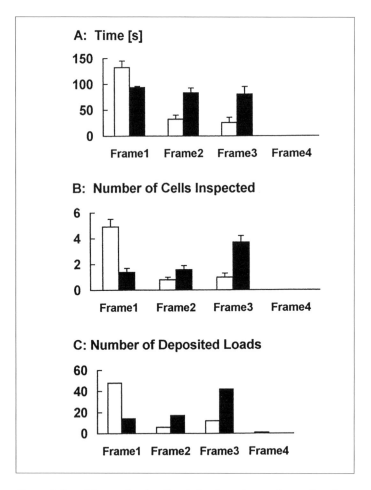

Figure 3 Searching and unloading behavior of returning pollen foragers, when the unsealed brood was either at the bottom of an observation hive on frame 1 (white bars) or in the upper part of the hive on frame 3 (black bars)

A total of 67 foragers were observed when the unsealed brood was on frame 1 (white bars), and 73 individuals were observed when the unsealed brood was on frame 3 (black bars). (A) The mean time ($\pm$SE) pollen foragers spent on each frame. When the unsealed brood was on frame 1, foragers spent more time there (t-test, $p < 0.05$); when the unsealed brood was on frame 3, pollen foragers spent more time on frames 2 (t-test, $p < 0.0001$) and 3 (t-test, $p < 0.01$). (B) The mean number ($\pm$SE) of cells inspected by pollen foragers on each frame. When the unsealed brood frame was in position 1, pollen foragers assessed more cells on that frame (t-test, $p < 0.0001$). When the unsealed brood frame was in position 3, more cells were inspected on frame 2 (t-test, $p < 0.05$) and third frame (t-test, $p < 0.0001$). (C) The total number of deposited pollen loads on each frame. Pollen foragers deposited their loads more frequently on the frame with unsealed brood, independent of its position (chi^2-test, $p < 0.0001$).

Interaction between genotypes and environmental stimuli

Workers from the high and low pollen hoarding strains of Hellmich et al. [5] responded to changes in the nest environment; however, they do not respond equally. Calderone and Page [8] cofostered (raised together in the same colonies) marked high and low strain workers developed by Hellmich et al. [5] in both high and low strain colonies. High strain colonies had significantly more stored pollen than did low strain colonies. High and low strain workers were both more likely to forage for pollen in low strain host colonies. However, the difference between host colony environments in the proportion of marked bees foraging for pollen was greatest for the low strain. These results demonstrated a significant genotype x host colony environment interaction and suggested that workers from the two strains had different response thresholds for pollen foraging stimuli.

Fewell and Page [37] showed that workers with different genotypes responded differently to the colony internal brood and pollen stimuli. They confined a colony in a cage and controlled the quality and quantity of food offered at artificial nectar (sugar syrup) and pollen feeders. They also controlled the quantities of brood and stored pollen contained in the hive. Cofostered workers derived from different genetic sources differentially became pollen or nectar specialists, or generalists. Brood, honey, and pollen stores were manipulated at the same time, and many foragers switched from nectar to pollen foraging when the pollen supply in the colony was low and the quantity of brood was relatively high. Under the same environmental conditions, workers from the genetically different sources differed in the likelihood that they would switch, indicating that genotype influences the threshold to react to changes of the internal colony environment.

External environmental stimuli, like food quality, also affect the foraging decisions of workers. However, genotype constrains those foraging decisions. In the same study, Fewell and Page [37] altered the quality of the food offered at feeders in the cages by either reducing the sugar concentration of the syrup offered, or mixing yeast with the pollen. Decreasing food quality resulted in less foraging, a result of some individuals switching resources and others reducing their rates of foraging. Individuals with different genotypes responded differently to the changes, demonstrating a genotype x foraging environment interaction effect.

Effects of development on foraging decisions

Worker honey bees perform different tasks as they age, a phenomenon called temporal polyethism (see [38] for a review). This change is mediated by increasing titres of juvenile hormone in the hemolymph [39, 40]. The most pronounced change in behavior occurs when bees are about 3 weeks old when they begin foraging [19, 41]. At this time, they cease performing most tasks within the nest and usually remain foragers for the rest of their lives. This age progression is not rigid. Developmental rates and trajectories are influenced by the nest environment. For example, some workers will forage precociously when colonies consist of all young bees. Likewise, foragers will revert to within nest tasks, like feeding larvae, if the younger bees are removed from the colony [40, 42–44].

It is known that a worker's genotype also affects her rate of behavioral development [7, 40, 42–45]. Page et al. [42] showed that differences in genotypes in their study were equivalent to at least 2–3 days of age development. However, genotype had no independent effect on the likelihood or rate of reverse development from forager to nurse bee after the young bees were removed from the colony. Calderone and Page [7, 45] showed that workers from the high pollen hoarding strain of Hellmich et al. [5] began foraging about 1 day earlier in life than did workers of the low strain. These data suggest that genotype and environment may interact with temporal development.

Another kind of behavioral development may be found as a consequence of behavioral canalization [46]. Behavioral canalization may occur as a consequence of the cumulative effects of early life experiences. For example, young workers may initially perform slightly different sets of tasks because, by chance, they differentially sample the stimulus environment of the hive. If previous experiences affect the likelihood of performing subsequent tasks, then individuals could become channelled into task trajectories that lead to foraging differentially for pollen or nectar. Calderone and Page [47] tested that hypothesis using high and low strain workers of Hellmich et al. [5]. Two cohorts of same aged high and low strain bees were marked and placed in screen cages in an incubator, 6 days apart. When the first group was 12 days old, a third cohort of high and low strain workers was marked, and all three cohorts were simultaneously introduced into an observation hive. At this time two groups had been deprived of early nest experiences for 6 and 12 days, respectively. The third group was not deprived. Observations were made of the age that each marked bee began foraging, and whether she foraged for pollen. Workers of the two strains initiated foraging at "normal" ages with the oldest foraging first. Also, high strain bees were much more likely to collect pollen, suggesting that early adult environmental experiences were not major determinants of foraging behavior.

Behavioral development may also be affected by preadult experiences. Calderone and Page [7, 8, 47] raised workers from high and low strain colonies in common "nursery" colonies until they were pupae. Pupae were transferred to an incubator until they emerged as adults, were marked, and placed into observation hives. Workers from the high and low strains still demonstrated strong preferences for foraging for pollen and nectar, respectively, even though they had common preadult experiences.

Physiological mechanisms

Robinson and Page [2] proposed a model for task specialization where individual workers vary for response thresholds to task specific stimuli. A worker performs a specific behavior such as foraging for nectar or pollen when the appropriate environmental stimuli exceed the threshold for that individual. Genotypic variation then results in persistent differences for response thresholds among individuals that share a common colony environment and, as a consequence, persistent differences in their tendencies to collect pollen or nectar. A result of variability in response thresholds may be a wider or more graded response to changes in the availability of resources.

Between the genotypes of individuals and their foraging behavior lie physiological mechanisms. The stimulus threshold model discussed above assumes that one of the physiological mechanisms couples stimuli perceived from the environment with the performance of specific foraging tasks. It is possible to test the perception and response of honeybees to one of the relevant foraging stimuli by using the proboscis extension reflex (PER) [48]. Bees reflexively extend their proboscis when a droplet of sugar solution of sufficient concentration is touched to the tip of the antenna. Returning pollen and nonpollen foragers (presumed to be foraging for nectar) were collected at the entrance of a colony and then tested for the PER with a series of sucrose solutions from 0 (water) up to 30% [49]. Returning pollen foragers had significantly lower response thresholds to sucrose and were more likely to respond to only water than were the nonpollen foragers.

Testing foragers of the high and low pollen hoarding strain revealed that returning high strain pollen foragers had lower response thresholds to sucrose and responded more to water than returning nonpollen foragers of the high strain. Returning high strain nonpollen foragers had lower sucrose response thresholds and higher response probabilities to water than did returning low strain nonpollen foragers. These results demonstrated that genotype affects sucrose response thresholds and responses to water as measured by PER, and that response thresholds for sucrose correlate with foraging behavior (pollen versus nonpollen foraging).

Based on these results Page et al. [49] predicted (i) that high strain returning nonpollen foragers should more often return with loads of water, (ii) that the nectar foraging high strain bees should accept lower concentrations of sugar in the nectar they collect, and (iii) low strain foragers should more often find no suitable nectar to collect and return empty. Measurements of the crop contents of returning nonpollen foragers revealed that PER indeed correlates with the kinds of materials collected by foragers, as predicted.

As a result of studies on sucrose response thresholds, the data of Hunt et al. [13] were reexamined [49]. In that study, the nectar loads of returning foragers were collected and analyzed for volume and concentration of sugars. Workers were derived from a hybrid queen who was mated to single drone from the high strain. As a consequence of male haploidy, workers varied in genotype at *pln1* and *pln2* only as a consequence of recombination in the queen. All individuals inherited a high strain allele from their father but could have either a high or low strain allele from their mother. Data were analyzed for all returning nectar foragers to determine whether nectar foragers inheriting high or low pollen hoarding alleles of *pln1* or *pln2* differed in the sugar concentrations of nectar they collected. The specific prediction was that individuals inheriting a high pollen hoarding allele should accept more dilute nectar which would be reflected by a lower average concentration of sugar associated with that allele. Individuals inheriting the high and low alleles at *pln1* did not differ in the concentration of the nectar collected, but those inheriting the high allele at *pln2* had significantly lower concentrations (Fig. 4). This demonstrated a plausible effect of *pln2* on the response thresholds for sucrose.

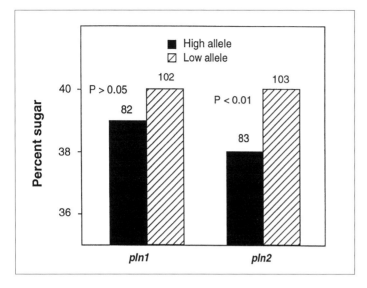

Figure 4 Median concentration of sucrose for returning nectar for-agers inheriting high or low alleles from their mother for pln1 *and* pln2
Probabilities given are results of Mann-Whitney U tests. Numbers on tops of bars are the numbers of individuals sampled.

Conclusion

Two quantitative genomic regions with major effects, *pln1* and *pln2*, influence the foraging decisions of individual workers in two ways: (1) both regions affect decisions to forage for pollen or nectar, while (2) *pln2* also appears to directly affect the nectar quality assessments of nectar foragers, presumably by affecting their sucrose response thresholds (Fig. 5). Together, genotype, age, development, and the stimulus environment affect individual foraging decisions of resource choice, foraging rate, load size, and recruitment. The collective activities of foragers determine the pollen intake rate of the colony, which influences two of the major stimulus factors determining pollen foraging behavior: the quantities of stored pollen, and brood. Stored pollen leads to a reduction in brood quantity when total nest space is a limiting factor [6], resulting in a decrease in the positive stimulus factor (brood) concurrently with an increase in the negative stimulus factor (stored pollen). The quality of the nectar and pollen available as forage are also part of the stimulus environment. Therefore, colony level selection

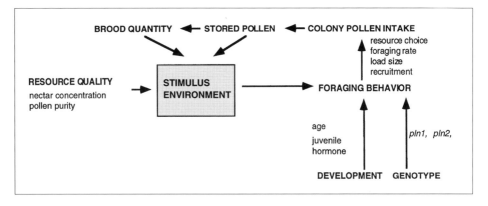

Figure 5
Regulatory mechanisms of pollen foraging activity in honey bees

results in allelic substitutions at genes that reside in individual workers that, after processes of development, affect foraging behavior. Individuals interact in a social context that results in the traits on which selection operates; therefore, there exist functional pathways from genes through neurophysiological processes to individual behavior that serve as the determinants for social behavior.

References

1   Darwin CR (1859) *On the origin of species*. 6th edn, Macmillan, New York [1962]

2   Robinson GE, Page RE (1989) Genetic basis for division of labor in insect societies. *In*: MD Breed, RE Page (eds): *The genetics of social evolution*. Westview Press, Boulder, CO, 61–80

3   Calderone NW, Robinson GE, Page RE (1989) Genetic structure and division of labor in the honey bee society. *Experientia* 45: 765–767

4   Dreller C, Fondrk MK, Page RE (1995) Genetic variability affects the behavior of foragers in a feral honeybee colony. *Naturwissenschaften* 82: 243–245

5   Hellmich RL, Kulincevic JM, Rothen-buhler WC (1985) Selection for high and low pollen-hoarding honey bees. *J Hered* 76: 155–158

6   Page RE, Fondrk MK (1995) The effects of colony-level selection on the social organization of honey bee (*Apis mellifera* L.) colonies: colony-level components of pollen hoarding. *Behav Ecol Sociobiol* 36: 135–144

7   Calderone NW, Page RE (1988) Genotypic variability in age polyethism and task specialization in the honey bee, *Apis mellifera* (Hymenoptera: Apidae). *Behav Ecol Sociobiol* 22: 17–25

8   Calderone NW, Page RE (1992) Effects of interactions among genotypically diverse nestmates on task specialization by

foraging honeybees (*Apis mellifera*). *Behav Ecol Sociobiol* 30: 219–226

9 Page RE, Waddington KD, Hunt GE, Fondrk MK (1995) Genetic determinants of honey bee foraging behaviour. *Anim Behav* 50: 1617–1625

10 Deng G (1996) *Foraging performance of honey bees* (Apis mellifera). PhD Dissertation, University of Miami

11 Waddington KD, Nelson M, Page RE (1998) Effects of pollen quality and genotype on the dance of foraging honey bees. *Anim Behav* 59: 35–39

12 Dreller C (1998) Division of labor between scouts and recruits: Genetic influence and mechanisms. *Behav Ecol Sociobiology* 43: 191–196

13 Hunt GE, Page RE, Fondrk MK, Dullum CJ (1995) Major quantitative trait loci affecting honey bee foraging behavior. *Genetics* 141: 1537–1545

14 Hunt GE, Page RE (1994) Linkage analysis of sex determination in the honey bee (*Apis mellifera*). *Mol Gen Genet* 244: 512–518

15 Barker RL (1971) The influence of food inside the hive on pollen collection by a honey bee colony. *J Apicultural Res* 10: 23–26

16 Free JB, Williams IH (1971) The effect of giving pollen and pollen supplement to honeybee colonies on the amount of pollen collected. *J Apicultural Res* 10: 87–90

17 Moeller FE (1972) Honey bee collection of corn pollen reduced by feeding pollen in the hive. *Amer Bee J* 112: 210–212

18 Fewell JH, Winston ML (1992) Colony state and regulation of pollen foraging in the honey bee, *Apis mellifera* L. *Behav Ecol Sociobiol* 30: 387–393

19 Lindauer M (1952) Ein Beitrag zur Frage der Arbeitsteilung im Bienenstaat. *Z Vergl Physiol* 34: 299–345

20 van Laere O, Martens N (1971) Influence d'une diminution artificielle de la provision de proteines sur l'activite de collete de la colonie d'abeilles. *Apidologie* 2: 197–204

21 Filmer RS (1932) Brood area and colony size as factors in activity of pollination units. *J Econ Entomol* 25: 336–343

22 Free JB (1967) Factors determining the collection of pollen by honey bee foragers. *Anim Behav* 15: 134–144

23 Cale GH (1968) Pollen gathering relationship to honey collection and egg laying in honey bees. *Am Bee J* 108: 8–9

24 Todd FE, Reed CB (1970) Brood measurement as a valid index to the value of honey bees as pollinators. *J Econ Entomol* 63: 148–149

25 Al-Tikrity WS, Benton AW, Hillman RC, Clarke WW (1972) The relationship between the amount of unsealed brood in honey bee colonies and their pollen collection. *J Apicultural Res* 11: 9–12

26 Eckert CD, Winston ML, Ydenberg RC (1994) The relationship between population size, amount of brood, and individual foraging behaviour in the honey bee, *Apis mellifera* L. *Oecologia* 97: 248–255

27 Seeley TD (1989) Social foraging in honey bees: how nectar foragers assess their colony's nutritional status. *Behav Ecol Sociobiol* 24: 181–199

28 Seeley TD (1992) The tremble dance of the honey bee — message and meanings. *Behav Ecol Sociobiol* 31/6: 375–383

29 Seeley TD, Tovey CA (1994) Why search time to find a food-storer bee accurately indicates the relative rates of nectar collecting and nectar processing in honey bee colonies. *Anim Behav* 47: 311–316

30 Camazine S (1993) The regulation of pollen foraging by honey bees: how foragers assess the colony's need for pollen. *Behav Ecol Sociobiol* 32: 265–272

31 Seeley TD (1983) The division of labor between scouts and recruits in honeybee foraging. *Behav Ecol Sociobiol* 12: 235–259

32 Seeley TD, Camazine S, Sneyd J (1991) Collective decision making in honey bees — how colonies choose among nectar sources. *Behav Ecol Sociobiol* 28/4: 277–290

33 Seeley TD, Towne WF (1992) Tactics of dance choice in honey bees — do foragers compare dances. *Behav Ecol Sociobiol* 30/1: 59–69

34 Esch H (1963) Über die Auswirkung der Futterplatzqualität auf die Schallerzeugung im Werbetanz der Honigbiene. *Verh Dtsch Zool Ges* 62: 302–309

35 von Frisch K (1967) *The dance language and orientation of honeybees*. Harvard University Press, Cambridge, MA

36 Waddington KD, Kirchner WH (1992) Acoustical and behavioral correlates of profitability of food sources in honey bee round dances. *Ethology* 92: 1–6

37 Fewell JH, Page RE (1993) Genotypic variation in foraging responses to environmental stimuli by honey bees, *Apis mellifera*. *Experientia* 49: 1106–1112

38 Robinson GE (1992) Regulation of division of labor in insect societies. *Annu Rev Entomol* 37: 637–665

39 Robinson GE (1987) Regulation of honey bee age polyethism by juvenile hormone. *Behav Ecol Sociobiol* 20: 329–338

40 Robinson GE, Page RE, Strambi C, Strambi A (1989) Hormonal and genetic control of behavioral integration in honey bee colonies. *Science* 246: 109–112

41 Seeley TD, Kolmes SA (1994) Age polytheism for hive duties in honey bees — illusion or reality. *Ethology* 87: 284–297

42 Page RE, Robinson GE, Britton DS, Fondrk MK (1992) Genotypic variability for rates of behavioral development in worker honeybees (*Apis mellifera*). *Behav Ecol Sociobiol* 4: 173–180

43 Robinson GE, Page RE, Strambi A, Strambi C (1992) Colony integration in honey bees: mechanisms of behavioral reversion. *Ethology* 90: 336–348

44 Tugrul G, Robinson GE (1994) Effects of intracolony variability in behavioral development on plasticity of division of labor in honey bee colonies. *Behav Ecol Sociobiol* 35: 13–20

45 Calderone NW, Page RE (1991) Evolutionary genetics of division of labor in colonies of the honey bee (*Apis mellifera*). *Amer Naturalist* 138: 69–92

46 Page RE, Robinson GE (1991) The genetics of division of labor in honey bee colonies. *Adv. Insect Physiol* 23: 117–169

47 Calderone NW, Page RE (1996) Age-based division of labour and the effect of pre-foraging experience on task selection by honey bee foragers. *Anim Behav* 51: 631–643

48 Menzel R, Greggers U, Hammer M (1993) Functional organization of appetitive learning and memory in a generalist pollinator, the honey bee. *In*: DR Papaj, AC Lewis (eds): *Insect learning ecological and evolutionary perspectives*. Chapman and Hall, New York, 79–125

49 Page RE, Erber J, Fondrk MK (1998) The effect of genotype on response thresholds to sucrose and foraging behavior of honey bees (*Apis mellifera* L.). *J Comp Physiol* 182: 489–500

Behavioral threshold variability: costs and benefits in insect societies

Robin F. A. Moritz and Robert E. Page Jr.

Introduction

Insect sociobiology is a rapidly developing field that spans many subdisciplines within biology. The past 5–10 years have seen a major shift in emphasis from questions dominated by ecology to those of genetics, physiology, development, and self-organization. Interindividual behavioral variation became a central theme following the publication of the proceedings of an IUSSI (International Union for the Study of Social Insects) symposium on the subject [1]. That volume dealt primarily with observations that workers within insect colonies demonstrate vastly variable behavior. However, recent studies have led to a better understanding of physiological processes underlying behavioral variation. Physiological processes have been linked to intrinsic and extrinsic factors that modulate individual development and are responsible for colony level changes in social organization. Testable models have been proposed for the regulation of division of labor in insect societies based either on nondevelopmental mechanisms, like the foraging for work hypothesis [2], or those that are developmentally based [3].

Shortly before and after publication of the symposium proceedings, evidence began accumulating that much of the observed variation among individuals is a result not only of development, age, and experience, but also of individual variation in genotype [4, 5]. Thus the field of sociogenetics was launched. Today, the field of sociogenetics has entered the age of sociogenomics. The honeybee genome has been mapped along with major genes that affect the determination of sex [6–9], foraging behavior [10], and defensive behavior [11]. Studies of behavioral quantitative trait loci (QTLs) suggest that observed variation in behavior results from the effects of these QTLs on the response thresholds of individuals to task-specific stimuli. Theoretical models of self-organizing processes have incorporated the response threshold-stimulus relationships and show how complex social behavior can emerge from the structure of the nest and

C. Detrain et al. (eds) Information Processing in Social Insects

simple behavioral rules of colony members [12, 13]. Due to local stimuli, colony members will or will not express a certain behavioral trait depending on its individual threshold. Members with similar sets of thresholds will tend to reveal similar behaviors and thus end up in behavioral castes.

Bauplan constraints and phenotypic plasticity

The presence of morphologically distinct castes within insect colonies is often striking. It is believed that specific castes evolved that facilitate the performance of particular tasks. The queen-worker dimorphism is usually most obvious with an anatomically specialized queen for egg laying. However, in some species of ants, worker morphology is also highly variable with minor and major caste workers easily discriminated. An extreme example has been presented for *Pheidologeton diversus*, where worker sizes range over several orders of magnitude [14]. If size has adaptive significance, we would expect task specialization to be associated with anatomical cast differentiation. Indeed, this is what we observe in many species of ants: majors more often serve as guards, whereas minors often are involved in brood care. If a specific mix of phenotypes is optimal to the colony fitness, how can natural selection possibly achieve such optimal phenotype compositions? There seem to be two principal nonexclusive ways of obtaining variable worker phenotypes in colonies. One is through strong environment-gene interactions. The genome must accommodate a large phenotypic plasticity in order to yield the various physical castes limited only by the constraints of the bauplan. The other is through behavioral threshold variability within a monomorphic worker caste.

Apparently trophic factors interact with the genome, resulting in differential gene expression leading to the development of distinct phenotypical castes. This route seems to be realized in many ant species, and colony demography is characterized by highly adaptive patterns [15]. Trophic factors have been found to be paramount for caste variability. Again, a good example comes from Attine ants [16]. Workers in incipient nests are small, which may result from shortage of food or other inadvertent conditions for larval development in the nest. With growing nest size, however, nutritional resources can be better exploited, and larger individuals can be reared. Finally, in full-size colonies the complete range of phenotypic castes can be expressed, most likely controlled through the local food availability within the nest.

Constraints of brood nest architecture

If there is an optimal mix of tasks performed by colonies, then we would expect natural selection for optimal caste ratios within specific environments. However, extreme worker caste differentiation with a wide array of different phenotypes is primarily observed in ants in the social Hymenoptera. The architecture of a nest may place constraints on caste production in other Hymenoptera. In wasps and bees, worker variability is much less expressed, and adults are very similar within a small size range. One apparent external constraint in larval development of some social wasps and bees is the use of uniform hexagonal cells for rearing the worker brood. This is not a constraint where the greatest examples of anatomical caste differentiation occur, the ants and termites. Although the shape and size of the "uniform" cells on combs have repeatedly been shown to be highly variable [17], cell sizes clearly set upper size limits on developing individuals and, therefore, reduce the degree of size variance in the adults. In spite of the reduced size and shape variability of these workers, we know from honeybees that division of labor exists and behavioral castes can be clearly identified.

Age polyethism and juvenile hormone

How can division of labor develop if there are no distinct morphological worker castes? The classic explanation is the age polyethism concept of Rösch [18, 19]. Workers change through various task cohorts as they mature in their adult life. It is not surprising to see young individuals engage in different tasks than older ones. Young worker bees do not have fully developed flight muscles, are poor flyers, and therefore participate primarily in intranest tasks. Natural selection may therefore have favored genotypes where young individuals have a low threshold for brood rearing and other intranest tasks. As soon as flight ability is reached, the animals leave the nest for various foraging tasks. In this stage, thresholds for intranest tasks are obviously increased, which forces the individuals to outdoor tasks.

Juvenile hormone (JH) has been shown to be a key factor in regulating age polyethism in honeybees [20–24]. However, the mechanisms by which JH is regulated are still unknown. One hypothesis, the activator/inhibitor model, proposes that individual honeybees are activated intrinsically to develop behaviorally. In this case, development occurs with increasing titres of JH in the hemolymph. Regulation of behavioral development then occurs through inhibition. Inhibitor(s) are extrinsic to individual bees. This model of behavioral development provides testable hypotheses for exploring mechanisms of age polyethism, but still lacks confirmation because the exact nature of the inhibitor(s) remains unknown. It is not clear to what extent the activator inhibitor model can explain the finer aspects of division of labor that are not age-linked such as differentiation of foragers into pollen, nectar, and water. There are no indications that forager specialization is due to JH titer variability, yet clear task specialization can be observed [25].

Genotypic variance and colony organization

In addition to age-dependent task specialization, genetic variance among individuals is a mechanism that generates behavioral variation and task specialization in those social insects where nest constraints, like uniform cell sizes, don't allow for morphological caste variability. Genotypic variability has repeatedly been shown to be important for interindividual behavioral differences in honeybees. It has been argued that colonies with greater genotypic variance can operate more efficiently due to better division of labor. If colonies with more genotypic variation are more fit than those with less variation, then we might expect natural selection to result in polygyny and polyandry. Indeed both polygyny and polyandry are common in social insects. Polygyny is typical for many ant and various wasp species, whereas polyandry has been found in vespine wasps and honeybees.

Costs of behavioral variability

Behavioral specialization may not be free of costs for the colony. In particular if colonies are small, the presence of too many different specialists might be detrimental. In solitary nest founding species we would expect the degree of worker specialization to be reduced in the incipient phase of colony development. Workers should respond to many different task situations but primarily those related to nest founding. In this phase it certainly would be best not to have a wide array of specialists but rather worker generalists with a small repertoire of tasks. We find many examples in ants where we can exactly observe this pattern. In *Solenopsis invicta* the minims caste is only present in the incipient nest phase [26] and disappears in later nest phases. The replacement of the minims by minors and in later stages major workers can be plausibly explained on the basis of ergonomic colony level selection.

If genotypic variability is important for variation in behavior, which is what we tend to believe, then colonies should adopt large numbers of nonrelated polyandrous queens. Such a strategy would not be free of costs. The problem lies in the loss of inclusive fitness for workers. Inclusive fitness theory predicts that it only pays to support the colony as long as benefits are larger than costs [27, 28]. Thus relatedness should not be expected to drop to zero, which might be the optimal solution for obtaining the best-operating behavioral mix. Different patterns of queen adoption are observed in ants. Polygynous *Formica truncorum* apparently recruit only related queens and maintain a high intranest relatedness ranging between 0.2 and 0.4 [29]. Yet looking at other empirical data in polygynous ants, intracolonial relatedness shows an extreme range, and can drop to values close to zero. In fact, in polygynous *F. aquilonia* and *F. lugubris* colonies, intranest relatedness was estimated less than zero [30]. These nests, however, are usually highly polydomous, and clusters of nests may actually represent one large colony rather than a population.

In monogynous colonies intranest relationships are usually higher than in polygynous colonies. Because all nestmates are at least half sibs, the intracolonial relatedness can never fall below $r = 0.25$. Looking at the most extreme cases of polyandry in honeybees, these extreme values are often realized. In the giant honeybee *Apis dorsata*, queens can mate with more than 55 drones, yielding intracolonial relationships close to 0.25 [31, 32]. Thus from the data it appears that in several highly eusocial species great efforts are taken to increase genotypic variance to the maximum possible value, in spite of the difficulties presented by a loss of high genetic relatedness among nestmates.

Hypotheses for the adaptive value of genotypic diversity

The very basic assumption common to all theories for the evolution of polygyny and/or polyandry is that there must be clear benefits at the level of colonial phenotypes [33–35]. One argument is that queens are limited by their capabilities to lay eggs, hence polygyny, or that males are limited in their ability to provide sufficient sperm to fertilize all of the eggs of queens, hence polyandry. These arguments are derived from observations that polygyny and polyandry seem to be more clearly expressed in species that form large, long-living colonies. However, sperm and egg-laying limitation hypotheses have been discarded by various authors [36, 37]. Single queens can establish gigantic colonies (e.g. in *Atta sexdens*) and males of polyandrous species usually have sufficient number of sperm to fully inseminate the queens (e.g. in *Apis mellifera*). Therefore, although the sperm limitation model cannot be completely ruled out for each and every species, hypotheses suggesting selective advantages for colonies with increased genetic diversity are currently more popular.

Genotypic diversity and control of the sex ratio

Some hypotheses focus on average intracolonial relatedness as the driving force for the evolution of intracolonial diversity. For example, sex ratio conflict between workers and queens is reduced in polyandrous mating systems [38, 39]. Yet this hypothesis finds little support if we look at the extreme cases of polyandry in honeybees. DNA fingerprinting techniques [31, 32, 40–42] have demonstrated that queens in some populations of African *A. mellifera* queens mate with up to 45 drones [43]. Intracolonial relatedness only marginally decreases once the number of effective matings exceeds 20. Assuming there are costs associated with mating number, fitness gains beyond 20 matings are difficult to explain. Therefore, models which focus on the impact of genotypic and behavioral variation seem to be more plausible.

Genotypic diversity and sex determination

Before polyandry can evolve as a consequence of its effect on genotypic diversity, two conditions must exist: (i) there must genetic variation for a trait or traits, and (ii) the trait(s) must have differential fitness associated with the different genotypes. As a consequence, there must be a mechanism to maintain variation in the population against the effects of natural selection favoring the best genotype [44].

Page [45] proposed that polyandry evolved in honeybees because of its effects on the distribution of diploid males produced as a consequence of homozygosity at the sex locus. Honeybees, like many species of Hymenoptera, have a single locus sex determination system where individuals that are diploid and heterozygous develop into females, whereas those that are haploid or homozygous at the sex locus develop into males. Diploid males are consumed by workers and are, therefore, effectively lethal. Rare sex alleles will be favored in populations because they are less likely to result in the production of diploid males and will, therefore, increase in frequency. Common sex alleles will be selected against because they are more likely to result in diploid male production. As a consequence, multiple sex alleles (genetic variation) are maintained in populations because of symmetrical overdominant selection.

The effect of the number of matings of queens is a change the variance in diploid male production among colonies. Colonies derived from queens with genotypes that result in more matings will have less variation for brood inviability resulting from diploid male production. As long as the relationship between queen fitness and the production of inviable diploid males is concave in shape (decreasing fitness differential with increasing brood viability), polyandrous mating will be selectively favored. Arguments have been given that the fitness relationship should be concave for species with populous colonies, like the honeybee, but there is little empirical evidence to support, or refute, those arguments [37, 44, 45].

Genotypic diversity and parasites and pathogens

The hypothesis of Page [45] proposed that polyandry evolved in honeybees as a consequence of factors intrinsic to them, their method of sex determination. However, Sherman et al. [46, 47] proposed a genotypic diversity hypothesis based on extrinsic factors, parasites and pathogens. According to their hypothesis colonies that have polyandrous queens are expected to have more genotypic diversity that increases their chances of surviving episodic outbreaks of diseases. The increased survival probability is a consequence of decreasing the variance in worker mortality resulting from disease. The relationship between worker mortality and survival is assumed to be either concave in shape, like that of Page [45], or a step function where all colonies below some critical threshold of worker survival perish. This hypothesis was criticized [44 but see also 47] because it lacks a mechanism for maintaining genetic variation in populations and it lacks any empirical evidence of the relationship between genotypic variation for disease resistance, and fitness.

Genotypic diversity and division of labor

There are two ways to view the possible effects of behavioral variation on colony fitness. Both assume that genetic variation is maintained in populations by colony level selection and that behavioral variation has a strong genetic component, as has been repeatedly demonstrated. One view is that more genetic variation leads to more behavioral specialization. Behavioral specialization is assumed to result in greater ergonomic efficiency. However, too many individuals performing one task may be detrimental, hence the necessity for more genotypic variation. Colonies with more and a broader base of specialists performing tasks are expected to better survive and reproduce. The other view is that genotypic variation results in colonies that are more average in their performance. By being average they avoid maladaptive colony organizational phenotypes.

The idea that the presence of specialists renders a colony more efficient was spelled out in principle by Oster and Wilson [15]. A more formal description of such a model is given by Fuchs and Moritz [48]. They show that through a process governed by intracolonial frequency-dependent selection, even very high degrees of polyandry can evolve in monogynous social Hymenoptera. The concept is based on the assumption that specialists need to be rare in a colony and the presence of too many specialists is detrimental. A complete lack of specialists, however, also has adverse effects on the colonial fitness (Fig. 1). Thus only in a small window of intracolonial specialist frequencies does the colony benefit from the presence of specialists. Data on fitness benefits of colonies of social insects as a function of genotypic diversity are, however, extremely rare and to our knowledge only available for honeybees.

An empirical example in support of the first view is given by the Cape honeybee *A. mellifera capensis*. Laying workers can establish themselves as pseudoqueens in colonies [49, 50] and significantly contribute to the population gene pool [51]. There is strong genetic variance for this trait [4], and only very few workers of a few subfamilies develop into pseudoqueens [52]. Colonies composed completely of specialist reproductive dominant workers are unable to reproduce because of a complete lack of workers rearing the brood [53]. In mixed colonies brood care is less strongly expressed, as in colonies with no dominant workers. The benefit to the colony of the pseudoqueen caste is the swift replacement of the queen after accidental queen loss. The price paid is that colonies lose brood-rearing capacity if too many dominant workers are present. This pattern exactly matches the hypothetical fitness function in Figure 1. Only few specialists in the colony are advantageous to the colonial phenotype, and colony fitness reduces with increasing number of specialists.

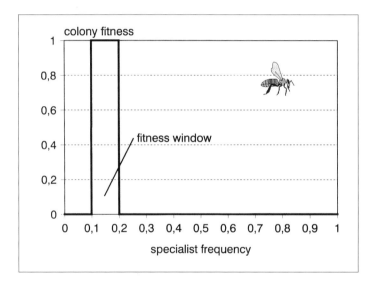

Figure 1 Fitness function depending on the frequency of specialists in a colony
Only within narrow frequency limits (fitness window) can the colony profit from the presence of specialists. Too few and too many specialists are disadvantageous to colony fitness.

Evidence for the second view was presented by Page and Fondrk [54]. They found that highly polyandrous honeybee colonies had a more average colonial phenotype than colonies with singly mated queens. If average colonies produce sufficient swarms for survival, the polyandrous colonies would reproduce. Colonies headed by queens with insufficient mating were highly variable, and a larger proportion of these would be more likely not to produce a swarm in the next season. The evolution of polyandry is still a central theme in social insect biology. Hypotheses abound and debates continue, but the jury is still out regarding its adaptive significance, if any.

Self-organization

Division of labor in insect societies can be explained by behavioral threshold variance among the members of a colony [12, 13]. As we have seen, variance in thresholds can result from differences in genotype, development, or experience. It is assumed that individuals respond to stimuli on the basis of their response thresholds. Members with similar sets of thresholds will tend to perform similar tasks and thus end up in the same behavioral roles, or subcastes. Even small differences in response thresholds among individuals may be sufficient to result in clear large differences in behavior through positive and negative feedback loops. For example, when individuals respond to a stimulus and perform a task, they can reduce stimulus levels. As a consequence, those with the lowest response thresholds repeatedly perform tasks and keep the stimulus level below that of other individuals with higher response thresholds.

Ten years ago the common view of insect social organization was one orchestrated by pheromones. Chemical signals in some way resulted in coordinated activities of colonies. However, today we are becoming increasingly skeptical about the role of pheromonal communication in relation to local decision-making behavior of colonial insects. Instead a new view is emerging where individuals have little information about the colony outside of their own realm of personal experience. There is no grand plan for task organization, only local decisions made with limited, local information. It appears that pheromones are reserved primarily for global reactions (like alarm or queen's presence). The clue to understanding the organization of insect societies lies in appreciating the simplicity of the individuals that make up a nest. An example of such an approach may be that of Page and Mitchell [55] for honeybees. By combining simple Boolean network models [56] with behavioral threshold variability, they could provide testable predictions for colonial organization. Although the sensory performance of social insects is remarkable, the more rewarding approach to understanding the complexity of insect societies may be to keep in mind a statement of Klaus Jaffé during the IUSSI congress in Bangalore 1990: "The more complex a society, the simpler the individual."

Acknowledgements

We are grateful to the Alexander von Humboldt Stiftung for enabling this cooperation.

References

1   Jeanne RL (1988) *Interindividual behavioral variability in social insects.* Westview Press, Boulder, CO

2   Franks NR, Tofts C (1994) Foraging for work: how tasks allocate workers. *Anim Behav* 48: 470–472

3   Huang ZY, Robinson GE (1999) Social control of division of labor in honey bee colonies. This volume, 165–186

4   Moritz RFA, Hillesheim E (1985) Inheritance of dominance in honeybees (*Apis mellifera capensis* Esch.) *Behav Ecol Sociobiol* 17: 87–89

5   Page RE, Robinson G (1991) The genetics of division of labour in honeybee colonies. *Adv Insect Physiol* 23: 117–169

6   Beye M, Epplen C, Moritz RFA (1994) Sex linkage in the honey bee *Apis mellifera* L. detected by multilocus DNA fingerprinting. *Naturwissenschaften* 81: 460–462

7   Beye M, Crozier RH, Crozier YC, Moritz RFA (1996) Mapping the sex locus of the honeybee (*Apis mellifera*). *Naturwissenschaften* 83: 424–426

8   Hunt GJ, Page RE (1995) Linkage map of the honey bee, *Apis mellifera*, based on RAPD markers. *Genetics* 139: 1371–1382

9   Hunt GJ, Page RE (1994) Linkage analysis of sex determination in the honey bee (*Apis mellifera* L.). *Mol Gen Genet* 85: 512–518

10 Hunt GJ, Page RE, Fondrk MK, Dullum CJ (1995) Major quantitative trait loci affecting honey bee foraging behavior. *Genetics* 141: 1537–1545

11 Hunt GJ, Guzmán-Novoa E, Fondrk MK, Page RE (1998) Quantitative trait loci for honey bee stinging behavior and body size. *Genetics* 148: 1203–1213

12 Bonabeau E, Theraulaz G (1999) Role and variability of response thresholds in the regulation of division of labor in insect societies. This volume, 141–164

13 Beshers SN, Robinson GE (1999) Response threshold and division of labor in insect colonies. This volume, 115–140

14 Moffet MW (1987) Sociobiology of the ants of the genus *Pheidologeton*. PhD dissertation, Harvard University, Cambridge, MA

15 Oster GF, Wilson EO (1978) *Caste and ecology in social insects.* Princeton University Press, Princeton

16 Wilson EO (1983) Caste and division of labor in leaf-cutter ants (Hymenoptera: Formicidae: *Atta*) IV: Colony ontogeny of *A. cephalotes. Behav Ecol Sociobiol* 14: 55–60

17 Hepburn HR (1986) *Honeybee wax.*

Springer, Berlin

18 Rösch GA (1927) Untersuchungen über die Arbeitsteilung im Bienenstaat. I. Die Tätigkeiten im normalen Bienenstaate und ihre Beziehungen zum Alter der Arbeitsbiene. *Z Vgl Physiol* 6: 264–298

19 Rösch GA (1930) Untersuchungen über die Arbeitsteilung im Bienenstaat. II. Die Tätigkeiten der Arbeitsbienen unter experimentell veränderten Bedingungen. *Z Vgl Physiol* 12: 1–17

20 Robinson GE, Page RE, Strambi C, Strambi A (1989) Hormonal and genetic control of behavioral integration in honeybee colonies. *Science* 246: 109–112

21 Robinson GE, Huang ZY (1998) Colony integration in honeybees: genetic endocrine and social control of division of labor. *Apidologie* 29: 159–170

22 Robinson GE, Ratnieks FW (1998) Induction of premature honey bee (Hymenoptera: Apidae) flight by juvenile hormone analogs administered orally or topically. *J Econ Entomol* 80: 784–787

23 Robinson GE (1987) Regulation of honey bee age polyethism by juvenile hormone. *Behav Ecol Sociobiol* 20: 329–338

24 Robinson GE (1992) Regulation of division of labor in insect societies. *Annu Rev Entomol* 37: 637–665

25 Dreller C, Page RE (1999) Genetic, developmental and environmental determinants of honeybee foraging behavior. This volume, 187–202

26 Porter SD, Tschinkel WR (1986) Adaptive value of nanitic workers in newly founded red imported fire ant colonies (Hymenoptera: Formicidae). *Ann Entmol Soc Amer* 79: 723–726

27 Hamilton WD (1964a) The genetical evolution of social behaviour I. *J Theor Biol* 7: 1–16

28 Hamilton WD (1964b) The genetical evolution of social behaviour II. *J Theor Biol* 7: 17–32

29 Sundström L (1989) Genetic relatedness and population structure in *Formica truncorum* Fabr. (Hymenoptera, Formicidae). *Actes Colloq Insect Soc* 5: 93–100

30 Pamilo P, Chautems D, Cherix D (1992) Genetic differentiation of disjunct populations of the ants *Formica aquilonia* and *Formica lugubris* in Europe. *Insect Soc* 39: 15–29

31 Moritz RFA, Kryger P, Koeniger G, Koeniger N, Estoup A, Tingek S (1995) High degree of polyandry in *Apis dorsata* queens detected by microsatellite variability. *Behav Ecol Sociobiol* 37: 357–363

32 Oldroyd BP, Smolenski AJ, Cornuet JM, Wongsiri S, Estoup A, Rinderer TE, Crozier RH (1996) *Ann Entomol Soc Amer* 89: 276–283

33 Keller L (1993) *Queen number and sociality in insects.* Oxford University Press, Oxford

34 Keller L (1995) Parasites, worker polymorphisms and queen number in social insects. *Amer Naturalist* 145: 842–847

35 Keller L, Reeve HK (1994) Genetic variability, queen number and polyandry in social Hymenoptera. *Evolution* 48: 694–704

36 Boomsma JJ, Ratnieks FLW (1996) Paternity in eusocial Hymenoptera. *Phil Trans R Soc Lond B*

37 Crozier RH, Page RE (1985) On being the right size: male contributions and multiple mating in social Hymenoptera. *Behav Ecol Sociobiol* 18: 105–115

38 Moritz RFA (1985) The effects of multiple mating on the worker-queen conflict in *Apis mellifera*. *Behav Ecol Sociobiol* 16: 375–377

39 Starr CK (1984) Sperm competition, kinship, and sociality in the aculeate Hymenoptera *In*: RL Smith (ed): *Sperm competition and the evolution of animal mating systems.* Academic Press, New York, pp 427–464

40 Moritz RFA, Meusel MS, Haberl M (1991) DNA Fingerprinting in the honey bee (*Apis mellifera* L.) with oligonucleotides. *Naturwissenschaften* 78: 42–428

41 Estoup A Solignac M, Cornuet J-M (1994) Precise assessment of the number of patrilines and of genetic relatedness in honeybee colonies. *Proc R Soc Lond B* 258: 1–7

42 Estoup A, Scholl A, Pouvreau A, Solignac M (1995) Monandry and polyandry in bumble bees (Hymenoptera: Bombinae) as evidenced by highly variable microsatellites. *Mol Ecol* 4: 89–93

43 Neumann P, Kryger P, Moritz RFA (1998) Variability of polyandry in the honeybee, *Apis mellifera* L. *Mol Ecol*; *in review*

44 Kraus B, Page RE (1998) Parasites, pathogens, and polyandry in the social insects. *Amer Naturalist* 151: 383–391

45 Page R (1980) The evolution of multiple mating behavior by honey bee queens (*Apis mellifera* L.). *Genetics* 96: 263–273

46 Sherman PW, Seeley TD, Reeve HK (1988) Parasites, pathogens, and polyandry in the social Hymenoptera. *Amer Naturalist* 131: 602–610

47 Sherman PW, Seeley TD, Reeve HK (1998) Parasites, pathogens, and polyandry in honey bees. *Amer Naturalist* 151: 392–396

48 Fuchs S, Moritz RFA (1998) Evolution of extreme polyandry in honeybees. *Behav Ecol Sociobiol* 45: 269–275

49 Onions GW (1912) South African "fertile worker bees". *Agric J Union S Afr* 3: 720–728

50 Anderson RH (1963) The laying worker in the Cape honeybee *Apis mellifera capensis*. J Apic Res 2: 85–92

51 Moritz RFA, Beye M, Hepburn RH (1998) Estimating the contribution of laying workers to population fitness in African honeybees (*Apis mellifera*) with molecular markers. *Insect Soc* 45: 277–287

52 Hillesheim E, Koeniger N, Moritz RFA (1989) Colony performance in honeybee (*Apis mellifera capensis* Esch.) depends on the proportion of subordinate and dominant workers. *Behav Ecol Sociobiol* 24: 291–296

53 Moritz RFA, Kryger P, Allsopp M (1996) Competition for royalty in bees. *Nature* 384: 31

54 Page RE, Fondrk MK (1995) The effects of colony-level selection on the social organization of honey bee (*Apis mellifera* L.) colonies: colony level components of pollen hoarding. *Behav Ecol Sociobiol* 36: 135–144

55 Page RE, Mitchell SD (1998) Self organization and the evolution of division of labor. *Apidologie* 29: 171–190

53 Kaufmann SA (1993) *The origins of order. Self organization and selection in evolution*. Oxford University Press, Oxford

Part 3 The individual at the core of information management

Individuality and colonial identity in ants: the emergence of the social representation concept

Alain Lenoir, Dominique Fresneau, Christine Errard and Abraham Hefetz

Summary

Colonial identity in social insects is based on nestmate recognition, which is mediated through cuticular substances. Although this is considered to be distinct from kin recognition, it is possible that through evolution the signal mediating kinship was replaced by the signal mediating "nestmateship". Cuticular hydrocarbons in *Cataglyphis niger* are responsible for modifying the ant's aggressive behavior and are considered to have a similar function in other ants species. In ants, the postpharyngeal gland (PPG) serves as a storage organ for these cues and functions as a "gestalt" organ, with the gestalt being permanently updated. Its content is constantly being exchanged with nestmates through trophallaxis and allogrooming. We hypothesize that already in the primitive ponerine ants the PPG evolved as a gestalt organ even without trophallaxis. We discuss two alternative primary selective pressures for the evolution of trophallaxis: facilitating food exchange versus exchanging recognition cues. Callow workers seem to be characterized by a "cuticular chemical insignificance" followed by a "chemical integration" period when they acquire the gestalt of the colony and learn the associated template. We hypothesize that the template has evolved from a simple personal chemical reference in primitive species with small colonies to an internal representation of the colonial identity in larger colonies.

Introduction

In this chapter we focus on nestmate recognition in ants and address questions pertaining to the nature of the signals (called labels or cues) at the basis of recognition, and their production and dissemination in adult ants and throughout onto-

C. Detrain et al. (eds) Information Processing in Social Insects
© 1999, Birkhäuser Verlag Basel/Switzerland

genesis. We discuss the mechanisms (decision rules) underlying the behavior of one individual facing another individual: whether amicable or aggressive behaviors follows comparison of the perceived signal with its own reference (called a template). We further discuss how this template is constructed. Finally, we introduce the notion of social representation: does the ant have an integrative representation of its social environment or does it react according to simple rules?

Nestmate versus kin recognition

Nestmate recognition is not to be confused with kin recognition. These two phenomena are indeed different. Nestmate recognition is typically manifested by rejecting alien intruders or preferentially transporting nestmate brood. Kin discrimination, on the other hand, is defined as the differential treatment of relatives according their level of relatedness so as to increase their fitness (nepotism) [1–3]. Since in most ant colonies relatedness is greater than zero, nestmate recognition can constitute a form of kin discrimination. However, the evidence that the template is acquired by exposure and learning at the callow stage suggests that these two phenomena are not necessarily linked. Indeed, there is only limited evidence of nepotism, the expressed behavior of kin discrimination in social insects. Demonstrating kin recognition is still hampered by many methodological difficulties, and the results obtained have been criticized [4]. In honeybees there is evidence for kin recognition [5, 1], but the degree of preference shown by workers for close kin is small, and whether it has an impact on the fitness of the larvae is still uncertain. In ants and wasps the few studies conducted could not find evidence for within-colony discrimination [6–9]. Recently, DeHeer and Ross [10] failed to demonstrate nepotism in multiple-queen colonies of the fire ant *Solenopsis invicta*. In agreement with Vander Meer and Morel [11] we consider that "virtually all recognition studies on ants involve nestmate recognition rather than kin recognition".

Although this dearth of evidence is due to the lack of necessary tools for measuring kin discrimination, we would nonetheless like to present an alternative explanation. The primary selective pressure for developing a recognition system may indeed have been the process of kin selection. However, when more complex societies evolved, that is, multiple mating by the queen and polygyny, selection pressures led to developing a nestmate recognition system that was not necessarily linked to kin. For example, Keller [3] hypothesized that polyandry which increases intracolony variability could explain the low level of nepotism in bees. It was parsimonious to adapt the system that had already developed for kin recognition for nestmate recognition, and therefore kinship became replaced by nestmateship (called also fellowship [12]). If this is indeed true, it means that these societies have lost the ability to recognize kin, which would explain many of the observed phenomena. For example, kin recognition is demonstrable in the primitively social bee *Lasioglossum zephyrum* [13] (and possibly honeybee), whereas in the ants all efforts to demonstrate this phenomenon have failed.

Signals of nestmate recognition: the role of cuticular hydrocarbons

It is a basic observation that when two ants encounter they may already recognize each other from a very short distance (1–2 cm), but generally physical contact is needed. The contact can be made anywhere on the partner's body, indicating that the signal is widely spread on the cuticle. Cuticular lipids, including hydrocarbons, have a primordial role in protecting the insects against desiccation and invasion of microorganisms or toxins. In social insects their hydrocarbon constituents may also have a determinant role in nestmate recognition [14, 15]. This point is largely controversial, as indicated in recent review papers [4, 11]. Two approaches have been developed to test this role of hydrocarbons: correlation studies and experimental studies.

Correlation studies

Cuticular hydrocarbon composition is highly diverse, and modern techniques permit the identification of more and more substances. For example, early analysis of cuticular lipids of *Myrmica incompleta* revealed 19 hydrocarbons [16], whereas further analyses resulted in the identification of 111 substances [17]. Recent analyses of the postpharyngeal gland secretion, representative of the composition of cuticular hydrocarbons (see below), in seven species of *Cataglyphis* resulted in the identification of a total of 242 different hydrocarbons [18]. Generally 30 to 60 substances are found in a given species. Similar to those found in other insects, they are species-specific, exhibiting qualitative variations, and also present intraspecific quantitative variations. They were accordingly used for chemosystematic studies [14, 18].

The hydrocarbon profile can also be characteristic of the population, rendering it a good index of speciation, as was studied, for example, in the genus *Cataglyphis* [18, 19]. More important is the fact that they are also colony-, caste-, and subcaste-specific [14, 20]. In all these studies the authors assessed the differences in hydrocarbon composition using a multivariate analysis, demonstrating that the colonies were well discriminated. A correlation was also found between the hydrocarbon pattern similarity and the closed nature of the colonies in *Cataglyphis cursor* [19]. All these findings do not provide unequivocal proof of the role of hydrocarbons, but they are indicative, taking into account the limitations of correlation studies.

Experimental studies

The removal and replacement of cuticular compounds have provided indications for their role in nestmate recognition. Hydrocarbons are efficiently collected by rinsing the ant with apolar solvents. Corpses of workers treated thus were consequently not recognized as nestmates but considered as neutral. Upon application of the extract to surrogates, these became considered as alien workers and were consequently aggressed (see [11]). These results are difficult to interpret, because a total body extraction contains additional exocrine products, as was observed in *Solenopsis invicta*, which contained large amounts of alkaloids. In this latter ant it was impossible to devise an appropriate bioassay for nestmate recognition, probably due to this high contamination [11]. A solvent rinse also includes lipids other than the hydrocarbons that are found on the cuticle. Therefore, this kind of experiment can at most lead to the conclusion that cuticular lipids are involved in recognition, but not necessarily hydrocarbons (fatty acids seem to be involved in the cuckoo ants [21]). Nevertheless, it does confirm that the substances present on the cuticle are involved in recognition, and hydrocarbons by virtue of their dominant presence are likely candidates for constituting the signal. Recognition pheromones are spread throughout the body surface, and even if they constitute a monomolecular layer on the epicuticle, there is a need for large quantities of material.

Chemical supplementation experiments constitute another approach and have been conducted in bees (see [4]) in which nestmate recognition was modified by applications of a C_{32} hydrocarbon. In the ant *Camponotus vagus*, applications of (Z)-9-tricosene induced some aggressive behaviors and intense antennations from the nontreated nestmates, indicating that the recognition cues had been modified [22]. These experiments are not conclusive, since many compounds that are applied exogenously may elicit aggression toward the ants.

The PPG

Recent studies on this gland (idiosyncratic to the Formicidae) have enabled considerable progress in our understanding of the elaboration of colonial odor in ants. The PPG contains hydrocarbons that are congruent with the cuticular ones [23–25]. This suggests a link between these two body parts, and that the PPG contains the recognition cues. Several studies using PPG content (and thereby avoiding contamination of secretions from other sources) have confirmed its role as a modifier of aggressive behavior. This was shown in *Cataglyphis niger* (Formicinae) and *Manica rubida* (Myrmicinae), where application of a nestmate's PPG on an alien ant reduced the aggression generally exhibited toward the latter, whereas application of an alien ant PPG secretion on a nestmate resulted in augmented aggression by her nestmates [26, 27]. The complementary results obtained with ants from two different subfamilies suggest that the role of the gland as a modifier of aggressive behavior may be a general phenomenon in ants.

Causative experiments confirming the role of hydrocarbons in nestmate recognition

To test the specific role of hydrocarbons in nestmate recognition, PPG secretion of *C. niger* was fractionated by column chromatography into hydrocarbons (hexane elution) and more polar lipids (chloroform:methanol elution) (S. Lahav et al., unpublished data). Of these two fractions, only the hydrocarbons modified the ant's aggressive behavior. Interestingly, augmented aggression toward a nestmate applied with an alien ant hydrocarbon fraction was much more pronounced than decreased aggression toward an alien ant applied with a nestmate's hydrocarbon complex. This may suggest that the mechanism of recognition relies on detecting differences rather than sameness.

A model of the production and dynamics of colonial odor: the gestalt model. A revisitation of the role of trophallaxis

The use of the PPG as a source for nestmate recognition cues is very adaptive. Since it opens to the mouth cavity the secretion can readily be applied onto the body surface by self-grooming. The position of the gland also facilitates exchange of substances between members of the nest, promoting the rapid distribution of the scent within the colony. Moreover, any new substance that is introduced into the colony can be quickly incorporated and distributed among the nest members and thus become a part of the recognition system. Thus the PPG can be regarded as a perfect gestalt organ.

The gestalt model was first proposed by Crozier and Dix from theoretical considerations [28], and later behaviorally demonstrated in various *Leptothorax* species [29, 30]. Recent studies using radioactive tracers have confirmed this role of the PPG. In *C. niger* the gland would appear to be only a place for storage, rather

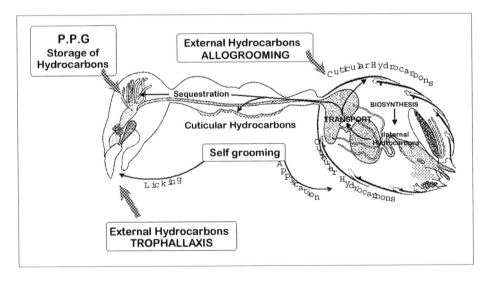

Figure 1 Intra- and interindividual hydrocarbon flow in the ants

than synthesis, of hydrocarbons [25]. According to these authors the hydrocarbons are synthesized elsewhere in the body, perhaps in the oenocytes as in other insects, and are either secreted directly to the epicuticle, or transported through the haemolymph to the PPG. Encounters between a prelabeled ant and an individual, or group of nonlabeled nestmates, further confirmed that hydrocarbons can be transferred from an individual during trophallaxis directly to the PPG of the partner(s) and then reapplied on its cuticle through autogrooming [25, 27]. By constructing encounters in which one or both of the ants had blocked mouthparts, these authors further showed that allogrooming also plays a role, albeit secondary, in the transfer of chemicals. Similar conclusions were reached by deposition of (Z)-9-tricosene on the cuticle of a *Camponotus vagus* worker and finding this substance on the cuticle and in the PPG of the nontreated nestmates [31].

As mentioned earlier, the use of the PPG as a gestalt organ facilitates the incorporation of new or external odors into the colony. It explains nicely how odors from environmental sources are used in nestmate recognition as found in fire ants [32], or how queen-derived odors can circulate within the colony (see below). Glandular exchanges also seem to be the key to the ability of artificially mixed species to cohabit [33, 34], and may reflect the situation in naturally mixed species colonies (e.g. parasite, slave maker; see later for additional discussion).

The finding that trophallaxis has a cardinal role in the creation of a homogenous colony odor sheds new light on the function of this behavior. Despite the early hypothesis of Wheeler [35] that social life evolved in insects through trophallaxis between adults and their progeny, there is little documentation on the possible role of this behavior in social evolution. It is not cited at all in the two recent books on social evolution in ants by Bourke and Franks, and Crozier and Pamilo [1, 36], and was not used as a factor in the evolutionary processes of ants discussed by Baroni Urbani [37]. It was recently suggested that food offered as part of an appeasement behavior may have selected for more elaborate trophallaxis in social Hymenoptera [38]. Having learned the role of trophallaxis in the exchange of semiochemicals, it was interesting to study how this process occurs in species that do not trophallax. Such a study was recently conducted with the ponerine ant *Pachycondyla apicalis*, which has a normal size PPG [39]. Using radioactive tracers and dyadic encounters, it was confirmed that the ants exchange hydrocarbons between PPG and cuticular surfaces. In contrast to *Cataglyphis niger,* the amount of transfer to the PPG was comparable irrespective of whether the donor had freed or blocked mouthparts. This indicates that all the transferred hydrocarbons to the donor were the result of allogrooming. Behavioral observations confirmed the absence of trophallaxis in this species. Cue exchange by allogrooming seems to be less effective than trophallaxis, as revealed by the finding that the magnitude of transfer in the 24-h dyadic encounters of species that employ trophallaxis such as *Manica rubida* or *C. niger* is more than threefold greater than in *P. apicalis*. Nonetheless, transfer by allogrooming may be sufficient for effective cue distribution in the small colonies of *P. apicalis* (80 individuals) [39–40]. Allogrooming is also probably involved in the chemical mimicry of the guest ant *Formicoxenus*, which licks its host *Myrmica* very frequently to obtain its cuticular hydrocarbons [41, 17].

If the case of *P. apicalis* can be considered as typical to ponerine ants, it would seem that the role of the PPG as a gestalt organ evolved earlier than trophallaxis. It can be postulated that as ant colonies became larger, recognition by individual odors became impossible, favoring the evolution of a colonial odor [28]. This in turn necessitated an effective means of odor sharing and the development of an exocrine gland in which the different odors could be admixed. The use of the PPG with its external opening to the mouth cavity was adaptive. Early signs of the role of trophallaxis in cue exchange are observed in *Ponera coarctata* [38], where droplets may comprise postpharyngeal gland secretion rather than food [39]. At a later stage, when colonies evolved to be even larger, transfer by allogrooming became inefficient, and transfer by trophallaxis became the major pathway for odor distribution within the colony. Whether trophallaxis evolution resulted from selective pressures to facilitate cue transfer in the colony, or primarily as a means of food transfer that was secondarily adapted for cue transfer, remains an open question.

The role of the queen in the composition of recognition cues

As mentioned above, ants from different colonies bear idiosyncratic compositions of hydrocarbons, even in cases in which they are reared under equal conditions and diet. This suggests a genetic component influencing their composition (see below). The question thus arises as to whose genes affect the colony odor: those of the queen or of the workers? In the case of monogynous and monandrous species, the genetic makeup of the colony is rather homogeneous, and unless the recognition cues constitute queen-specific products, both the queen or the workers can equally contribute to colony odor. In polyandrous monogynous species the genetic makeup of the colony is more heterogeneous, and the question regarding the impact of the queen becomes more acute. It is definitely relevant in polygynous species. The solution to this biological problem may not be uniform. The impact of the queen may differ in different species with different social structures.

Of the few studies that have addressed this question, two possible models emerge. The first, based on behavioral experiments in *Camponotus floridanus*, suggests that the queen is the origin of the gestalt [42]. The second, based on biochemical studies in *C. niger*, suggests that the queen acquires the average colony odor and therefore positions herself in the center of the gestalt, rather than being its origin (43). These models may not be mutually exclusive, especially in view of the fact that *C. floridanus* is monogynous whereas *C. niger* is polygynous.

The basic hypothesis behind the experiments done with *C. floridanus* was that in a mixed species group consisting of an unequal number of workers from each species and a queen of the minority species, if the workers are responsible for the colony odor then this should have the odor of the majority species. Through a series of elegant studies with such mixed colonies it was demonstrated that whatever the respective number of workers, the colony still bears the odor of the species to which the queen belongs.

The study with *C. niger* utilized a more direct approach to address this question [43]. Through a series of dyadic and group encounters it was shown that the queen receives more PPG secretion that she tends to give away. It was further shown that when given the choice, workers preferred to transfer PPG material to the queen rather than to their nestmates. The end result is that the queen possesses three times as much secretion in her gland and, moreover, that she always has an average colony odor. This central position of the queen is also suggested from the experiment in which polygynous colonies were split into several daughter monogynous colonies. After several months of separation the hydrocarbon profiles of the daughter colonies diverged (see below), but the queen was always at the average of each respective colony (although the workers in the colony were not necessarily her descendants; S. Lahav et al., unpublished data).

Genetic and environmental origins of the hydrocarbons

Many studies have shown that colonial recognition is at least partly determined by genetic factors [4, 11]). There is also evidence that hydrocarbon composition is genetically determined. Intermediary cuticular profiles that have been found in areas cohabited by two *Solenopsis* species are apparently the result of hybridization between these two species [44, 45]. In honeybees different patrilines have discriminable cuticular profiles, emphasizing again the genetic components delineating hydrocarbon composition [5].

An environmental origin of the colonial odor must also be considered, because the lipid layer of cuticle is a veritable trap for all odors. Some myrmecophiles rub their body against the walls of the nest to acquire the odor of the host colony as camouflage. In bees and wasps, nesting material was described to influence colonial recognition [46, 47], whereas in attine ants the quality of food leaves influences the nest odor [48]. In *Leptothorax nylanderi*, which nests in sticks on the forest ground, nestmate recognition is largely influenced by the origin of the stick, being either pine or oak [49]. Ants living in pinewood probably incorporate and use the characteristic conifer odors trapped by their cuticle.

Ontogeny of the PPG secretion and changes of hydrocarbon profile with age

C. niger callow ants have a visibly structured, but not yet fully developed PPG. There is a clear structural as well as chemical age-dependent development. The epithelial thickness increases with age, reaching its maximum within the first week after emergence, and also showing a progressive development of microvilli and occurrence of lamellar inclusions. Concomitantly there is an increase in the quantity of hydrocarbons present in the gland [50]. A similar chemical ontogeny is also described for *C. iberica* [51]. This glandular ontogeny corroborates nicely with the studies pertaining to the ontogeny of nestmate recognition. Since the early work of Fielde [52], who hypothesized the existence of a progressive odor change with age in ants, other authors have confirmed that there is an age-dependent quantitative variation in the production and release of the cues implied in the formation of colonial odor (see [11, 51]). In workers of *Manica rubida* and *Formica selysi*, chemical analysis showed that both species already had their own specific cuticular profiles within 48 h of emergence, and that their total amount increased with age until they stabilized at the age of 1 month [53, 54]. This quantitative variation in the production of chemical cues is also observed in *C. iberica*, where worker chemical profiles are distinctive according to age group and gradually converge from a "callow profile" with specific hydrocarbons to a profile characteristic of mature workers [51].

Callow workers are more likely to be accepted into alien colonies than a 1-week-old worker [29, 55–58], which may be linked to weak aggressive behavior, but also due to the weak signal they possess on their cuticle. The lack of external chemical characteristics, which we term "cuticular chemical insignificance", also seems to be the basis of the successful formation of artificially mixed species groups of *M. rubida* and *F. selysi*. This is followed by a period of "chemical integration" that results in a pseudosocial colony, and is possible only with callow workers [53], thus supporting the observations that at this stage the PPG is rather undeveloped (A. Hefetz and C. Errard, unpublished data). It is interesting that in more phylogenetically remote species, workers must be younger in order to achieve a successful mixed species group [59]. We can speculate that in these cases the qualitative differences between their respective hydrocarbons are high, and the small amount of hydrocarbons that the callow workers possess is enough to signify the differences between the individual species in question. On the other hand, when the species are more related phylogenetically, qualitative differences in the recognition cues are smaller and there is a need for higher quantities to detect the differences between the species. A cuticular chemical insignificance phenomenon may also be the basis of the successful usurpation by slave maker and parasitic ants. We would like to emphasize that the processes of cuticular chemical insignificance followed by chemical integration that we observe in artificially mixed colonies do not necessarily reflect the mechanisms employed by naturally mixed societies. In artificially mixed societies, the ants employ their predisposed behavioral repertoire once mutual tolerance has been established (the cuticular chemical insignificance phase), that is, cue exchange by trophallaxis and allogrooming, and achieve a uniform colonial odor. Slave makers and parasites, on the other hand, have seemingly coevolved with their host, and any of various mechanisms, from chemical camouflage to the maintenance of independent chemical signatures, could have been selected for.

Template acquisition

The template is an "internal representation of the labels expected in kin within the mind of the discriminating individual" [36]. Many ethological observations have shown the existence of a learning period for template acquisition. In all the species studied so far, nestmate recognition cues appear to be learned shortly after eclosion into the adult stage even in species utilizing cues that are primarily genetically determined, as in sweat bees *Lasioglossum zephyrum* [2, 13]. Workers reared in mixed groups learn and memorize the allospecific chemical cues during their early social experience, incorporating them into their template and keeping them for at least 1 year [53, 60]. The supposition of an early learning period does not exclude the possibility of template updating (see below).

The decision rules

When encountering another ant, the individual has to compare his own template with the received cues and choose between amicable or aggressive reactions. The response can be evoked according to a threshold rule or can be graded (these are by no means mutually exclusive). In some species (e.g. *Cataglyphis cursor*, *Myrmica rubra*) ants exhibit scaled aggression from complete rejection to acceptance [19, 26, 27, 32, 55]. On the other hand, in *Cataglyphis iberica*, allocolonial workers are always immediately rejected [61], suggesting that the ants react according to a threshold. A behavioral response according to threshold does not mean that discrimination does not follow a gradual scale. In *C. iberica*, young workers at the onset of hibernation possess a PPG composition that diverges from the colonial odor [62]. This difference is apparently recognized by older workers, since the young workers are preferentially transported to other satellite nests. Since the difference in glandular composition does not elicit aggression, we can assume that they are below the threshold. Thus in this ant we have at least three levels of discrimination with a parallel gradation of dissimilarity in cuticular hydrocarbons: alien (behavioral response expressed as aggression), young (adult transport), and old (various amicable behaviors such as trophallaxis, allogrooming, etc.). Although the role of hydrocarbons in nestmate recognition was not demonstrated in this species, we draw a parallelism from the experiments with *C. niger*.

Additional evidence for a graded model in discrimination is the longer antennation time observed when the signature differs slightly from the template. This was shown in various *Solenopsis* and *Leptothorax* [63, 64] in which nestmates are recognized immediately when encountered, but any slight difference induces longer inspections and recognition is more difficult. Longer antennation, before actual recognition, presumably the result of slightly diverging cuticular compositions, was also observed in adult *C. iberica* that were reunited after a period of separation [65]. In a colony composed of non-related workers, mutual antennation was longer than normal [66, 67], which was not the case when the workers constituted different matrilines [7].

The conclusion arising from all these studies is that the gestalt colonial odor has a dynamic component and is modified with time according to the composition of the group and/or food, as has been observed in other ants [63, 64]. It is further emphasized that there is a graded response to odor/template matching up to a threshold which is variable according to the species.

A new insight into the graded response was achieved in studies that tested the reaction of ants from mixed species colonies that were treated with postpharyngeal secretions under various contexts. Accordingly, it was hypothesized that deciphering the signal in the recognition process may be hierarchical and the reaction inverse to the familiarity of the signal. It was further demonstrated that the template to which the signal is compared is acquired, and since the label is dynamic, template plasticity must follow. Thus the template is not fixed after the sensitive period, but has to be reinforced via constant perception of nestmate odors [29, 68]. In some highly polygynous and pandemic species, ants generally accept introduced alien workers. It is possible that the workers are habituated to be confronted with various odors due to low relatedness (the lowest known case is $r = 0.02$ in *Linepithema humile* [69]) and more variable genetic cues [70]. They learn a less restrictive template, and consequently their threshold response is very weak, or even entirely absent in pandemic tramp species [71].

Discussion and perspectives

Nestmate recognition is undoubtedly one of the major selective forces in social evolution, and has provided a multidisciplinary subject for research. Although it was a natural theoretical outcome from kin selection considerations, when it came to elucidating the proximal mechanisms, it proved to be a complex question. In order to progress in our understanding of the system, we have on the one hand to differentiate between the label and the template, but at the same time we also have to appreciate that the mechanisms controlling them are interlinked. Our understanding of the nature of the label and its origin has improved considerably in recent years, but we are far from understanding these features regarding the template.

Concerning the label, the theoretical suggestion that a gestalt odor must exist in complex social insects was experimentally substantiated in all the species studied so far. While the theory of the gestalt provided an excellent explanation for establishing of a colonial odor, biochemical experiments have outlined the processes involved, including the involvement of hydrocarbons as a part of the recognition system, and mutual exchanges as a means for creating the gestalt.

A basic feature of the label is its dynamic character and the ability of the ants to update the cues through their mutual sharing with a subsequent refreshment on the cuticular surface.

There are still open questions regarding the evolution of the label, its glandular origin, and its mode of dissipation within the colony, in particular in very populous colonies. Was there a functional evolution of the PPG and comparable glands? Is there a cuticular lipid storing gland in solitary species, or has it evolved specifically in ants? An indication for this kind of convergent evolution comes from a recent study on the biosynthesis of the hydrocarbon sex pheromone of the tiger moth. It appears that the pheromone is biosynthesized in association with the abdominal integument and carried by a specific lipophorin to the pheromone-disseminating gland in the abdomen [72].

The chemical gestalt present in the ants can be compared to a network. The gestalt is based on a constant flow of "information" and is naturally influenced by the number of participants. While optimizing the system so as to create uniformity, it is obvious that as the system becomes more complex, it is vulnerable to more breakdowns. To what extent, then, can the gestalt odor function in large colonies? Does it break down or are there alternative methods for achieving a cohesive closed colony?

The idea of a template has been around for a long time (e.g. hypotheses pertaining to learning and sensitive period), but little experimental progress has been achieved. This is not surprising, since the experimental paradigms for testing the nature of the template are extremely difficult and multidimensional. We still do not understand how nestmate recognition cues are perceived, and at what levels the information is processed. At the behavioral level we know that the manifestation of the tested chemical comprises a complex behavior that is largely dependent on the context of their perception. For example, in the ponerine ant *Pachycondyla apicalis*, the outcome of a dyadic encounter depends on whether the ants are resident or intruders in accordance with the theory of games [40, D. Fresneau et al., unpublished data).

Another largely unsolved problem regarding the template is the impact of memory on the integration of the perceived signals. Adoption experiments point to the existence of a learning sensitive period during which the sensory imaging of the colony is formed. But this memory evidently has to be plastic rather than fixed, since the experimental evidence demonstrates that the label is continuously updated. Therefore, an efficient system has to possess a comparably flexible template system. Returning to the case exhibited by the ponerine ants, following experimental manipulation, ants that are adopted into an alien colony participate after a while in the defense of their adoptive nest. This means that they have substituted their original template for a new one. There are two alternative hypotheses to explain this phenomenon: the template is memorized, but if it is not reinforced or the stimulus permanently changes, the old memory is erased by a new one; alternatively, the template is not neural but chemical. The ants constantly compare the odor on their body surface (or part of it, e.g. antennae) to the odor perceived at any given moment [73]. This latter system is simpler, since the information is already filtered at the peripheral system without a need to refer to memory embedded within the brain. These systems are not mutually exclusive. We hypothesize that the simpler chemical reference could have been selected for in the primitive social species with small colonies. In larger colonies the complexity of the systems increases exponentially and may have also required an internal representation of the colonial identity. We hope that with advances in insect neuroscience answers to such questions will be resolved.

Acknowledgments

We thank Ms. Nomi Paz for the English revision of the text.

References

1 Bourke AFG, Franks NR (1995) *Social evolution in ants*. Princeton University Press, Princeton

2 Hölldobler B, Wilson EO (1990) *The ants*. Springer, Berlin

3 Keller L (1997) Indiscriminate altruism: unduly nice parents and siblings. *TREE* 12: 99–103

4 Breed MD (1997) Chemical cues in kin recognition: criteria for identification, experimental approaches, and the honey bee as an example. *In*: RK Vander Meer, MD Breed, M Winston, C Espelie (eds): *Pheromone communication in social insects: ants, wasps, bees and termites*. Westview Press, Boulder, CO, 57–78

5 Arnold G, Quenet B, Cornuet J-M, Masson C, De Schepper B, Estoup A, Gasqui P (1996) Kin recognition in honeybees. *Nature* 379: 498

6 Bernasconi G, Keller L (1996) Reproductive conflicts in cooperative associations of fire ant queens (*Solenopsis invicta*). *Proc R Soc Lond B* 263: 509–513

7 Carlin NF, Reeve HK, Cover SP (1993) Kin discrimination and division of labor among matrilines in the polygynous carpenter ant, *Camponotus planatus*. *In*: L Keller (ed): *Queen number and sociality in insects*. Oxford University Press, Oxford, 362–401

8 Queller DC, Hughes CR, Strassmann JE (1990) Wasps fail to make distinctions. *Nature* 344: 388

9 Snyder LE (1993) Non-random behavioural interactions among genetic subgroups in a polygynous ant. *Anim Behav* 46: 431–439

10 DeHeer CJ, Ross KG (1997) Lack of detectable nepotism in multiple-queen colonies of the fire ant *Solenopsis invicta* (Hymenoptera: Formicidae). *Behav Ecol Sociobiol* 40: 27–33

11 Vander Meer RK, Morel L (1998) Nest-mate recognition in ants. *In*: RK Vander Meer, MD Breed, M Winston, C Espelie (eds): *Pheromone communication in social insects: ants, wasps, bees and termites*. Westview Press, Boulder, CO, 79–103

12 Jaisson P (1991) Kinship and fellowship in ants and social wasps. *In*: PG Hepper (ed): *Kin recognition*. Cambridge University Press, Cambridge, 60–93

13 Greenberg L (1979) Genetic component of bee odor in kin recognition. *Science* 206: 1095–1097

14 Howard RW (1993) Cuticular hydrocarbons and chemical communication. *In*: DW Stanley-Samuelson, DR Nelson (eds): *Insect lipids: chemistry, biochemistry and biology* University of Nebraska Press, Omaha, 179–226

15 Lorenzi MC, Bagnères A-G, Clément J-L (1996) The role of cuticular hydrocarbons in social insects: is it the same in paper wasps? *In*: S Turillazzi, MJ West-Eberhard (eds): *Natural history and evolution of paper wasps*. Oxford University Press, Oxford, 178–189

16 Howard RW, Stanley-Samuelson DW, Akre RD (1990) Biosynthesis and chemical mimicry of cuticular hydrocarbons from the obligate predator, *Microdon albicomatus* Novak (Diptera: Syrphidae) and its ant prey, *Myrmica incompleta* Provancher (Hymenoptera: Formicidae). *J Kans Entomol Soc* 63: 437–443

17 Lenoir A, Malosse C, Yamaoka R (1997) Chemical mimicry between parasitic ants of the genus *Formicoxenus* and their host *Myrmica* (Hymenoptera, Formicidae). *Biochem System Ecol* 25: 379–389

18 Dahbi A, Lenoir A, Tinaut A, Taghizadeh T, Francke W, Hefetz A (1996) Chemistry of the postpharyngeal gland secretion and its implication for the phylogeny of Iberian *Cataglyphis* species (Hymenoptera, Formi-

cidae). *Chemoecology* 7: 163–171

19 Nowbahari E, Lenoir A, Clément J-L, Lange C, Bagnères A-G, Joulie C (1990) Individual, geographical and experimental variation of cuticular hydrocarbons of the ant *Cataglyphis cursor* (Hymenoptera: Formicidae): their use in nest and subspecies recognition. *Biochem System Ecol* 18: 63–73

20 Bonavita-Cougourdan A, Clément J-L, Lange C (1993) Functional subcaste discrimination (foragers and brood-tenders) in the ant *Camponotus vagus* Scop. polymorphism of cuticular hydrocarbon patterns. *J Chem Ecol* 19: 1461–1477

21 Franks NR, Blum MS, Smith R, Allies AB (1990) Behaviour and chemical disguise of cuckoo ant *Leptothorax kutteri* in relation to its host *Leptothorax acervorum*. *J Chem Ecol* 16: 1431–1444

22 Meskali M, Provost E, Bonavita-Cougourdan A, Clément J-L (1995) Behavioural effects of an experimental change in the chemical signature of the ant *Camponotus vagus* (Scop.). *Insectes Soc* 42: 347–358

23 Bagnères A-G, Morgan ED (1991) The postpharyngeal glands and the cuticle of Formicidae contain the same hydrocarbons. *Experientia* 47: 106–111

24 Do Nascimento RR, Billen J, Morgan ED (1993) The exocrine secretions of the jumping ant *Harpegnathos saltator*. *Comp Biochem Physiol B Comp Bioch* 104: 505–508

25 Soroker V, Vienne C, Hefetz A (1995) Hydrocarbon dynamics within and between nestmates in *Cataglyphis niger* (Hymenoptera, Formicidae). *J Chem Ecol* 21: 365–378

26 Hefetz A, Errard C, Chambris A, Le Négrate A (1996) Postpharyngeal gland secretion as a modifier of aggressive behavior in the myrmicine ant *Manica rubida*. *J Insect Behav* 9: 709–717

27 Soroker V, Vienne C, Hefetz A, Nowbahari E (1994) The postpharyngeal gland as a "gestalt" organ for nestmate recognition in the ant *Cataglyphis niger*. *Naturwissenschaften* 81: 510–513

28 Crozier RH, Dix MW (1979) Analysis of two genetic models for the innate components of colony odor in social Hymenoptera. *Behav Ecol Sociobiol* 4: 217–224

29 Alloway TM (1997) The role of workers and queens in the colony-member recognition systems of ants. Are there any differences that predispose some kinds of ants to social parasitism? *In*: G Greenberg, E Tobach (eds): *Comparative psychology of invertebrates. The field and laboratory study of insect behavior*. Garland, New York, 193–219

30 Stuart RJ (1988) Collective cues as a basis for nestmate recognition in polygynous Leptothoracine ants. *Proc Natl Acad Sci USA* 85: 4572–4575

31 Meskali M, Bonavita-Cougourdan A, Provost E, Bagnères A-G, Dusticier G, Clément J-L (1995) Mechanism underlying cuticular hydrocarbon homogeneity in the ant *Camponotus vagus* (Scop.) (Hymenoptera: Formicidae): role of postpharyngeal glands. *J Chem Ecol* 21: 1127–1148

32 Obin MS, Vander Meer RK (1988) Sources of nestmate recognition cues in the imported fire ant *Solenopsis invicta* Buren (Hymenoptera: Formicidae). *Anim Behav* 36: 1361–1370

33 Hefetz A, Errard C, Cocojaru M (1992) Heterospecific substances in the postpharyngeal gland secretion of ants in mixed groups. *Naturwissenschaften* 79: 417–420

34 Vienne C, Soroker V, Hefetz A (1995) Congruency of hydrocarbon patterns in heterospecific groups of ants: transfer and/or biosynthesis? *Insectes Soc* 42: 267–277

35 Wheeler WM (1928) *The social insects: their origin and evolution*. Kegan Paul,

Treanch, Trubner and Co, London

36 Crozier RH, Pamilo P (1996) *Evolution of social insect colonies, sex allocation and kin selection.* Oxford University Press, Oxford

37 Baroni Urbani C (1989) Phylogeny and behavioural evolution in ants, with a discussion of the role of behaviour in evolutionary processes. *Ethol Ecol Evol* 1: 137–168

38 Liebig J, Heinze J, Hölldobler B (1997) Trophallaxis and aggression in the Ponerine ant, *Ponera coarctata*: implications for the evolution of liquid food exchange in the Hymenoptera. *Ethology* 103: 707–722

39 Soroker V, Fresneau D, Hefetz A (1998) The formation of colony odor in the ponerine ant *Pachycondyla apicalis* (Formicinae: Ponerinae). *J Chem Ecol* 24: 1077–1090

40 Fresneau D (1994) Biologie et comportement social d'une fourmi ponérine néotropicale (*Pachycondyla apicalis*). Thesis Doctorat d'État, État Université Paris Nord

41 Errard C, Fresneau D, Heinze J, Francoeur A, Lenoir A (1997) Social organization in the guest-ant *Formicoxenus provancheri*. *Ethology* 103: 149–159

42 Carlin NF, Hölldobler B (1987) The kin recognition system of carpenter ants (*Camponotus* spp) II. Larger colonies. *Behav Ecol Sociobiol* 20: 209–217

43 Lahav S, Hefetz A, Vander Meer RK (1998) Nestmate recognition in the ant cataglyphis niger: do queen matter? *Behav Ecol Sociobiology* 43: 203–212

44 Ross KG, Vander Meer RK, Fletcher DJC, Vargo EL (1987) Biochemical phenotypic and genetic studies of two introduced fire ants and their hybrids (Hymenoptera: Formicidae). *Evolution* 41: 280–293

45 Vander Meer RK, Lofgren CS, Alvarez FM (1985) Biochemical evidence for hybridization in fire ants. *Fla Entomol* 68: 501–506

46 Breed MD, Garry MF, Pearce AN, Hibbard BE, Page REJ (1995) The role of wax comb in honey bee nestmate recognition. *Anim Behav* 50: 489–496

47 Gamboa GJ (1996) Kin recognition in social wasps. *In*: S Turillazzi, MJ West-Eberhard (eds): *Natural history and evolution of paper wasps*. Oxford Univ Press, Oxford, 161–177

48 Jutsum AT, Saunders E, Cherrett JM (1979) Intraspecific agression in leaf-cutting ant *Acromyrmex octospinosus*. *Anim Behav* 27: 839–844

49 Heinze J, Foitzik S, Hippert A, Hölldobler B (1996) Apparent dear-enemy phenomenon and environment-based recognition cues in the ant *Leptothorax nylanderi*. *Ethology* 102: 510–522

50 Soroker V, Hefetz A, Cocojaru M, Billen JPJ, Francke J, Francke W (1995) Structural and chemical ontogeny of the postpharyngeal gland in the desert ant *Cataglyphis niger*. *Physiol Entomol* 20: 323–329

51 Dahbi A, Cerdá X, Lenoir A (1999) Ontogeny of colonial hydrocarbon label in callow workers of the ant *Cataglyphis iberica*. *C R Acad Sci Paris* 321: 395–402

52 Fielde AM (1905) The progressive odor of ants. *Biol Bull Marine Lab, Woods Hole* 10: 1–16

53 Errard C (1994) Development of interspecific recognition behavior in the ants *Manica rubida* and *Formica selysi* (Hymenoptera: Formicidae) reared in mixed-species groups. *J Insect Behav* 7: 83–99

54 Errard C, Jaisson P (1991) Les premières étapes de la reconnaissance interspécifique chez les fourmis, *Manica rubida* et *Formica selysi* (Hymenoptera: Formicidae) élevées en colonies mixtes. *C R Acad Sci Paris* 313, sér III: 73–80

55 Carlin NF, Hölldobler B (1986) The kin

recognition system of carpenter ants (*Camponotus* spp) I. Hierarchical cues in small colonies. *Behav Ecol Sociobiol* 19:123–134

56 Morel L, Vander Meer RK, Lavine BK (1988) Ontogeny of nestmate recognition cues in the red carpenter ant (*Camponotus floridanus*). *Behav Ecol Sociobiol* 22: 175–183

57 Nowbahari E, Lenoir A (1989) Age related changes in aggression in ant *Cataglyphis cursor* (Hymenoptera, Formicidae). *Behav Processes* 18: 173–181

58 Stuart RJ (1992) Nestmate recognition and the ontogeny of acceptability in the ant, *Leptothorax curvispinosus*. *Behav Ecol Sociobiol* 30: 403–408

59 Errard C, Jaisson P (1984) Etude des relations sociales dans les colonies mixtes hétérospécifiques chez les fourmis (Hymenoptera, Formicidae). *Folia Entomol Mexicana* 61: 135–146

60 Errard C (1994) Long-term memory involved in nestmate recognition in ants. *Anim Behav* 48: 263–271

61 Dahbi A, Cerdá X, Hefetz A, Lenoir A (1996) Social closure, aggressive behavior, and cuticular hydrocarbon profiles in the polydomous ant *Cataglyphis iberica* (Hymenoptera, Formicidae). *J Chem Ecol* 22: 2173–2186

62 Dahbi A, Cerdá X, Hefetz A, Lenoir A (1997) Adult transport in the ant *Cataglyphis iberica*: a means to maintain a uniform colonial odour in a species with mutiple nests. *Physiol Entomol* 22: 13–19

63 Provost E, Rivière G, Roux M, Morgan ED, Bagnères A-G (1993) Change in the chemical signature of the ant *Leptothorax lichtensteini* Bondroit with time. *Insect Biochem Mol Biol* 23: 945–957

64 Vander Meer RK, Saliwanchik D, Lavine BK (1989) Temporal changes in colony

cuticular hydrocarbon patterns of *Solenopsis invicta*: implications for nestmate recognition. *J Chem Ecol* 15: 2115–2125

65 Dahbi A, Lenoir A (1998) Nest separation and the dynamics of the Gestalt odor in the polydomous ant *Cataglyphis iberica* (Hymenoptera, Formicidae). *Behav Ecol Sociobiol* 42: 349–355

66 Carlin NF, Hölldobler B, Gladstein DS (1987) The kin recognition system of carpenter ants (*Camponotus* spp) III. Within colony discrimination. *Behav Ecol Sociol* 20: 219–227

67 Kaufmann B, Passera L (1991) Première approche du problème de la reconnaissance coloniale chez *Iridomyrmex humilis* (Formicidae, Dolichoderinae). *Actes Coll Insectes Soc* 7: 75–82

68 Errard C, Hefetz A (1997) Label familiarity and discriminatory ability of ants in mixed groups. *Insectes Soc* 44: 189–198

69 Kaufmann B, Boomsma JJ, Passera L, Petersen KN (1992) Relatedness and inbreeding in a French population of the unicolonial ant *Iridomyrmex humilis* (Mayr). *Insectes Soc* 39: 195–200

70 Keller L (1995) Social life: the paradox of multiple-queen colonies. *TREE* 10: 355–360

71 Passera L (1994) Characteristics of tramp species. *In*: DF Williams (ed): *Exotic ants. Biology, Impact and control of introduced species*. Westview Press, Boulder, CO, 23–43

72 Schal C, Sevala VL, Cardé RT (1997) Synthesis and transport through the hemolymph of a hydrocarbon pheromone in tiger moth (Lepidoptera: Arctiidae). *Abstr Int Conf Chemical Ecology*, Vancouver, July 1997

73 Fresneau D, Errard C (1994) L'identité coloniale et sa "représentation" chez les fourmis. *Intellectica* 19: 91–115

Key individuals and the organisation of labor in ants

Simon K. Robson and James F.A. Traniello

Summary

In this chapter we examine the organisation of group behaviour at the level of the individual, and discuss the extent and significance of individual behavioural specialisation to group success. Studies of ants in which the behaviour of individuals as well as the group is recorded often reveal a high degree of individual specialisation in the absence of either age or morphological caste differences. Typically referred to as examples of elitism, idiosyncracy or specialisation, we incorporate these examples into a new classification, the key individual concept. This concept defines specialised individuals on the basis of their functional relationship to the behaviour of the group as a whole and indicates that a number of different organisational principles can underlie group behaviour. Some key individuals (catalysts) act to increase the activity level of other group members, whereas other key individuals (organisers) serve to ensure group cohesion and task completion. The demonstration that even large-scale process such as nest emigration can be organised by a subgroup of highly active individuals suggests that studies of group action might benefit from simultaneous analysis at both the individual and group level. Individuals may not always be behaviourally interchangeable, and individual behavioural specialisation, in the absence of age or caste differences, can be a significant component of group organisation.

Introduction

Social insect colonies are comprised of numerous workers that differ in the manner in which they contribute to colony fitness. This behavioural specialisation, division of labor, is a defining feature of social insect colonies [1]. Much of the variance in individual behaviour within a colony can be explained by the existence of physical castes or age polyethism, and the evolution of division of labor due to morphological or age differences has attracted considerable attention [2–4]. Ergonomic studies seeking to explain the mechanism of colony-level efficiency have concentrated on how the total work effort required by a colony can be best divided among the available temporal and physical castes. This is because colony efficiency appears to reflect behavioural variability between the various groups that comprise the colony, rather than individual variability within any one group. It is the series-parallel actions of numerous relatively simple individuals that appear to contribute most to colony success, not interindividual variance within a particular group *per se* [3, 5]. This approach has recently been extended by the application of models and techniques such as self-organisation, cellular automata and neural networks, that allow the relationship between simple individual and emergent group-level behaviours to be examined [6–8]. These approaches focus on the behaviour of groups: studies of collective behaviour typically minimise the role of individual behavioural variance (reviewed in [9–11]).

The study of collective behaviour has emphasised individual similarity. However, individuals of the same temporal or physical caste can differ significantly in the type and frequency of tasks they perform [12]. Nested within the influence of age and morphology, a variety of factors such as genetic predisposition, individual experience and flexible developmental pathways are also known to influence individual behaviour (reviewed in [10, 13, 14, 16–19]). Cases in which the variance in individual behaviour appears extreme have been described as representing elitism, idiosyncratic behaviour or behavioural specialisation [3, 4]. While the mechanisms underlying these phenomena remain unclear, the demonstration of significant behavioural variance within individuals of similar temporal or physical caste suggest that modes of organisation of labour other than those related to age or size exist. Not all social insect behaviours may be based on relatively simple and similar individuals, and it appears likely that multiple organisational principles may apply to different aspects of division of labor.

In this chapter we review studies of the behaviour of ants to reappraise the existence and significance of extreme individual behavioural variability. After reviewing previous descriptions and explanations for behavioural specialisation and its underlying processes, we suggest an alternative framework, the key individual concept, for defining and interpreting this behaviour, and discuss its implications for studies of behavioural organisation.

The extent and significance of individual behavioural specialisation

Patterns of individual behavioural specialisation

A list of studies reporting extreme individual behavioural specialisation is presented in Table 1. We define an individual as behaviourally specialised if, within its age or physical caste, it engages in a particular behaviour far more frequently than expected, based on the activity of other similar individuals. These are individuals whose frequency of behaviour lies well outside the activity distribution of other members of its age or physical caste. Behavioural specialisation is thus based on the pattern of behaviour relative to other individuals of the same age or caste group and the available task space and is independent of both the functional role of this behaviour and the mechanism or process that underlies it. This "individual-based" definition is distinct from the more general sense in which specialisation has been used in discussions of colony design, that is, a "group-level" definition referring to behavioural differences at the temporal or physical caste level.

There are fewer than 20 examples of extreme behavioural specialisation in Table 1, but we suspect this is more indicative of the technical difficulty of studying individually marked workers than the presence of the phenomenon. Typically, studies of behavioural organisation in ants are relatively brief and use techniques that allow the description of individual behaviour only at the caste or age level, without reference to the identities of the individuals involved.

Cases of specialised individual behaviours appear to be taxonomically widespread and contextually variable (Tab. 1). Examples occur in at least 4 of the 16 formicine subfamilies and involve a variety of behaviours in several social contexts that range from within-nest activities such as nest construction and brood retrieval to outside-nest activities such as prey retrieval. The duration of specialisation can also be highly variable, and the degree of specialisation can be extreme. The duration of individual specialisation ranges from less than 20 min for *Formica schaufussi* workers retrieving prey to at least 3 months in *Myrmica rubra* workers involved in nest emigration. In some cases it is possible to infer the functional correlates of these specialisations. Nest emigration in *Formica sanguinea* and *Camponotus sericeus*, for example, appears to be dependent on

241

Table 1. Examples of individual behavioral specialisation in ants

Group behavior	Subfamily	Species	% active*	duration of specialization†	Selected References
Construction	Formicinae	Formica fusca	20	–	[22]
Brood retrieval	Dolichoderinae	Tapinoma erraticum	10	10 days	[57]
	Dolichoderinae	Tapinoma erraticum	–	6 weeks	[36]
	Formicinae	Lasius niger	14	3 weeks	[58]
	Formicinae	Lasius niger	4	–	[35]
Excavation	Formicidae	Camponotus japonicus	–	7 months	[27, 28]
Freeing nest entrance	Formicinae	Lasius niger	12‡	–	[33]
Nest emigration social carrying§	Dolichoderinae	Tapinoma erraticum	70	–	[59]
	Myrmicinae	Myrmica rubra	–	5 days	[25]
	Myrmicinae	Myrmica rubra	18–34||	8–10 days	[48]
	Myrmicinae	Myrmica rubra	–	3 months¶	[34]
	Formicinae	Formica sanguinea#	11**	2 months	[20]
	Formicinae	Formica fusca††	–	–	*[41]*
	Formicinae	Cataglyphis iberica	6	–	[24]
tandem running	Formicinae	Camponotus sericeus#	6**	2 months	[20]
carry and tand. run	Ponerinae	Diacamma rugosum	5–25	35 days	[60]
Prey retrieval	Ponerinae	Pachycondyla caffraria‡‡	–	–	[61]
	Formicinae	Formica schaufuss§§	–	<20 mins	[21]

* Methods used to estimate the percentage of active individuals varied between experiments.

† Minimum estimates based on experimental duration (though see¶).

‡ Perform 91% of the work.

§ Includes adult and brood transport.

‖ "All carrying acts were effected by only 18–34% of the workers".

¶ Individuals were observed over an 8-month period.

Known to be performers.

** Were active in 80% of all nest moving experiments.

†† "Only a few workers…. were actively involved in transport"; "individual recruiters frequently performed more than 200 carrying acts".

‡‡ "Stingers" versus "transporters".

§§ Known to be organizers.

groups of specialised individuals. Over a 2-month period, small groups of individuals (11% and 6% of all *Formica sanguinea* and *Camponotus sericeus* workers, respectively) performed over 80% of the tandem running during nest emigration. Removal of these active workers resulted in a significant increase in the time required for the colony to emigrate, and when the replacement group of workers was removed, "the colony was almost unable to achieve an organised nest emigration, even when their nest conditions became very unfavourable" [20, p. 231]. Cooperative prey retrieval in *Formica schaufussi* is also dependent on specialised individuals. Successful prey retrieval requires the presence of the individual that first located the prey and initiated recruitment from the nest. Removal of this individual alone, even after the recruits have reached the prey, significantly decreases retrieval success [21]. While we cannot yet assess the adaptive significance of a colony nest emigration or prey retrieval strategy based on specialist versus generalist individuals, it does appear likely that the behaviour of these specialised workers has a significant impact on colony operations.

The comprehensive theory of caste evolution given by Oster and Wilson [3] represents the first attempt at synthesising examples of within-subcaste individual behavioural variability. We use their definitions as a reference for discussion, but note that the overlapping, nonexclusive and often changing nature of these classifications has made it difficult to develop adaptive explanations for elitism, specialisation and idiosyncrasy. Oster and Wilson defined elitism as "the existence of exceptionally active or entrepreneurial individuals within age-size cohorts" [3, p. 155]. Combes [22], who appears to be the first to use the term "elitism", considered the activity levels of elites to be more persistent than those of other workers. Workers were also noninterchangeable: the role of elites could not be taken over by other less active workers. More recent studies have added a functional component to the definition, considering elites to be those individuals "who do almost all the work" [23, p. 420). While Oster and Wilson [3] did not formally define the concept of specialisation, they distinguished specialised from elite workers based on the range of tasks they carried out and the duration of their highly active behaviour. Elite individuals can be very active in a number of tasks, but specialists concentrate their activities on a subset of tasks and persist in this task over a longer period than elites. Later studies have also built upon this definition by considering that specialised workers are not only persistent but more efficient at their tasks [e.g. 24]. Idiosyncratic individuals were considered by Oster and Wilson [3] to display an element of "peculiarity". This is consistent with an earlier definition of idiosyncrasy as the observation of "individual behavioural characteristics" [25, p. 17], but differs from later studies that have variously defined idiosyncracy as "different intrinsic individual predispositions" for a particular behaviour [14, p. 5], or "tendencies to simultaneously display a single level of activity for several behaviours" [26, p. 58]. Given the varying usage of the terms "elitism", "specialisation" and "idiosyncracy", it is not surprising that these terms have been typically used only as descriptors, without reference to an underlying mechanism or social process. While these definitions may lack clarity, the phenomena they describe all suggest that factors other than age and morphology can produce significant behavioural variation. As noted by Oster and Wilson, these studies emphasise that "individual worker ants do not fit the popular image of invariant replicas that perform like parts in a machine" [3, p. 153].

Processes underlying individual behavioural specialisation

Explanations for the processes underlying these patterns of individual specialisation are as varied as the manner in which they have been described, often reflecting the prevailing view of sociality. Chen [27, 28] interpreted the dynamics of nest excavation within the contexts of social facilitation and dominance hierarchies, a theoretical framework that now receives only limited application (e.g. [29–31]). More recent studies have interpreted extreme within-cohort behavioural variance within the context of such factors as individual differences in genotype or experience [13, 14], task fixation [32], social context [21] or even stochastic processes alone [33, 34]. The studies of Lenoir and Ataya [33] and Abraham et al. [34] appear to have anticipated the growing field of self-organisation by suggesting that some examples thought to illustrate specialisation may be scale-dependent and an unavoidable consequence of the organisation of the groups themselves. When viewed over longer time periods, the differences in work effort between individuals can become less pronounced. It is likely that many factors influence the existence and detection of an individual's behavioural predisposition. Can a cohesive theory be developed to explain the pattern of behaviours collectively referred to as elitism, idiosyncrasy or specialisation? Many of the examples listed in Table 1 were detected with different experimental protocols and appear to represent a disparate group of behaviours. They share in common a demonstration of extreme behavioural variance, but the different systems and descriptive techniques used make interspecific comparisons and generalisations difficult. It has been difficult to distinguish between the pattern, the underlying process and the role of behavioural specialisation. This may also explain why the organisation of these behaviours has typically not been examined within an ergonomic framework. Indeed, the efficiency of the labour of specialised workers has received little attention.

Resolving the proximate mechanisms or processes underlying behavioural specialisation requires knowledge of at least three aspects of the pattern of the behaviour. These include knowledge of what tasks an individual engages in relative to the available tasks it could potentially carry out. Is the individual particularly active in a number of tasks, or does it concentrate its efforts on a single task only (elitism versus specialisation *sensu* Oster and Wilson [3])? Next, we need to know what tasks an individual engages in relative to the tasks of other individuals. This is a critical component as task specialisation is itself a relative concept, yet previous descriptions of specialisation differ in the manner in which individual behaviour is integrated within group action. An extremely active individual can only be described as an elite with reference to the behaviour of other members of the group through a comparison of activity levels. The example of an individual repeatedly foraging in the same area, however, can be termed idiosyncratic. Knowledge of the foraging activity of nestmates is not required. Finally, knowledge of the duration of individual specialisation is required. The identity of the highly active *Lasius niger* workers involved in brood retrieval can fluctuate on a day-to-day basis [35], whereas in *Tapinoma erraticum* individuals involved in brood retrieval can remain active over at least a 6-week period [36]. The mechanism underlying the relatively short-lived, 20-minute foraging specialisation in *Formica schaufussi* may not be the same as the mechanism producing the comparatively longer 3-month specialisation shown by *Myrmica rubra* workers involved in nest emigration (Tab. 1). Rather than attempting to define the phenomena precisely and resolve the processes underlying the specialisation phenomena illustrated by these examples, and in the absence of clear patterns that allow comparisons between taxa, we present an alternative approach to the study of extreme individual variance and group-level behaviour: the key individual concept.

Key individuals and the organisation of labor

Understanding the evolution of organisational principles within the social insects requires a comparative approach that provides consistent descriptions of the behaviour of interest in a variety of taxa, in order to reveal the historical factors and ecological correlates [4]. As a first step in this direction, we propose a novel interpretation and classification of behavioural specialisation in groups comprised of similar temporal or physical caste individuals. This classification recognises the existence of key individuals that have a specific function within a group, and it defines an individual's behaviour on the basis of its interactions and functional relationships with other group members. Such a definition emphasises that an individual's behaviour must be interpreted within the context of group behaviour, and hence that behaviour at both the individual and group level must be examined simultaneously. In turn, group behaviour can be defined based on the functional relationships of the individuals that comprise it. Unlike the concepts of elitism, idiosyncrasy and specialisation, the classification based on key individuals (i) describes individual behaviour in the context of the social group; (ii) provides testable hypotheses about the function of key individuals; and (iii) provides a first step in attempting to elucidate the mechanisms underlying this process. It may ultimately be possible to develop evolutionary interpretations of the distribution of these organisational principles for various social insect behaviours. Determining the functional role of key individuals is a necessary step in testing adaptational hypotheses.

A key individual has a specific function in a group, and it is the nature of this function that defines the individual's role. Based on the examples provided in Table 1, we have detected at least three ways in which key individuals can be functionally related to the group, but additional relationships are possible. These three relationships are summarised in Table 2 and are defined below.

Catalysts

Key individuals that function as catalysts stimulate greater activity in group members. Their presence is not required for a process to proceed, but they act to increase the rate at which the group behaviour proceeds. The organisation of excavation groups in *Camponotus japonicus* could be based on key individuals

Table 2 A summary of the key individual concept

| | Key individual category | | | |
	Organiser	Catalyst	Performer	Example
Necessary for completion of labor	(+)*	–	+	Nest emigration in *Camponotus sericeus* [20]
Stimulate or otherwise affect rate of labor	–	+	–	Nest excavation in *Camponotus japonicus* [27, 28]
Maintain group cohesion	+	–	–	Group retrieval in *Formica schaufussi* [21]
Effect of removal of key individual on group activity	Stops	Slows	Stops	

(+)* Required for group success by maintaining group cohesion, not by performing all the labor.

acting as catalysts. Individuals involved in excavation can be defined as either rapid, median or slow workers. All individuals can excavate in isolation, but the presence of active individuals (leaders) results in a greater work effort by a group of inactive workers (followers) [27, 28]. This behaviour was originally interpreted by Chen as an example of social facilitation (an increase in individual response "merely from the sight or sound of others making the same movements" [37]), and later as elitism [3, 38]. In the classification presented here, the active individuals or leaders are key individuals because they act as catalysts: their presence stimulates greater excavation activity in other colony members. Removing a catalyst results in a reduction in the rate of labor provided by the group (Tab. 2).

Performers

Unlike catalysts, performers alone complete a task. Other individuals may be involved, but the majority of the work is carried out by performers, which do not stimulate others to assist. The individuals involved in nest emigration in *F. sanguinea* and *C. sericeus* fall within this category [20]. A subgroup of individuals perform the majority of work, but their presence is required for the task to be completed. If individuals denoted as "transport specialists" are removed, the time required for nest emigration increases significantly, and if the few workers that take their place are in turn removed, nest emigration stops. These performers complete the majority of work associated with emigration: their removal ends the process entirely (Tab. 2).

Organisers

As is true for performers, the presence of organisers is required for a task to be completed, but organisers differ in that they do not carry out the work themselves. Their presence is required to maintain group cohesion, which entails keeping other individuals involved in a task until it is completed. Studies of group retrieval in *F. schaufussi* illustrate this type of key individual. In this species, individuals locating large prey items return to the colony and initiate the recruitment of nestmates which cooperatively retrieve the prey [39, 40]. An individual that locates prey initiates prey retrieval and maintains the cohesion of the retrieval group as it leaves the colony, travels to the prey and transports the prey to the nest. The removal of the prey retrieval organiser will result in the dispersal of the retrieval group members, even if they reach the prey. Ultimately, prey retrieval fails [21]. Although the individuals who first locate the prey may assist in prey transport and can initiate additional recruitment from the colony during prey retrieval, their principle function appears to be in organising the task of group retrieval. Like performers, organisers are not interchangeable or redundant. Removing an organiser results in the failure of the group to complete its task (Tab. 2).

The highly active individuals involved in nest emigration in *F. sanguinea* and *C. sericeus* [41] and *M. rubra* [25] have been described as possessing an organising role. These authors appear to have used the term to describe the general importance of these individuals to the success of the group process, rather than the specific functional relationship we imply here. We consider that nest emigration in *F. sanguinea* and *C. sericeus* represents examples involving performers, rather than organisers.

Implications for the study of behavioural organisation

Specialist versus generalist individuals

The demonstration that group behaviour can depend on subgroups of specialised individuals (Tab. 1) seems at odds with the notion that group action in the social insects is optimised when based on a series-parallel organisation. The theory of why series-parallel operations optimise group success is well developed, and many studies have emphasised how the collective actions of numerous relatively simple and redundant individuals can underlie group-level behaviour [3, 5, 11, 42]. Individual specialisation may be expected to be more likely in smaller groups, but the examples listed in Table 1 indicate this may not always be true. Large-scale processes such as nest construction and brood retrieval can be dependent on the behaviour of specialised individuals. There is no obvious "scale-dependent" relationship between the degree of individual specialisation and group size in these examples, and *a priori* assumptions that minimise between-individual behavioural variance may be problematic. While the application of decentralised models that minimise individual variance to study group behaviour has been productive [11], future studies may need to confirm the pattern of individual behavioural variance on which the models are based.

Levels of analysis and the organisation of group behaviour

The key individual concept emphasises the need to examine the organisation of social insect behaviour at the individual level. Yet much of our understanding of the flexible organisation of social insect behaviour is based on the actions of groups of individuals and their responses to colony needs [10, 11]. These studies suggest that the behaviour of individuals within a group is distributed in such a way that it can be interpreted in terms of population measures such as the mean and variance. Techniques are employed that detect behavioural differences between groups of individuals, such as providing different groups with unique marks, but these techniques preclude detecting any within-group individual variability that might contribute to group behaviour. It is possible that many examples of behavioural flexibility and organisation reflect a subset of specialised individuals within a particular group responding to task needs, rather than equal participation by all group members. The process of mass recruitment is typically viewed as being independent of the actions of specialised individuals, but the few studies of mass-recruiting species that also observed behaviour at the individual level revealed a significant variability in individual recruitment capabilities. Individual *Solenopsis geminata* [43] and *Lasius niger* workers [44] differ in both the probability and intensity of trail laying, whereas *Tetramorium caespitum* workers also differ in their trail-following capabilities [45]. Up to one-third of *T. caespitum* workers recruited from the colony left the recruitment trail and did not reach the food. It is possible that even in these examples of mass communication, a subset of individuals may perform most of the recruitment effort. The importance of individual worker identity to group dynamics was noted by Chen as early as 1937 [27, 28]. Chen reported that the degree of behavioural specialisation found in experimental groups of workers was dependent on the specific individuals placed together. While overall patterns of behaviour of social insects may be consistent with a model based on generalist individuals, this assumption needs further examination.

The key individual concept can be applied to other aspects of behavioural organisation, such as the formation of teams. While developing a theoretical basis for the organisation of caste, Oster and Wilson [3, p. 151] noted that social insects can recognise individuals as belonging to a particular caste, but "do not appear to recognise each other as individuals." The consequence of this "lower grade of discrimination is that members of colonies do not form cliques and teams", and their organisation with reference to each other is essentially random. This view appears to suggest limitations on the nature of interactions at the individual level. However, Franks [46] demonstrated that prey retrieval groups in the army ant *Eciton burchelli* possess a nonrandom caste distribution, and interpreted this as evidence of teams. While prey retrieval groups typically involve individuals from two caste groups (one individual from the submajor caste and at least one individual from another caste group, either medias or minors), retrieval groups tend to include only a single submajor, irrespective of retrieval group size. The nonrandom distribution of castes was therefore interpreted to represent teams. Although this team structure seems at odds with the individual-orientated concept of Oster and Wilson [3], Hölldobler and Wilson [4] later defined a team as "members of different castes that come together for highly coordinated activity in the performance of a particular task", seemingly accepting Franks's interpretation. Recognition and cooperation at the individual level, once critical for a team, now characterised "cliques" [4, p. 343]. The organisation of cooperative prey retrieval groups in *E. burchelli* could be described with the concept of key individuals. Teams (*sensu* Hölldobler and Wilson [4]) can be thought of as cooperative groups that require more than one organiser. The first submajor arriving at the prey may assume an organising role, with its presence preventing additional submajors from joining the retrieval group. Individuals from the remaining castes involved in prey transport (medias and minors) are also required for successful prey transport; but their presence may not directly influence the arrival of additional individuals of the same caste.

Detecting and interpreting the behaviour of key individuals

Critical to understanding the mechanisms underlying the key individual phenomenon is the determination of the duration of individual behavioural specialisation and the distribution of relative work effort among individuals. Designing experimental protocols in the absence of any *a priori* information can be difficult, and information on the duration over which individuals act as key individuals has been determined for only two of the examples listed in Table 1. Prey retrieval in *F. schaufussi* involves only a single key individual that acts in this role for the duration of only a single retrieval event (approximately 20 min) and appears to become a key individual in response to the phase, or social context, at which it enters the retrieval process [21]. In contrast, nest emigration in *M. rubra* involves more than a single individual that maintains a role for up to 3 months. The reasons individuals adopt such a specialised behaviour remain unknown [34].

Studies of nest emigration in *M. rubra* demonstrate how sampling effort influences our ability to detect and interpret the behaviour of key individuals. Experiments conducted over an 8-month period were required to determine that the individuals initially described as being actively involved in nest emigration over at least a 5–10-day period [25, 47, 48] could remain active for up to 3 months [34]. When examined over this longer period, the distribution of work effort was continuously distributed, suggesting that individual specialisation reflected more of a continuum in activity levels than a discrete subgroup of active individuals. The distribution of individual work effort followed a log normal relationship with rank, a relationship confirmed in a number of other studies of behavioural specialisation [13, 33]. While summing the total work efforts of individuals over the entire course of study does not reveal patterns of individual temporal variability or how it may be related to the efforts of other group members, these studies emphasise that experiments must continue for a duration greater than the task itself and involve a sufficient number of individuals in order to understand how labor is divided among individuals. Determining the duration of individual specialisations represents a first step in understanding the underlying mechanisms.

Key individuals and social context

The processes underlying the relationships between individual function and the identity of group members remain almost entirely unknown. In the prey retrieval behaviour of *F. schaufussi* we have some idea of the context that leads to individuals entering into an "organising" relationship with the group [21]. In this monomorphic species, an individual's ability to act as a key individual is independent of its age or morphology and is determined by the "social context" or phase at which it enters the sequence of prey retrieval events. The first individual to locate the prey acts as an organiser, and within the context of this single retrieval event, no other individual is able to perform this role. While individuals can act as either organisers or recruits in subsequent prey retrieval events, they are almost unable to switch roles within a single retrieval effort. The social context prevailing at the stage appears to constrain their behaviour during the remainder of the prey retrieval event.

The demonstration that social context can determine an individual's role represents an unexpected source of rapid and flexible behavioural variability. Although we do not yet understand the mechanisms underlying the process by which social context influences individual roles, these observations provide a step toward understanding the physiological basis underlying task specialisation. Individual *F. schaufussi* can rapidly change their behaviour in response to conspecifics in a certain behavioural state, and enter a behavioural sequence or role. Behavioural predispositions in the honeybee are known to be influenced in the long term by juvenile hormones mediated by social interactions [49], and it is possible that the rapid secretion of biogenic amines in response to immediate stimuli may also underlie short-term behavioural trajectories in these and other social insects. Serotonin and octopamine levels, for example, can change rapidly in response to stress and other environmental factors, and are known to modulate neuronal activity and behaviour in a number of insect species [50, 51]. An individual *F. schaufussi* worker locating a large prey may be influenced by a different combination of amines than an individual recruited from the nest. The level of biogenic amines may not only determine the role an individual performs, but also act to constraint individuals from switching roles within the context of a single retrieval effort. Perhaps amine levels can be elevated more rapidly than they can be reduced?

The adaptive significance of key individuals

Understanding the adaptive significance of individual specialisation requires information on the types of group behaviours that involve key individuals, the species in which they occur and their fitness consequences. Combining phylogenetic analysis with comparative behavioural studies has been a fundamental approach to understanding the evolution and organisation of group behaviour in social insects [1, 4, 52–55], and promises to be a fruitful approach to the study of key individuals. Are similar group-level behaviours in different taxa based on homologous organisational principles? Have particular types of key individuals evolved independently and repeatedly? How have phylogenetic and ecological factors interacted to influence group organisation? Preliminary data suggest that the types of key individuals underlying group behaviour are highly variable both within and between taxonomic groups. Similar behaviours can be based on different organisational principles, and similar organisational principles can underlie different behaviours. Group digging appears to involve catalysts in *C. japonicus* [27, 28] and *F. subsericea* [31] but not in *F. fusca* [38] and *F. lemani* [56]. And while nest emigration in both *F. sanguinea* and *C. sericeus* is based on the existence of performers, the type of behaviour involved (social carrying versus recruitment via tandem running), are different (Tab. 1). Combining phylogenetic, ecological and behavioural studies should help resolve the issue of how frequently key individuals underlie group behaviour and how these organisational principles may have evolved within the ants. This analysis may be facilitated by a classification scheme such as the key individual concept, which compares group behaviours based on similarity in the underlying organisational principles, rather than apparent similarity in the behaviour of the group itself. Combining examples according to the type of group-level behaviour alone may have masked differences in the underlying organisational principles.

Conclusion

Understanding the nature of individual variability is critically important to understanding the integration and organisation of social insect colonies. This review suggests that individual behavioural variability and specialisation are significant in a variety of group-level behaviours. Group action may be based on behavioural specialisation at the level of the individual to a greater degree than suspected, and it is likely that multiple organisational principles exist to account for the various phenomena that encompass individual specialisation. The frequency of occurrence of behavioural specialisations strongly suggests they are more than just "noise" in the system. Ergonomic and adaptive interpretations of the sometimes rapid, flexible and short-lived individual variance in task performance still await more detailed descriptions of these processes. Researchers are just beginning to adequately describe the organisational and functional relationships underlying these group behaviours, and it would be premature to initiate a discussion of their adaptiveness. Nevertheless, it is clear that short-term individual specialisation can be a significant component of group action. Future studies of group behaviour in ants and other social insects may benefit from study at the level of both the individual and the group, and the recognition of the social context as a potential determinant of flexible, context-specific behavioural specialisation.

Acknowledgements

We thank Claire Detrain, Jean-Louis Deneubourg and Jacques M. Pasteels for inviting our contribution and an anonymous referee for helpful comments. We also acknowledge the financial support of the National Science Foundation and the Whitehall Foundation (J.F.A.T), the Fulbright Foundation, Boston University, James Cook University and the Australian Research Council (S.K.R.).

References

1 Wilson EO (1971) *The insect societies.* Belknap Press of Harvard University Press, Cambridge, MA

2 Wilson EO (1976) Behavioral discretization and the number of castes in an ant species. *Behav Ecol Sociobiol* 1: 141–154

3 Oster GF, Wilson EO (1978) *Caste and ecology in the social insects.* Princeton University Press, Princeton

4 Hölldobler B, Wilson EO (1990) *The ants.* Belknap Press of Harvard University Press, Cambridge, MA

5 Herbers JM (1981) Reliability theory and foraging by ants. *J Theor Biol* 89: 175–189

6 Deneubourg JL, Goss S (1989) Collective patterns and decision-making. *Ethol Ecol Evol* 1: 295–311

7 Karsai I, Penzes Z (1993) Comb building in social wasps: self-organization and stig-

mergic script. *J Theor Biol* 161: 505–525

8 Cole BJ, Cheshire D (1996) Mobile cellular automata models of ant behavior: movement activity of *Leptothorax allardycei. Amer Naturalist* 148: 1–15

9 Traniello JFA, Robson SK (1995) Trail and territorial communication in social insects. *In*: WJ Bell, R Cardé (eds): *The chemical ecology of insects,* vol 2. Chapman and Hall, New York, 241–286

10 Gordon DM (1996) The organization of work in social insect colonies. *Nature* 380: 121–124

11 Bonabeau E, Theraulaz G, Deneubourg JL, Aron S, Camazine S (1997) Self-organization in social insects. *Trends Ecol Evol* 12: 188–193

12 Fresneau D, Lachaud JP, Jaisson P (1987) Individual behaviour and polyethism. *In*: J

Eder, H Rembold (ed): *Chemistry and biology of social insects*. Peperny, Munich, 126–127

13 Lenoir A (1987) Factors determining polyethism in social insects. *In*: JM Pasteels, JL Deneubourg (eds): *From individual to collective behavior in social insects*. Birkhäuser, Basel, 219–240

14 Jaisson P, Fresneau D, Lachaud JP (1988) Individual traits of social behavior in ants. *In*: RL Jeanne (ed): *Interindividual behavioral variability in social insects*. Westview Press, Boulder, CO, 1–51

15 Jeanne RL (ed) (1988) *Interindividual behavioral variability in social insects*. Westview Press, Boulder, CO

16 Robinson GE (1992) Regulation of division of labor in insect societies. *Annu Rev Entomol* 37: 637–665

17 Stuart RJ (1997) Division of labor in social insect colonies. Self-organization and recent revelations regarding age, size and genetic differences. *In*: G Greenberg, E Tobach (eds): *Comparative psychology of invertebrates. The field and laboratory study of insect behavior*. Garland Publishing, New York, 135–155

18 Beshers SN, Robinson GE, Mittenthal J (1998) Response thresholds and division of labor in insect colonies. *In*: C Detrain, JL Deneubourg, JM Pasteels (eds): *Information processing in social insects*. Birkäuser, Basel, 115–140

19 Bonabeau E, Theraulaz G (1998) Role and variability or response thresholds in the regulation of division of labor in insect societies. *In*: C Detrain, JL Deneubourg, JM Pasteels (eds): *Information processing in Social Insects*. Birkäuser, Basel, 141–164

20 Möglich M, Hölldobler B (1974) Social carrying behavior and division of labor during nest moving in ants. *Psyche* 81: 219–236

21 Robson SK, Traniello JFA (1998) Social

context: A novel source of behavioural specialisation in ants; *in review*

22 Combes M (1937) Existence probable d'une élite non differenciée d'aspect constituant les véritable ouvrières chez les Formica. *C R Acad Sci* 204: 1674–1675

23 Plowright RC, Plowright CMS (1988) Elitism in social insects: a positive feedback model. *In*: RL Jeanne (ed): *Interindividual behavioral variability in social insects*. Westview Press, Boulder, CO, 419–431

24 Cerda X, Retana J (1992) A behavioural study of transporter workers in *Cataglyphis iberica* ant colonies (Hymenoptera Formicidae). *Ethol Ecol Evol* 4: 359–374

25 Abraham M (1979) Comportement individuel lors de déménagements successifs chez *Myrmica rubra* L. *In*: *Ecologie des insectes sociaux*. Compte Rendu Colloque Annuel U.I.E.I.S., Lausanne, 17–19

26 Bonavita-Cougourdan A, Morel L (1988) Interindividual variability and idiosyncrasy in social behavior in the ant *Camponotus vagus* Scop. *Ethology* 77: 58–66

27 Chen SC (1937) The leaders and followers among the ants in nest-building. *Physiol Zool* 10: 437–455

28 Chen SC (1937) Social modification of the activity of ants in nest-building. *Physiol Zool* 10: 420–436

29 Clayton DA (1978) Socially facilitated behavior. *Quart Rev Biol* 53: 373–392

30 Imamura S (1982) Social modifications of work efficiency in digging by the ant *Formica* (*Formica*) *yessensis* Forel. *J Fac Sc Hokkaido Univ Ser 6* 23: 128–142

31 Klotz JH (1986) Social facilitation among digging ants (*Formica subsericea*). *J Kans Entomol Soc* 59: 537–541

32 Forsyth AB (1978) *Studies on the behavioral ecology of polygynous social wasps*. PhD thesis, Harvard University

33 Lenoir A, Ataya H (1983) Polyéthisme et répartition des niveaux d' activité chez la

fourmi *Lasius niger* L. *Z Tierpsychol* 63: 213–232

34 Abraham M, Deneubourg JL, Pasteels JM (1984) Idiosynchrasie lors du déménagement do *Myrmica rubra* L. (Hymenoptera, Formicidae). *Actes Coll Insect Soc* 1: 19–25

35 Lenoir A (1981) Brood retrieving in the ant *Lasius niger* L. *Sociobiology* 6: 153–178

36 Meudec M, Lenoir A (1982) Social responses to variation in food supply and nest suitability in ants (*Tapinoma erraticum*). *Anim Behav* 30: 284–292

37 Allport FH (1924) *Social psychology.* Houghton Mifflin, Boston

38 Sakagami SF, Hayashida K (1962) Work efficiency in heterospecific ant groups composed of hosts and their labor parasites. *Anim Behav* 10: 96–104

39 Traniello JFA, Beshers SN (1991) Maximization of foraging efficiency and resource defense by cooperative retrieval in the ant *Formica schaufussi. Behav Ecol Sociobiol* 29: 283–289

40 Robson SK, Traniello JFA (1998) Resource assessment, recruitment behavior and the organization of cooperative prey retrieval in the ant *Formica schaufussi. J Insect Behav* 11: 1–22

41 Möglich M, Hölldobler B (1975) Communication and orientation during foraging and emigration in the ant *Formica fusca. J Comp Physiol* 101: 275–288

42 Wilson EO, Hölldobler B (1988) Dense heterarchies and mass communication as the basis of organization in ant colonies. *Trends Ecol Evol* 3: 65–68

43 Hangartner W (1969) Structure and variability of the individual odor trail in *Solenopsis geminata* Fabr. (Hymenoptera, Formicidae). *Z Vergl Physiol* 62: 111–120

44 Beckers R, Deneubourg JL, Goss S (1992) Trail laying behaviour during food recruitment in the ant *Lasius niger* (L.) *Insect Soc* 39: 59–72

45 Pasteels JM, Deneubourg JL, Goss S (1987) Self-organization in ant societies (I): Trail recruitment to newly discovered food sources. *In*: JM Pasteels, JL Deneubourg (eds): *From individual to collective behavior in social insects.* Birkhäuser, Basel, 155–175

46 Franks NR (1986) Teams in social insects: group retrieval of prey by army ants (*Eciton burchelli*, Hymenoptera: Formicidae). *Behav Ecol Sociobiol* 18: 425–429

47 Abraham M, Pasteels JM (1977) Nest-moving behavior in the ant *Myrmica rubra. Proc 8th Int Congress IUSSI*, Wageningen, Netherlands, 286

48 Abraham M, Pasteels JM (1980) Social behavior during nest-moving in the ant *Myrmica rubra* L. (Hymenoptera, Formicidae). *Insect Soc* 27: 127–147

49 Huang ZY, Robinson GE (1992) Honeybee colony integration: worker-worker interactions mediate hormonally regulated plasticity in division of labor. *Proc Natl Acad Sci USA* 89: 11 726–11 729

50 Harris JW, Woodring J (1992) Effects of stress, age, season, and source colony on levels of octopamine, dopamine and serotonin in the honey bee (*Apis mellifera* L.) brain. *J Insect Physiol* 38: 29–35

51 Hirashima A, Eto M (1993) Effect of stress on levels of octopamine, dopamine and serotonin in the American cockroach (*Periplaneta americana* L.). *Comp Biochem Physiol C Comp Pharmacol Toxicol* 105: 279–284

52 Wheeler WM (1910) *Ants. Their structure, development and behavior.* Columbia University Press, New York

53 Jeanne RL (1980) Evolution of social behavior in the vespidae. *Annu Rev Entomol* 25: 371–396

54 Baroni-Urbani C (1989) Phylogeny and behavioural evolution in ants, with a discussion of the role of behavior in evolutionary processes. *Ethol Ecol Evol* 1:

137–168

55 Wenzel JW (1992) Behavioural homology and phylogeny. *Annu Rev Ecol Syst* 23: 361–381

56 Sudd JH (1972) The absences of social enhancement of digging in pairs of ants (*Formica lemani* Bondroit). *Anim Behav* 20: 813–819

57 Meudec M (1973) Note sur les variations individuelles du comportement de transport du couvain chez les ouvrières de *Tapinoma erraticum* Latr. *C R Acad Sci, Paris, ser D* 277: 357–360

58 Verron H (1976) Note sur la stabilité de certains traits éthologiques chez les ouvrières de *Lasius niger*. *C R Acad Sci,*

Paris, D 283: 671–674

59 Meudec M (1977) Le comportement de transport du couvain lors d'une perturbation du nid chez *Tapinoma erraticum* (Dolichoderinae): rôle de l'individu. *Insect Soc* 24: 345–352

61 Agbogba C, Howse PE (1992) Division of labour between foraging workers of the ponerine ant *Pachycondyla caffraria* (Smith) (Hymenoptera: Formicidae). *Insect Soc* 39: 455–458

60 Fukumoto Y, Abe T (1983) Social organization of colony movement in the tropical ponerine ant, *Diacamma rugosum* (Le Guillou). *J Ethol* 1: 101–108

Temporal information in social insects

Vincent J.L. Fourcassié, Bertrand Schatz and Guy Beugnon

Summary

Social insects are able to regulate their daily patterns of activity according to the periodic variations of the biotic and abiotic factors of their environment. The ability to control temporal information in social insects is important at both the individual and collective level because it allows to minimize or, conversely, maximize interactions with other species and to synchronize collective activities within a colony. The fact that some species are able to achieve temporal learning, that is, to memorize a specific time of day, allows in addition a certain degree of flexibility in the tuning of their activity pattern to that of other species. Since the basis of the control of temporal information lies in the existence of activity rhythms, we begin our paper with a review of the studies that have investigated the biological nature of activity rhythms in social insects. These rhythms are either exogenous, that is, entirely dependent on the variations of an external signal, or endogenous, that is, based on an internal clock which is periodically entrained by an external signal called a Zeitgeber. We then proceed by reviewing the studies dealing with temporal learning, with a particular emphasis on time-place learning process. Future lines of research on the subject are discussed in the conclusion.

Introduction

Along with food type, habitat and space, time is an essential dimension of the ecological niche of a species. To minimize competition organisms need to control temporal in addition to spatial information. For example, the coexistence of sympatric species sharing the same resources is allowed by temporal niche partitioning, both on a daily and seasonal time scale. Time is also of considerable

C. Detrain et al. (eds) Information Processing in Social Insects

importance from a behavioural point of view, both at the individual and social level. At the individual level, the control of temporal information allows an organism to adapt to the time structure of its environment by partitioning its time budget in period of activity and inactivity, by anticipating the timing of regular events or by fixing the timing of specific behaviours [1]. In addition, time underlies almost all foraging decisions that involve the measurement of food patch or food source profitability [2–4]. At the social level, the control of temporal information plays also an important role by allowing the synchronization of individual activities, for example during group foraging or mating flight in social insects [5]. Finally, from an evolutionary point of view the control of temporal information provides the basis for some important processes. For example, sympatric speciation may emerge from the differential timing of reproductive activities among individuals of the same species and coevolution from the synchronization of the activity rhythms of organisms belonging to different species.

The existence of natural activity rhythms in animals provides the basis for the control of temporal information. A rhythm can be exogenous, that is, entirely dependent on the variations of an external signal, or endogenous, that is, based on a pacemaker that functions as an internal clock. This internal clock is periodically entrained by an external signal called a Zeitgeber, which allows an animal to continuously monitor the passage of time and can depend on abiotic factors, for example ambient luminosity or temperature, or on biotic factors, for example the time of a food source availability or the activity of other animals of the same or of different trophic levels. Due to the rotation of the earth on its axis, abiotic factors are generally characterized by a 24-h-periodicity, and circadian rhythms are thus prevalent in the animal kingdom. The first part of this chapter will be devoted to a review of the question of the biological nature of activity rhythms in social insects.

That animals are able to control temporal information is best evidenced by the existence of temporal learning processes and time memory. According to Gallistel [6], one can distinguish two types of time memory that are based on radically different temporal information processing systems. The first form of time memory is the memory for a time of day (phase sense). It allows an animal to memorize the time of occurrence of a daily event. This is the type of time memory which has by far received the greater attention in social insects. The second form of time memory is the memory for an interval of time (interval sense). It allows an animal to monitor the interval of time elapsed since the onset of an event, that is, to measure the duration of an event. It also allows to measure the interval of time elapsed between two occurrences of the same event, and is thus the first step in measuring the rate of occurrence of an event. We will restrict our review in the second part of this chapter to the studies dealing with the phase sense of social insects.

The biological nature of activity rhythms in social insects

Every species has its own distinctive foraging schedule, and the study of sponta-neous activity patterns is thus part of every basic field research [7]. A distinctive rhythm exists when a periodicity can be revealed by the analysis of the activity pattern. Assumptions on the mechanisms underlying the periodicity observed are generally drawn from the correlations between the pattern of activity and the pat-tern of change of the abiotic factors of the environment [8]. Most studies, how-ever, stop at this stage and thus remain purely descriptive. The two standard methods of chronobiology (the free-running and phase-shift experiments [8, 9]) have been applied to only a limited number of species of social insects. The first method consists in placing the animal in environmental conditions that are as constant as possible. One can determine in this way whether a rhythm has an exogenous or an endogenous origin and, if the latter is true, bracket the value of its period. The second method consists in shifting the phase of a rhythm by a cer-tain amount of time, for example by transcontinental travels. It is used to deter-mine the nature of the Zeitgebers that entrain an endogenous rhythm. Both meth-ods have been employed to investigate the physiological basis of activity rhythms in honeybees and ants.

Although honeybees can occasionally forage on clear nights, in natural con-ditions their outdoor activity is essentially restricted to the diurnal part of the day; bees that stay in the hive at night show physiological changes that are akin to the sleep of vertebrates [10, 11]. Their pattern of activity is thus characterized by a clear alternance of activity and rest that matches the light/dark cycle of the nat-ural photoperiod. By rearing bees for several days in an artificial chamber in fully controlled conditions (constant light, temperature and humidity and control of the electrical conductivity of the atmosphere), Beling [12] was the first to point to the endogenous origin of the bee rhythm: bees trained to visit a feeding dish at a certain time of the day in the artificial chamber persisted in doing so after several days, even in the absence of food. Under these conditions bees are said to be placed in self-selected light/dark conditions (LD) because they are able to choose the time they spend resting in the darkness of the hive and the time they spend foraging in the constantly illuminated chamber. In order to make sure that bees could not use any unknown fluctuating local cues, Renner (13) trained bees to visit a feeding dish at a certain time of the day under self-selected LD in an artificial room located in Paris. After several days of training the hive was transferred to a similar room in New York. Bees proved to be unaffected by the translocation and continued to visit the feeding place at the usual Parisian training time. Other isolation experiments can be found elsewhere in the literature [10, 14–29]. Although different methods have been applied to measure the activity of the bees (activity at the feeding dish [12–14, 20, 25]; activity at the hive entrance [22, 23, 28]; walking activity [21]; oxygen consumption [27, 29]), all these experiments show that in constant darkness conditions (DD) the circadian clock of honeybees tends to free-run with a period slightly shorter than 24 h, whereas in constant light (LL) or self-selected conditions, the free-running period is generally slightly above 24 h. A free-running circadian rhythm is always observed, whether the activity is measured on whole colonies [22, 23] small groups of workers [27] or, although to a lesser extent, fully isolated individuals (10, 18, 21, 28). In the latter case, caste differences were found in the free-running period [18]. Differences in free-running rhythms among honeybee races have also been reported [15–17, 23, 30].

An interesting point is that the free-running period of individual worker bees has a much greater variance than that of whole colonies [28], which suggests that individual bees synchronize their activity rhythm within the colony. The search for the signal underlying this synchronization has long been a challenge to investigators, and the question has been addressed repeatedly [24, 27–29, 31]. Medugorac and Lindauer [31] have shown that individual bees can adopt the rhythm of their social group very rapidly. When a small group of workers trained to visit a feeding dish at a certain time of the day was transferred to another colony trained to forage at a different time, the transplanted workers showed a peak of activity both at their usual training time and at the training time of their colony of adoption. Conversely, in self-selected LD conditions, the free running rhythm of isolated workers was completely out of phase from that of their colony of origin after 12 days of separation [28]. Tactile contacts seem to be important in the synchronization process [24], but the coordination of individual rhythms

can also be achieved through the synchronization of intragroup temperature fluctuations [29]. Although the queen is not indispensable for the synchronization to
occur, the process may nonetheless be accelerated by its presence [27]. Finally,
periodic restricted daily feeding can also act as a social Zeitgeber [9, 23]. The
primary Zeitgeber of the honeybee activity rhythm in natural conditions, however, is provided by the daily change of light level: whole colonies or individual
bees can be readily entrained in environmental chamber to different square-wave
LD cycles [16, 18, 21, 32, 33]. The importance of light as a Zeitgeber is also
shown by a translocation experiment performed by Renner [34]. A colony of
honeybees was trained to forage outdoors at a certain time of the day on the East
Coast of the United States. It was then flown overnight to the West Coast. Under
these conditions, the searching activity of the displaced honeybees was progressively shifted over 3 days from the usual training local time in the East to the
local time in the West that would have corresponded nominally to the training
time on the East coast, showing that the internal clock of the honeybee was progressively reset. The role of other external cues as potential Zeitgebers has also
been investigated. Thus, Moore and Rankin [33] have shown that temperature
can act as an external Zeitgeber on individual bees reared in constant DD conditions. A long argument was initiated by Martin et al. [35] and Martin and Martin
[25] who claimed that bees could use the daily fluctuations of the earth's magnetic field as an additional external Zeitgeber. This hypothesis, however, was definitely dismissed by a series of experiments performed by Neumann [26] (see
also discussion in [32]).

Most species of ants are characterized by a specific daily foraging activity
pattern: some ants are active day and night, others during either the whole or a
limited part of the photo- or scotophase [7, 36] and unimodality, bi- or pluri-
modality can be observed in their activity schedule. Ants also show a great ver-
satility in their foraging activity pattern. They are flexible enough to adapt their
foraging schedule to the modifications of environmental conditions, whether
these modifications occur rapidly or progressively. The question of whether the
periodicity observed in the ants' activity pattern is the result of an endogenously
driven rhythm or depends entirely on the fluctuations of external factors has been
addressed by a few authors [5, 36–45].

De Bruyn and Kruk-de Bruin [38], Rosengren [39], Rosengren and Fortelius
[40] and North [41, 42] studied the chronobiology of several species of the redwood ant group. In natural conditions these species are mostly diurnal. However,

depending on the season and on meteorological conditions, an important nocturnal activity can sometimes be observed. Working with whole colonies of *Formica
polyctena*, Rosengren [39] and Rosengren and Fortelius [40] have reported that
the periodicity in worker activity in LD conditions disappeared when they were
placed in self-selected LD conditions. The periodicity resumed when environmental fluctuations were reestablished, either by a temperature or a luminosity
cycle [39]. Using the same method as Moore and Rankin [21] for honeybees,
North [41, 42] succeeded in measuring the period of the free-running rhythm of
isolated workers in *F. rufa*. He found a mean period of 22.54 ± 0.45 h for ants
placed in DD and of 24.51 ± 0.33 h for ants placed in LL. In DD a clear self-sus-
tained periodicity has also been observed in the activity of males of *Camponotus
clarithorax* [5] and in the thermal sensitivity of the workers of the species *C. mus*
and *C. rufipes* [43–45]. Caste differences in the activity rhythm of several species
of ants placed in DD have also been noted [37, 46]. Altogether, these results suggest that at least some species of ants possess an endogenously driven oscillator.

As in most insects, the primary Zeitgeber in ants seems to be provided by the
alternance of LD period. North [41] found that the rhythms of isolated workers
of *F. rufa* could be readily entrained by a LD cycle. However, the rhythms were
not synchronized, and there was a lot of variability among workers. A formal
description of the mechanisms by which the patterns of activity of individual
workers can be synchronized in whole colonies has been given by Cole ([47] and
this volume). A clear rhythm has also been observed in LD conditions with several other species [5, 36, 37, 40]. North [42] working on individually isolated *F.
rufa* workers and Rosengren [39] on whole colonies of *F. polyctena* have also
shown that the activity rhythm could be maintained in DD or LL conditions by a
temperature cycle. Moreover, Roces and Nunez [45] have provided some evidence that cyclic temperature changes alone are able to entrain the daily bimodal
thermal sensitivity rhythm of the nurse workers of *C. mus*.

Temporal learning in social insects (phase sense)

With a few exceptions [48], studies on temporal learning in social insects have
been restricted to the Apidae and Formicidae, and we will thus focus our review
on these two taxons. One can distinguish different levels of complexity in temporal learning. The first level of complexity is the memorization of single versus

multiple circadian temporal reinforcements. The second level of complexity is the association between a time of day and one or several signals involving different sensory modalities. The third level of complexity is the memorization of several times of day, each associated with a different signal. In the following text we will give several examples of these three types of temporal learning in honeybees and ants. Finally, the fourth level of complexity may be described by the learning of a temporal reinforcement with a noncircadian periodicity. All experiments that have attempted to test this latter type of learning, however, have failed [12, 14], and there are hitherto no reports of such a learning achievement in the literature on social insects.

Forel [49] and Buttel-Reepen [50] were the first to report that honeybees can adjust their daily foraging schedule according to the temporal availability of the food sources they visit. It was Beling [12], however, who first showed experimentally that honeybees can be trained to visit a feeding dish at a definite time of day. Bees tended to arrive a little earlier at the feeding dish, an observation which was subsequently made by several authors [13, 20, 22, 23, 25, 26, 31, 32, 34]. As noted by von Frisch [51] this phenomenon should not be interpreted as indicating imprecision in the time sense of the bees; rather, it makes sense ecologically because such anticipation prevents other competitors from preempting the food source. Wahl [14] repeated Beling's experiments successfully and showed that after 6 to 8 days bees could be trained to visit a feeding dish at up to six different times of day. The biological significance of this learning achievement has been discussed by Kleber [52].

Koltermann [53], using a method slightly different from Wahl's, found that individually trained bees are able to link a particular scent to up to nine different times of day after only one day of training, and Wahl [54] demonstrated that honeybees are able to link different times of day to different food quality. The question of whether bees are also able to link different stimuli to different times of day was addressed in further experiments by Koltermann [53]. He succeeded in training bees to visit a food source associated with a different odor or a different color at two different times of day. In fact, Bogdany [55] showed that honeybees link time, color and scent together when these signals are presented simultaneously during training.

One could consider that in all temporal learning experiments spatiotemporal learning is implicit since animals are able to associate one to several times of day with the spatial information that characterize the location of the food source visited. In most of these experiments, however, the location of the food source remained unchanged throughout the day. A spatiotemporal learning of a higher degree of complexity would be achieved if the location of the food source was changed at each period of reinforcement. Gould [56] reported a learning performance of this type in honeybees: using a passive avoidance learning paradigm, he showed that foragers are able to store information on how to land on different parts of an artificial flower according to the time of day. An operant learning paradigm, however, would be closer to the kind of learning observed in natural conditions. Such a paradigm was used by Wahl [14] in a series of experiments performed more than 60 years ago. Wahl trained bees during 7 days to visit a food source A in the evening and a food source B in the morning. On the test day without rewards, bees visited source B more frequently in the morning and source A more frequently in the evening. Wahl claimed therefore that bees had learned to associate a particular time of day with a particular location. A close examination of his data, however, do not support unequivocal evidence for the existence of true time-place learning. First, when one considers only the category of bees that flew by or over the food locations, one comes to the conclusion that bees were almost equally likely to fly by at site A and at site B in the morning and in the afternoon: at each period of reinforcement, honeybees were actually commuting between the two training locations. Second, when one considers only the bees that landed at the food source, an analysis of Wahl's data at the individual level reveals that only 7 individuals out of a sample of 34 trained bees (21%) landed at the two sites at the right time. Most bees (51%) were spotted at each period of reinforcement, but landed at both sites within each period. Finally, some bees (21%) were spotted at only one period of reinforcement whereas still others visited neither of the two locations (6%). If one takes into account only the bees that were spotted at the two periods of reinforcement, only 29% (7 out of 25) of the bees landed at the right site only at each period of reinforcement, that is, on site B in the morning and on site A in the evening. This figure is not significantly different from what would have been observed if the bees had chosen their landing site randomly at each period of reinforcement (25%). Finally, there is a third additional problem in Wahl's experiment which comes from the fact that he did not fully control for the effect of recruitment on visiting bees. One could argue that the seven individuals that visited each site at the right time had simply been recruited by a small group of nestmates that had specialized on a single period of reinforcement. For example, a small group of bees that had specialized on the morning period could be able to entrain their nesmates to visit this site in the morning, and this same group of bees could in turn be itself entrained in the afternoon by another group of bees that had specialized on the afternoon period. This objection can be tempered in the light of a series of carefully conducted experiments by Moore et al. [32]. These authors trained two groups of individually marked bees of the same hive to visit two separate food locations at two different times of day. Their results show that such training led to two separate and noninteracting groups of foragers: on the test day without rewards the group of foragers trained in the afternoon was not entrained in the morning by the group of foragers trained in the morning and *vice versa*. Moore et al. also monitored the positions of the bees inside the hive: they showed that each group of conditioned bees form a dense cluster on the dance floor of the hive before the onset of their training time; this cluster persists until shortly after the end of the training time and disperses thereafter (see also Körner [57] for similar observations). Most important, although bees clustered on the dance floor as their training time drew close, they did not perform any recruitment waggle dance in the absence of food rewards. Menzel et al. [58] recently carried out a time-place experiment using an experimental protocol which seems to exclude any of the objections raised previously against Wahl's experiment. They showed that bees that were trained at two different locations at two different times of day were able to link each reinforced time of day with a specific compass direction: individual bees captured while departing from the hive at either training time and released at an unknown location flew in a compass direction equivalent to the direction from the hive of the location of the feeder which was reinforced at the corresponding training time.

Reichle [59], Dobrzanski [60] and Rosengren [39] attempted to condition several species of temperate ants to visit a food source at a specific time of day. Despite extensive training, none of the 16 species they tested were able to achieve such a task. To this date temporal learning has only been shown in two species of tropical Ponerinae: the giant tropical ant *Paraponera clavata* [61] and the species *Ectatomma ruidum* [62]. The common characteristic of these two species is that they rely heavily on nectar as a source of carbohydrate food. In contrast, the 15 species of temperate ants studied by Reichle, Dobrzanski and Rosengren mainly collect Homopteran honeydew to fulfill their carbohydrate requirements. Along with the experiments on honeybees, this suggests that, as in other animals [63], nectarivory may be a prerequisite for the evolution of temporal learning in social insects.

Harrison and Breed [61] trained workers of five colonies of *Paraponera clavata* to feed at the same time and location during a 5-day period. The results show that in absence of rewards on the 6th day, most ants arrived at the usual food location within 30 min of the training time. This experiment thus shows that in the same way as honeybees, ants are able to learn the temporal pattern of food availability and to anticipate the time of daily peak production of the sources of nectar. Given this similarity with honeybees, one would expect ants also to be able to link temporal information with other types of information. This in fact was shown in *E. ruidum* [62]. In its natural environment this species is characterized by a triphasic rhythm of foraging activity which appears to be correlated with the rhythm of nectar production of the extrafloral nectaries they visit [64]. This daily rhythm of foraging activity is maintained in the laboratory when ants are offered food *ad libitum*. Schatz et al. [62] tested whether in these conditions ants would be able to associate up to three different times of day with three different spatial locations. Honey was thus available at three different periods of the day, each time at a different location. After 22 days of training, individually marked ants were tested in the absence of food reinforcements at all sites. The results show that the peak of frequentation at each site occurred just during the period when it was usually reinforced. The majority of ants did not commute between the three sites but stayed at each site during a length of time that roughly corresponded to the duration of food availability experienced during training. This suggests that ants are able to form expectancy as to the duration of food availability at the three locations. Twenty-six out of 30 foragers (86%) specialized in the collection of honey visited all three sites at the right time.

Temporal information is linked to spatial information in another context, namely celestial orientation. As mentioned above, using only celestial cues, honeybees are able to learn to link a particular compass direction to a particular time of day [34, 58]. In addition, as many insects, bees and ants are able to compensate the apparent daily movement of the sun across the sky. Temporal learning, however, seems to play only a minor role in shaping this capability. Experiments in the ant *Cataglyphis fortis* [65] and in the bee *Apis mellifera* [66] suggest that these insects possess an innate template of the general pattern of solar movement. Coupled with their internal clock, this template allows them to infer the position of the sun relative to terrestrial landmarks at any time of the day.

Conclusion

Our review of studies on the nature of biological rhythm in social insects clearly shows that we are still a long way from fully understanding the mechanisms underlying the periodicity of activity patterns observed in these insects, whether at the individual or at the colony level. Carefully conducted experiments using the standard methods of chronobiology are lacking in ants, whereas whole groups of insects (wasps, bumblebees, stingless bees, termites) remain to be investigated. Another remark is that most studies of temporal patterns in social insects have hitherto focused on the spontaneous locomotor activity of workers belonging to the forager subcaste. Recent studies in ants [44, 45], however, have revealed the existence of rhythms also in nurse workers. Attention should now be turned to other subcastes and other possible rhythms, for example rhythm in physiological processes, in production of pheromones or in responsiveness to various stimuli (light, pheromone, odor, temperature, gravity) should also be more systematically explored.

The control of temporal information in social insects has been illustrated in this review by the phase sense displayed by the honeybee *A. mellifera* and two species of ants. It would be worth exploring whether such a time sense could be found in other species of social insects with similar diets, for example stingless bees, bumblebees and many species of tropical ants relying on extrafloral nectaries for their alimentation. The control of temporal information associated with spatial information at the individual level may have important consequences at the colony level. Through recruitment, a single individual that has achieved spatiotemporal learning may in fact be able to entrain a whole group of nestmates. This will lead to a subdivision of the forager force in different groups, each with its own spatiotemporal specialization. As for the interval sense, studies on foraging strategies [67] and social regulation [68, 69] provide some clues that social insects are indeed endowed with the capability of measuring and comparing time intervals and rate of events. However, there are still no indications whatsoever about the mechanisms underlying these processes.

Another area which deserves to be explored concerns the neurobiological, physiological and pharmacological basis of temporal learning. Attempts have been made in the past to disturb the internal clock of honeybees by using cold [13], narcosis [31] or drugs interfering with metabolism [13, 70], but none of these experiments have yet been replicated. The same is true for studies on the neuronal substrate of the time sense in honeybees [20, 25]. Given the extent of knowledge accumulated over the last 20 years on the neurobiological and pharmacological basis of other types of learning in honeybees [71], this species should constitute a biological model particularly suited for chrononeurobiologists.

The ecological and evolutionary significance of temporal learning should be more systematically investigated by comparing the performances of different species. An aspect of temporal learning which deserves to be examined is the dynamic of extinction of spatiotemporal memory. For example, Greggers and Menzel [72] in the honeybee and Chittka et al. [73] in bumble bees have shown that insects are indeed able to develop expectations about the specific properties of food sources and to subsequently use these expectations to guide their choice among different types of food sources. One could hypothesize that there should be an optimal forgetting time that could be correlated with the stability of the food sources in the natural environment. In a temperate climate, for example, nectar production is more dependent on changing weather conditions and could cease temporarily at one location. In that case, insects would not benefit by forgetting the location of a food source after only 1 or 2 days of unrewarded visits. In a tropical climate, on the other hand, nectar production should be much more predictable: once nectar production has ceased, it should be definitive. Insects should thus give up visiting the food location after only a few unrewarded visits. The longer forgetting time observed in the bee *A. mellifera* [14] as opposed to the tropical ant *Paraponera clavata* [61] seems to support this hypothesis.

References

1 Daan S (1981) Adaptive daily strategies in behavior. *In*: J Aschoff (ed): *Handbook of behavioral neurobiology*, Plenum Press, New York, 275–298

2 Krebs JR, Kacelnik A (1984) Time horizons of foraging animals. *In*: J Gibbon, LG Allen (eds): *Timing and time perception*. New York Academy of Sciences, New York 423: 278–291

3 Kacelnik A, Brunner D, Gibbon D (1990) Timing mechanisms in optimal foraging: some applications of scalar expectancy theory. *In*: RN Hughes (ed): *Behavioural*

mechanisms of food selection, NATO ASI Series, vol G20. Springer, Berlin

4 Gibbon J, Church RM (1990) Representation of time. *Cognition* 37: 23–54

5 McCluskey ES (1965) Circadian rhythms in male ants of five diverse species. *Science* 150: 1037–1038

6 Gallistel CR (1990) *The organization of learning*. MIT Press, Cambridge, MA

7 Hölldobler B, Wilson EO (1990) *The ants*. The Belknap Press of Harvard University Press, Cambridge, MA

8 Saunders DS (1976) *Insect clocks*.

Pergamon Press, Oxford

9 Aschoff J (1981) Freerunning and entrained circadian rhythms. *In*: J Aschoff (ed): *Handbook of behavioural neurobiology*. Plenum Press, New York, 81–93

10 Kaiser W, Steiner-Kaiser J (1983) Neuronal correlates of sleep, wakefulness and arousal in a diurnal insect. *Nature* 301: 707–709

11 Kaiser W (1988) Busy bees need rest, too: behavioural and electromyographic sleep signs in honeybees. *J Comp Physiol* 163: 565–584

12 Beling I (1929) Über das Zeitgedächtnis der Bienen. *Z Vergl Physiol* 9: 259–338

13 Renner M (1957) Neue Versuche über den Zeitsinn der Honigbiene. *Z Vergl Physiol* 40: 85–118

14 Wahl O (1932) Neue Untersuchungen über das Zeitgedächtnis der Bienen. *Z Vergl Physiol* 16: 529–589

15 Bennet MF, Renner M (1963) The collecting performance of honey bees under laboratory conditions. *Biol Bull* 125: 416–430

16 Beier W (1968) Beeinflussung der inneren Uhr der Bienen durch Phasenverschiebung des Licht-Dunkel-Zeitgebers. *Z Bienenforschung* 9: 356–378

17 Beier W, Lindauer M (1970) Der Sonnenstand als Zeitgeber für die Biene. *Apidologie* 1: 5–28

18 Spangler HG (1972) Daily activity rhythms of individual worker and drone honey bees. *Ann Entomol Soc Amer* 65: 1073–1075

19 Spangler HG (1973) Role of light in altering the circadian oscillations of the honey bee. *Ann Entomol Soc Amer* 66: 449–451

20 Martin U, Martin H, Lindauer M (1978) Transplantation of a time-signal in honeybees. *J Comp Physiol* 124: 193–201

21 Moore D, Rankin MA (1985) Circadian locomotory rhythms in individual honeybees. *Physiol Entomol* 10: 191–197

22 Aschoff J (1986) Anticipation of a daily meal: a process of "learning" due to entrainment. *Monit Zool Italian* 20: 195–219

23 Frisch B, Aschoff J (1987) Circadian rhythms in honeybees: entrainment by feeding cycles. *Physiol Entomol* 12: 41–49

24 Southwick EE, Moritz RFA (1987) Social synchronization of circadian rhythms of metabolism in honeybees (*Apis mellifera*). *Physiol Entomol* 12: 209–212

25 Martin H, Martin U (1987) Transfer of a time-signal isochronous with local time in translocation experiments to the geographical longitude. *J Comp Physiol A* 160: 3–9

26 Neumann MF (1988) Is there any influence of magnetic or astrophysical fields on circadian rhythm of honeybees? *Behav Ecol Sociobiol* 23: 389–393

27 Moritz RFA, Sakofski F (1991) The role of the queen in circadian rhythms of honeybees (*Apis mellifera* L.). *Behav Ecol Sociobiol* 29: 361–365

28 Frisch B, Koeniger N (1994) Social synchronization of the activity rhythms of the honeybees within a colony. *Behav Ecol Sociobiol* 35: 91–98

29 Moritz RFA, Kryger P (1994) Self-organization of circadian rhythms in groups of honeybees (*Apis mellifera* L.) *Behav Ecol Sociobiol* 34: 211–215

30 Kefuss JA, Nye WP (1970) The influence of photoperiod on the flight activity of honey-bees. *J Apicultural Res*, 9: 133–139

31 Medugorac I, Lindauer M (1967) Der Einfluss der CO_2-Narkose auf das Zeitgedächtnis der Bienen. *Z Vergl Physiol* 55: 450–474

32 Moore D, Siegfried D, Wilson R, Rankin MA (1989) The influence of time of day on the foraging behavior of the honeybee, *Apis mellifera*. *J Biol Rhythms* 4: 305–325

33 Moore D, Rankin MA (1993) Light and temperature entrainment of a locomotor rhythm in honeybees. *Physiol Entomol* 18: 271–278

34 Renner M (1959) Über ein weiteres Versetzungsexperiment zur Analyse des Zeitsinns und der Sonnenorientierung des Honigbiene. *Z Vergl Physiol* 42: 449–483

35 Martin H, Lindauer M, Martin U (1983) "Zeitsinn" und Aktivitätsrhythmus der Honigbiene – endogen oder exogen gesteuert? *In: Bayrische Akademie der Wissenschaften. Mathematisch-Naturwissenschaften Klasse. Sonderdruck 1 aus der Sitzungsberichten,* 1–41

36 McCluskey ES, Soong S-MA (1979) Rhythm variables as taxonomic factors in ants. *Psyche* 86: 91–102

37 McCluskey ES, Brown WL (1972) Rhythms and other biology of the giant tropical ant *Paraponera. Psyche* 79: 335–347

38 de Bruyn GJ, Kruk-de Bruin M (1972) The diurnal rhythm in a population of *Formica polyctena* Forst. *Ekologia Polska* 20: 1–127

39 Rosengren R (1977) Foraging strategy of wood ants (*Formica rufa* group). II. Nocturnal orientation and diel periodicity. *Acta Zool Fenn* 150: 1–30

40 Rosengren R, Fortelius W (1986) Light:dark induced activity rhythms in *Formica* ants (Hymenoptera: Formicidae). *Entomol Gen* 11: 221–228

41 North RD (1987) Circadian rhythm of locomotor activity in individual workers of the wood ant *Formica rufa. Physiol Entomol* 12: 445–454

42 North RD (1993) Entrainment of the circadian rhythm of locomotor activity in wood ants by temperature. *Anim Behav* 45: 393–397

43 Roces F (1995) Variable thermal sensitivity as output of a circadian clock controlling the bimodal rhythm of temperature choice in the ant *Camponotus mus. J Comp Physiol A* 177: 637–643

44 Roces F, Nunez JA (1995) Thermal sensitivity during brood care in workers of two *Camponotus* ant species: circadian variation and its ecological correlates. *J Insect Physiol* 41: 659–669

45 Roces F, Nunez JA (1996) A circadian rhythm of thermal preference in the ant *Camponotus mus*: masking and entrainment by temperature cycles. *Physiol Entomol* 21: 138–142

46 McCluskey ES (1963) Rhythms and clock in harvester and Argentine ants. *Physiol Zool* 36: 273–292

47 Cole B (1991) Short-term activity cycles in ants: generation of periodicity by worker interaction. *Amer Naturalist* 137: 244–259

48 Grabensberger W (1933) Untersuchungen über das Zeitgedächtnis der Ameisen und Termiten. *Z Vergl Physiol* 20: 1–54

49 Forel A (1910) *Das Sinnesleben der Insekten.* E. Reinhardt, Munich

50 Buttel-Reepen H von (1900) *Sind die Bienen Reflexmaschinen?* Leipzig

51 Frisch K von (1967) *The dance language and orientation of bees.* Belknap Press of Harvard University Press, Cambridge, MA

52 Kleber E (1935) Hat das Zeitgedächtnis der Bienen einen biologischen Bedeutung? *Z Vergl Physiol* 22: 221–262

53 Koltermann R (1971) 24-Std-Periodik in der Langzeiterrinerung an Duft- und Farbsignale bei der Honigbiene. *Z Vergl Physiol* 75: 49–68

54 Wahl O (1933) Beitrag zur Frage des biologischen Bedeutung des Zeitgedächtnisses der Bienen. *Z Vergl Physiol* 18: 709–717

55 Bogdany FJ (1978) Linking of the learning signals in honeybee orientation. *Behav Ecol Sociobiol* 3: 323–336

56 Gould JL (1987) Honey bees store learned flower-landing behaviour according to time of day. *Anim Behav* 35: 1579–1581

57 Körner I (1939) Zeitgedächtnis und Alarmierung bei den Bienen. *Z Vergl Physiol* 27: 445–459

58 Menzel R, Geiger K, Chittka L, Joerges J,

Kunze J, Müller U (1996) The knowledge base of bee navigation. *J Exp Biol* 199: 141–146

59 Reichle F (1943) Untersuchungen über Frenquentzrhythmen bei Ameisen. *Z Vergl Physiol* 30: 227–256

60 Dobrzanski J (1956) Badania nad zmysten czasu u mrowek. *Folia Biologica Krakow* 4: 385–397

61 Harrison JM, Breed M (1987) Temporal learning in the giant tropical ant, *Paraponera clavata*. *Physiol Entomol* 12: 317–320

62 Schatz B, Beugnon G, Lachaud JP (1994) Time-place learning by an invertebrate, the ant *Ectatomma ruidum* Roger. *Anim Behav* 8: 236

63 Falk H, Biebach H, Krebs JR (1992) Learning a time-place pattern of food availability: a comparison between an insectivorous and a granivorous weaver species (*Ploceus bicolor* and *Euplectes hordeaceus*). *Behav Ecol Sociobiol* 31: 9–15

64 Passera L, Lachaud JP, Gomel L (1994) Individual food source fidelity in the neotropical ponerine ant *Ectatomma ruidum* Roger (Hymenoptera, Formicidae). *Ethol Ecol Evol* 6: 13–21

65 Wehner R, Müller M (1993) How do ants acquire their celestial ephemeris function? *Naturwissenschaften* 80: 331–333

66 Dyer FC, Dickinson JA (1994) Development of sun compensation by honeybees: how partially experienced bees estimate the sun's course. *Proc Natl Acad Sci USA* 91: 4471–4474

67 Seeley TD (1985) The information-center strategy of honeybee foraging. *In*: B Hölldobler, M Lindauer (eds): *Experimental behavioral ecology*. G Fischer Verlag, Stuttgart, New York, 75–91

68 Seeley TD (1989) Social foraging in the honeybee: how nectar foragers assess their colony's nutritional status. *Behav Ecol Sociobiol* 24: 181–198

69 Gordon D, Richard EP, Thorpe K (1993) What is the function of encounter pattern in ant societies? *Anim Behav* 45: 1083–1100

70 Werner G (1954) Tänze und Zeitempfinden der Honigbiene in Anhängigkeit vom Stoffwechsel. *Z Vergl Physiol* 36: 464–487

71 Menzel R, Müller U (1996) Learning and memory in honeybees: from behaviour to neural substrates. *Annu Rev Neurosci* 19: 379–404

72 Greggers U, Menzel R (1993) Memory dynamics and foraging strategies of honeybees. *Behav Ecol Sociobiol* 32: 17–29

73 Chittka L, Gumbert A, Kunze J (1997) Foraging dynamics of bumble bees: correlates of movements within and between plant species. *Behav Ecol* 8: 239–249

The individual at the core of information management

Bernhard Ronacher and Rüdiger Wehner

Most of the chapters in this book treat colonies of social insects as superstructures whose inner workings are governed by self-organizing processes. It is also in the abiotic world that many patterns, such as windblown ripples of sand, the form of liquids in motion, spirals, spots and stripes are woven by self-organization through simple, local interactions among their component parts. In all these cases global properties are not imposed upon the system by overarching blueprints, templates or other forms of instruction from outside, but emerge solely from numerous lower-level interactions. However, if we move from physical to biological systems, an extra dimension is added. Now the superstructures become "adaptive"; the component parts and the rules governing their interactions are finely tuned by natural selection. Whereas the patterns formed by, say, Bénard convection cells can be regarded as merely incidental by-products of the interactions of their elementary subunits [1], biological superstructures are the product of group-level adaptations [2, 3]. Natural selection is the driving agent of the interactions among the subunits of a biological system. In evolutionary terms, a functional higher-order unit can be viewed as a vehicle built by the replicating genes to improve their survival and reproduction [4].

This building of biological superstructures through nested hierarchies of lower-level units [5] has led to an ever-increasing complexity already at the lower level of the subunits. In insect societies—the uppermost biological superstructures—the subunits are highly sophisticated multicellular organisms that display awe-inspiring behavioural repertoires [6–9]. Furthermore, there is not only complexity but also heterogeneity among the component parts of biological superstructures. The genotype of an insect colony is distributed among hundreds or thousands of genetically unique individual workers [10]. In fact, it might have been this very genetic diversity that led originally to social organizations characterized by pronounced functional differentiations among the individual group members. Computer simulations show that even if there are only slight differences in individual response thresholds with respect to, say, pollen or nectar for-aging, a strong division of labour will emerge as a self-organized property of the group [11]. These theoretical inferences are supported by empirical evidence. If, for example, strictly solitary carpenter bees, *Ceratina flavipes*, are coerced to cofound a nest within a cage containing only a single nesting site, there is an almost spontaneous reproductive division of labour, with one individual laying the eggs and guarding the entrance, and the other collecting the food [12]. Colony foundation in *Ceratina* bees can also be induced by crowding, that is, by liberating numerous females in a cage with only few potential nesting sites. Under these experimental conditions, colonies are formed more readily with females of different sizes than with females of about the same size. Similarly, a pronounced division of labour emerges in pairs of young queens of desert ants, *Pogonomyrmex barbatus*, if the queens are forced into foundress associations by being placed in small breeding chambers [11]. One foundress doing the majori-ty of the digging becomes the nest-excavating specialist. She is also the one who, in her solitary stage, has spent more time digging than has her cofoundress. In conclusion, individual predispositions favour the "spontaneous" occurrence of the division of labour even within small groups of interacting conspecifics.

Evolution of eusociality might have proceeded along the same lines. In halic-tine bees, for instance, most species are either solitary or eusocial. This and other examples provide increasing evidence that the eusocial way of life might have arisen without intervening (subsocial or quasisocial, semisocial) stages directly within hitherto solitary populations of bees [13].

In summary, biological superstructures differ from physical ones by having natural selection as their driving force in building increasingly complex subunits which interact physiologically and behaviourally in increasingly complex ways. Part 3 of *Information Processing in Social Insects* focuses on the performances of individual workers. It asks what degree of complexity is required at this indi-vidual level to generate the superstructural complexity we observe at the group level. Of course, self-organizing principles still play the decisive role in building insect-society superstructures, but as the subunits of biological systems (say, individual bees and ants) differ fundamentally from their physical counterparts (say, sand grains and liquid molecules), the study of their interactions becomes a diverse and multifaceted enterprise.

The biological status of the organismic subunits has important consequences. Contrary to what happens in a physical self-organizing system, in a colony of social insects—and in any other biological system as well—the idiosyncratic interests of the subunits must be mutually counterbalanced. For example, hymenopteran workers should prefer to rear their own sons rather than their brothers, to which they are related only by 0.25. Hence, they should have a common interest in revolting against the queen in laying their own eggs. Furthermore, they should cooperate in their revolt, because their fitness will increase by rearing male eggs laid by nestmate workers rather than rearing male eggs laid by the queen. (A worker bee is related to her nephews by 0.375.) Such, however, is the case only if the queen has mated only once. Actually, a honeybee queen mates with several (10–20) males so that the workers become half-sisters. Now things change. The workers should cease to prefer rearing each other's sons (to which they are now related by only 0.125) and instead rear their mother's sons. In fact, as shown by Ratnieks and Visscher [14], they preferentially eat each other's eggs and feed the queen-laid male eggs. How do they discriminate between male-destined eggs laid by either workers or queens? Most probably, this is done by a pheromone produced in the queen's Dufour gland and applied to queen-laid eggs [15]. Furthermore, how does the queen assure mating with several males? These are just two of a set of proximate questions to be asked at the individual level of complexity. And these are the kinds of questions addressed in Part 3 of this book.

In this context, let us return to one of the most important inherent properties of insect societies, the allocation of different physiological and behavioural tasks to different groups of individuals. Here again, self-organization comes to the fore. It is crystal clear that a single individidual (the queen) as central processing and command unit would be unable to oversee, govern and steer the spatial and temporal foraging activities of hundreds of workers, and to record the ways in which food is deposited and extracted from storage compartments within the complex three-dimensional interior space of an insect colony. Instead, the orderly spatial distribution of brood, pollen and nectar within a honeybee colony can be fully explained by the operation of a few behavioural rules which describe the spatial and temporal aspects of egg deposition by the queen and of storing and extracting pollen and nectar by the workers [16]. Computer simulations show that there is no need to refer to any more sophisticated knowledge of the individuals about the overall distribution of brood and food within the colony. Nor must a honeybee forager be informed about the locations and profitabilites of hundreds of food sources within the 100-km^2 foraging area surrounding the colony. The necessary information is conveyed to unemployed foragers by the dance activities of a large number of successful foragers. This information-transfer system results in a highly efficient allocation of foragers to the nectar and pollen sites, with the numbers of foragers being adjusted to the respective profitabilities of the sources and the needs of the colony. The latter can be read off from the rates with which the returning foragers get unloaded by food-storer bees. These rates, in turn, depend on the ratios of empty to filled cells inside the hive. Similar simple rules seem to hold for the foraging behaviour and nest construction of ants [7, 17, 18].

The upshot of all these examples is that, in the course of self-organization, sets of rather simple rules lead to emergent properties of unexpected complexity. The limited precision of the simple "rules of thumb" is balanced by the statistics of large numbers. "Simple", however, is an attribute that, in the context of biological superstructures, should not be overemphasized. The tasks accomplished by the individual group members and the rules of how these members interact can be much more intricate and sophisticated than the few examples mentioned above might let one assume. In the most highly advanced eusocial communities there are elements of small-chain hierarchical control. Wilson and Hölldobler [19] speak of "heterarchical" (rather than purely hierarchical) organizations characterized by dense interconnections and feedback loops between different subgroups. Hence, in the discussion of biological self-organizing principles there is one often neglected point we would like to stress here: it is the individual worker that is at the core of information management, and it is its behavioural repertoire that sets the stage for complex superstructures to arise. In fact, as noted by Bonabeau et al. [20], discussing evolutionary issues of self-organized group behaviour without understanding how individual behaviours are actually implemented, and by what parameters they are influenced, might become a dangerous exercise.

The contribution of Lenoir et al. deals with the question of how the coherence of large groups and subgroups of individuals can be maintained. Of course, a basic prerequisite for cooperation of individuals in social insect colonies is nestmate recognition. Several lines of evidence indicate that in this context cuticular hydrocarbon profiles play a major role. The cuticular hydrocarbon composition has been found to be characteristic not only for species, populations and colonies, but in some instances even for castes and subcastes of patrilines [21]. Certainly, there are genetic influences upon the colony-specific hydrocarbon composition, and these influences are mediated by the queen. In the case of multiple queens (and also multiple matings), this might result in a higher diversity of recognition cues as compared with monogynous colonies. This, in turn, may lead to a lower rejection rate of non-nestmates [22]. In addition, the colony-specific odour blends are prone to be influenced by food and nesting material [23], since the cuticular lipids act as an odour trap. Central to the formation of colony-specific odour in ants are the postpharyngeal gland (PPG) and the behavioural trait of trophallaxis, by which the contents of the PPGs can be distributed rapidly within the colony. This mutual exchange guarantees a spatially uniform colony-specific odour blend ("odour gestalt"). The content of the PPG is distributed over the cuticle by auto- and allogrooming, thereby allowing identification of colony members or aliens by antennation on various parts of the body. An interesting hypothesis in this context has been put forward by Keller [24]: the fast spreading of chemical recognition labels amongst colony members may also serve to reduce or eliminate information about kinship of subgroups of individuals, and thus may help to reduce conflicts and to suppress nepotism, which probably would lead to a reduction in colony efficiency (for reviews on these aspects see [24], see also [25]).

How are individuals "imprinted" on the colony-specific "odour gestalt", and how do changes of the colony hydrocarbon profile get incorporated into the recognition system? An interesting concept, put forward by Lenoir et al. in this volume, is that a chemical template—unlike the acoustic templates of song birds —is based simply on the bearer's own hydrocarbon profile rather than on some kind of neuronal representation of it. In such a "chemical reference" system the discrimination and recognition would rely more on detecting differences than on evaluating the familiar. This notion gains some support from experiments in which ants have been treated with PPG extracts from other colonies or species (Lenoir et al., this volume). It is also in accord with the observation that the label is dynamic and continually updated.

Robson and Traniello take a different point of view. They focus on the possible role of individual specializations in the organization of labour and on the allocation of individual ants to different tasks. In their terminology, task specialization is used in a more restricted way than meaning merely specializations based on physical castes (as they occur in ants and termites) or on age-dependent temporal polyethism. In line with prior classifications such as "elitism", "specializations" and "idiosyncrasy", these authors entertain the concept of the "key individual". These key individuals are further subdivided into organizers, catalysts and performers (depending on the effects of their removal on the performance of the group; see also [26, 27]).

In this context, however, one must be careful. First, it will be important to distinguish between "true" individual specializations and apparent ones, as they may result from the timing of behavioural actions. Of course, this distinction will prove to be difficult and will likely be biased by experimental difficulties of observing individually marked workers for a sufficiently long period of time. Second, given that individuals indeed show degrees of specializations that can reliably be separated from the mean and variance of the group or cohort behaviour, then the question arises whether these specializations are based on genetic differences between the individual workers (due to patrilines or polygyny, e.g. [28–31]), or whether such variations can be explained sufficiently by epigenetic factors such as differences in physiological maturation, prior experience, social regulation and so forth (see e.g. [32, 33]). Finally, the analysis should focus on the ultimate consequences of such individual differences. Do these differences lead to new emergent properties at the colony level? How is the success of a colony related to this kind of task specialization, and what are the selective forces that may act on maintaining this trait (see e.g. [28, 34–39]):

As can be deduced from the evolutionarily stable strategy (ESS) approach, the division of labour between two tasks may be accomplished in one of two ways: (i) A certain proportion of individuals, say 20%, show specialization on task A, while the remaining 80% focus on task B. (ii) Alternatively, each single individual could switch between tasks A and B with appropriate probabilities. Solution (i) may be advantageous when a premium is put on maximum efficiency of task performance. Difficulties arise, however, if different and varying numbers of specialists must be allocated rapidly to particular tasks. In this respect, solution (ii) may be more flexible, especially under highly variable ecological conditions. An important test could consist in forming experimental groups that are composed exclusively of presumed specialists (e.g. organizers in Robson's and Traniello's parlance), and to observe whether some of these individuals then switch to a performer role.

Individual variation may enhance the diversity of possible task performances above the level set by fixed task specializations. Another possible function of individual variation might be deduced from the phenomenon of stochastic resonance: individual variation may play a role similar to a (moderate) level of noise in stochastic resonance events. It could facilitate or speed up the transition from one state of activity to another. As self-organization depends on the amplification of local fluctuations [20], one important role of individual specializations may be to generate a sufficient amount of fluctuation. Similar effects might have contributed to the occurrence of multiple matings and the evolution of polygynous colonies.

An often underrated factor influencing the spectrum of possible emergent states is the mere number of colony members. The limited number of individuals in small colonies should have serious consequences for the kind and amount of information transfer, and for the degree of division of labour. Indeed, in small colonies of polistine wasps direct physical attacks by the dominant (reproducing) female seem to be the main means of controlling monogyny [40]. This mechanism obviously delimits the size of the colony. In fact, colonies of independently founding polistine wasps commonly comprise less than 200 adult individuals. Furthermore, in such small colonies the foundresses perform several tasks in addition to oviposition, whereas in the larger colonies of swarming polistine species all social tasks except oviposition are performed by the workers. Finally, large insect societies have lost the ability to recognize kin (i.e. members of different matrilines or patrilines; [41]), as can be found in primitively eusocial forms (*sensu* [42]; see [43]). In all ant species, for example, kin recognition has been replaced by nestmate recognition via a common colony odour (see above).

With respect to the relation of task specialization and colony size comparative studies are highly recommended. For example, even if the same behavioural rules were present in two species characterized by different colony sizes, one would expect that the same local rules resulted in different global patterns of behaviour. It would be especially interesting to investigate what individual rules already apply to small colonies at the verge of eusociality, where the numbers of individuals are small, and hence the balancing effects of statistics are weak.

Fourcassié et al. review the management of temporal information in social insects. The capacity to handle temporal information has been demonstrated by recording spontaneous activity patterns as well as by learning experiments. In the context of this review it is interesting to note that the free-running periods of individual worker bees show larger variances than those of whole colonies. Tactile contacts and intragroup temperature fluctuations seem to be important for this kind of social synchronization. Another point worth mentioning is the ability of individual workers to associate temporal information with spatial information, for example to learn that particular food sources are available at particular locations only at particular times of day [44, 45].

Here, however, as in so many other cases, individual behavioural complexity cannot be integrated yet into the overall pattern of group performance. This is a wide and more promising field of proximate research than the perpetual formal claim of self-organization processes governing the workings of insect societies could ever be.

References

1 Bak P (1996) *How nature works: the science of self-organized criticality.* Springer, New York
2 Williams GC (1966) *Adaptation and natural selection.* Princeton University Press, Princeton
3 Seeley TD (1995) *The wisdom of the hive. The social physiology of honey bee colonies.* Harvard University Press, Cambridge, MA
4 Dawkins R (1982) *The extended phenotype.* WH Freeman, San Francisco
5 Bronowski J (1974) New concepts in the evolution of complexity. *In*: RJ Seeger, RS Cohen (eds): *Philosophical foundations of science.* Reidel, Dordrecht, 133–151
6 Frisch K von (1967) *The dance language and orientation of bees.* Harvard University Press, Cambridge, MA
7 Hölldobler B, Wilson EO (1990) *The ants.* Belknap, Harvard University Press, Cambridge, MA
8 Seeley TD (1995) *The wisdom of the hive. The social physiology of honey bee colonies.* Harvard University Press, Cambridge, MA
9 Wehner R, Lehrer M, Harvey WR (1996) Navigation. *J Exp Biol* 199: 125–162, 225–261
10 Gadagkar R (1997) *Survival strategies. Cooperation and conflict in animal societies.* Harvard University Press, Cambridge, MA
11 Page RE Jr (1997) The evolution of insect societies. *Endeavour* 21: 114–120
12 Sakagami SF, Maeta Y (1987) Sociality, induced and/or natural, in the basically solitary small carpenter bees (*Ceratina*). *In*: Y Ito, JL Brown, J Kikkawa (eds): *Animal societies: theories and facts.* Japan Scientific Society Press, Tokyo, 1–16
13 Michener C (1985) From solitary to eusocial: need there be a series of intervening species? *In*: B Hölldobler, M Lindauer (eds): *Experimental behavioural ecology and sociobiology.* G. Fischer, Stuttgart, 293–305
14 Ratnieks FLW, Visscher PK (1989) Worker policing in the honey bee. *Nature* 342: 796–798
15 Ratnieks FLW (1995) Evidence for a queen-produced egg-marking pheromone and its use in worker policing in the honey bee. *J Apicult Res* 34: 31–37
16 Camazine S (1991) Self-organizing pattern formation on the combs of honeybee colonies. *Behav Ecol Sociobiol* 28: 61–76
17 Wehner R (1987) Spatial organization of foraging behavior in individually searching desert ants, *Cataglyphis* (Sahara Desert) and *Ocymyrmex* (Namib Desert). *In*: JM Pasteels, J-L Deneubourg (eds): *From individual to collective behavior in social insects.* Birkhäuser, Basel, 15–42
18 Franks NR, Deneubourg JL (1997) Self-organizing nest construction in ants: individual worker behaviour and the nest's dynamics. *Anim Behav* 54: 779–796

19 Wilson EO, Hölldobler B (1988) Dense heterarchies and mass communication as the basis of organization in ant colonies. *Trends Ecol Evol* 3: 65–68

20 Bonabeau E, Theraulaz G, Deneubourg JL, Aron S, Camazine S (1997) Self-organization in social insects. *Trends Ecol Evol* 12: 188–193

21 Arnold G, Quenet B, Cornut J, Masson C, De Schepper B, Estoup A, Gasqui P (1996) Kin recognition in honeybees. *Nature* 379: 498

22 Morel L, van der Meer RK, Lofgren CS (1990) Comparison of nestmate recognition between monogyne and polygyne populations of *Solenopsis invicta* (Hymenoptera: Formicidae). *Ann Entomol Soc Amer* 83: 642–647

23 Breed MD, Leger EA, Pearce AN, Wang YJ (1998) Comb wax effects on the ontogeny of honey bee nestmate recognition. *Anim Behav* 55: 13–20

24 Keller L (1997) Indiscriminate altruism: unduly nice parents and siblings. *Trends Ecol Evol* 12: 99–103

25 Carlin NF, Reeve HK, Cover SP (1993) Kin discrimination and division of labour among matrilines in the polygynous carpenter ant *Camponotus planatus. In*: L Keller (ed): *Queen number and sociality in insects*. Oxford University, Oxford, 362–401

26 Möglich M, Hölldobler B (1975) Communication and orientation during foraging and emigration in the ant *Formica fusca. J Comp Physiol* 101: 275–288

27 Abraham M, Pasteels JM (1980) Social behaviour during nest-moving in the ant *Myrmica rubra L*. (Hymenoptera, Formicidae). *Insect Soc* 27: 127–147

28 Moritz RFA (1989) Colony level and within colony level selection in honeybees. *Behav Ecol Sociobiol* 25: 437–444

29 Page RE Jr, Robinson GE (1991) The genetics of division of labour in honey bee colonies. Adv. *Insect Physiol* 23: 117–169

30 Giray T, Robinson GE (1994) Effects of intracolony variability in behavioral development on plasticity of division of labor in honey bee colonies. *Behav Ecol Sociobiol* 35: 13–20

31 Calderone NW, Page RE Jr (1996) Temporal polyethism and behavioural canalization in the honey bee, *Apis mellifera. Anim Behav* 51: 631–643

32 Jaisson P, Fresnau D, Lachaud JP (1988) Individual traits of social behavior in ants. *In*: RL Jeanne (ed): *Interindividual behavioral variability in social insects*. Westview Press, Boulder, 1–51

33 Gordon DM (1996) The organization of work in social insect colonies. *Nature* 380: 121–124

34 Fewell JH, Winston ML (1992) Colony state and regulation of pollen foraging in the honey bee, *Apis mellifera L. Behav Ecol Sociobiol* 30: 387–393

35 Fuchs S, Schade V (1994) Lower performance in honeybee colonies of uniform paternity. *Apidologie* 25: 155–168

36 Page RE Jr, Fondrk MK (1995) The effects of colony-level selection on the social organization of honey bee (*Apis mellifera L.*) colonies: colony-level components of pollen hoarding. *Behav Ecol Sociobiol* 36: 135–144

37 Page RE Jr, Robinson GE, Fondrk MK, Nasr ME (1995) Effects of worker genotypic diversity on honey bee colony development and behavior (*Apis mellifera L.*). *Behav Ecol Sociobiol* 36: 387–396

38 Page RE Jr, Waddington KD, Hunt GJ, Fondrk MK (1995) Genetic determinants of honey bee foraging behaviour. *Anim Behav* 50: 1617–1625

39 O'Donnell S (1998) Genetic effects on task performance, but not on age polyethism, in a swarm-founding eusocial wasp. *Anim Behav* 55: 417–426

40 Jeanne RL (1980) Evolution of social

behaviour in the Vespidae. *Annu Rev Entomol* 25: 371–396

41 Jaisson P (1991) Kinship and fellowship in ants and social wasps. *In*: PG Hepper (ed): *Kin recognition*. Cambridge University Press, Cambridge, 60–93

42 Michener C (1974) *The social behaviour of the bees. A comparative study*. Belknap, Harvard University Press, Cambridge, MA

43 Greenberg L (1979) Genetic component of bee odor in kin recognition. *Science* 206: 1095–1097

44 Bogdany FJ (1978) Linkage of learning signals in honey bee orientation. *Behav Ecol Sociobiol* 3: 323–336

45 Menzel R, Geiger K, Jorges J, Müller U, Chittka L (1998) Bees travel novel homeward routes by integrating separately acquired vector memories. *Anim Behav* 55: 139–152

Part 4 Amplification of information and emergence of collective patterns

Activity cycles in ant colonies: worker interactions and decentralized control

Blaine J. Cole and Franc I. Trampus

Summary

In this chapter we look at the phenomenon of periodic activity cycles in ant colonies. Although most of the information on activity cycles in ant colonies is available for *Leptothorax*, it seems to be a phenomenon that occurs widely. Cycles of activity that last approximately one-half hour occur in colonies, with their occurrence being a function of the number of individuals that are in an aggregate or a colony. The presence of brood has an important influence on activity cycles, increasing the degree of periodicity, though brood are neither necessary nor sufficient for the production of periodic activity. We discuss models for the production of activity cycles and divide the models into two basic groups, those that require global colonywide variables that influence each colony member and those which are based on local interactions with no global connections among workers. We find that the local models, which have explicit spatial structure, are more realistic and have greater predictive power even though they are often more cumbersome to use. Finally, we consider the functional basis for periodic activity in ant colonies. There is essentially no information about the function or possible adaptive significance of this widespread, colony-level phenomenon.

Introduction

One of the most intriguing questions about the organization of social insects is the means by which the actions of individuals produce colonywide phenomena. Successful integration of individual effort enables efficient foraging, nest construction and brood care. How do the independent actions of workers produce such large-scale effects? What sort of coordination and control mechanisms

C. Detrain et al. (eds) Information Processing in Social Insects
© 1999, Birkhäuser Verlag Basel/Switzerland

result in the emergence of collective behavior? At one end of the continuum of possibilities a centralized control mechanism directs the production of an outcome. At the other extreme collective patterns emerge from a decentralized control mechanism by dynamic interactions among individuals. In this chapter we will discuss the production of activity patterns in ant colonies. We will argue that decentralized control causes the self-organization of activity cycles.

Colony activity, defined here as movement of any type within the colony, is the subject of this review. We regard activity patterns as important for study for two basic reasons. First, activity itself is a fundamental property of behavior. It is difficult to conceive of behavior without movement. If activity has temporal pattern, then other behaviors such as foraging, brood care, nest construction and so on must also be affected. We also see the study of colony activity levels as a model system for the general question of how colony level phenomena are produced by the actions of individuals. Activity has the useful property that it can be objectively and automatically quantified. This gives considerable flexibility to data collection.

This chapter consists of three parts. In the first we discuss the empirical evidence concerning colony activity cycles, the activity of individuals and the interactions among them. In the second section we discuss models that have been developed to describe this phenomenon, the data that pertain to each model and the relationship among the models. In the third section we discuss the significance of activity cycles in ant colonies.

The phenomenon of activity cycles

Although observers of ant behavior have long been aware that activity in ant colonies was not constant, this phenomenon has only been quantified relatively recently [1, 2]. For long periods of time, all of the workers under observation may be completely immobile. One ant initially becomes active, and then activity appears to spread to neighboring workers. Eventually, all of the workers under observation may be actively moving about. This activity reaches a peak and gradually dies out, and every worker may be quiescent again (Fig. 1a). These episodes of activity last approximately 30 min. We shall first describe the cycles of activity in intact colonies of ants, emphasizing work using *Leptothorax allardycei* and *L. acervorum*, about which the most is known. Next, we shall summarize exper-

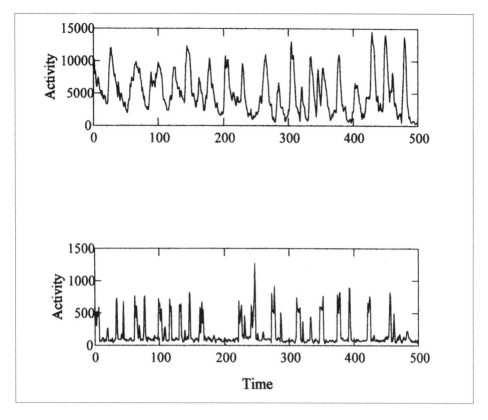

Figure 1 The activity pattern of an intact colony of Leptothorax allardycei *(top) and an isolated worker of* L. allardycei *(bottom)*
The y-axis is activity measured in pixel changes between successive images [1, 3], and the x-axis is time in 30-s intervals.

iments that illustrate the relationship between the activity of individual workers and the activity patterns of colonies. We shall discuss the mechanism of interaction among individuals, and other important correlates of activity cycles such as colony hunger, worker age and the presence of brood.

Periodic activity of colonies

The pattern of activity in colonies of *Leptothorax* ants has a periodic component. In some cases the periodic signature is extremely strong. Fourier analysis is used to analyze periodicity in the data record. Any temporal data record can be decomposed into Fourier components that describe the contribution of sine waves of period (1, 2,..., n/2) time intervals where n is the number of measurements that are made in the record. When added together, the Fourier components reconstruct the data record. Because the sum of the coefficients is related to the variance of the data record, a Fourier analysis is similar to an analysis of variance where the size of the coefficients of the Fourier decomposition is proportional to the fraction of variance in the data record that is due to oscillations at that frequency. If a data record consists of the sum of several perfectly sinusoidal components, the coefficients of the Fourier decomposition that correspond to these components will be quite large. If there is oscillation with noise, there will be large Fourier components that correspond to the frequency of the signal, but also components that correspond to the pattern of noise. If there is a sloppy oscillation, with a tendency to drift in frequency, there may be one or more peaks of power, and the peaks will probably be broader. All biological oscillations, from heartbeats to circadian rhythms, have noise to one extent or another, and some may drift in period or amplitude.

Strong periodic components occur in the colony activity rhythms of ants, but they are never purely periodic [1–5]. How periodic are the activity rhythms? There is variation among colonies and preparations. The degree of periodicity can be measured by looking at the fraction of the variation in the data record that is due to periodic oscillations at the largest Fourier component. Cole and Cheshire [3] and Cole and Hoeg [4] report that in intact colonies about 33% of the variation is due to a single frequency component.

While most of the data has been collected from *Leptothorax* species, ants from a number of other genera have also been observed to measure activity patterns. Colonies of *Camponotus planatus* (Fig. 2a), *Tapinoma littorale* (Fig. 2b), *Pseudomyrmex cubaensis* (Fig. 2c) and *Monomorium floricola* (Fig. 2d) all show periodic components in activity. Additionally, periodic activity has been found in *Leptothorax pastinifer*, *L. isabellae*, *Pseudomyrmex elongatus*, *P.* sp., *Zacryptocerus varians*, *Solenopsis picta* and *Xenomyrmex floridanus*. The broad representation of species from at least four subfamilies indicates that periodic activity cycles are not restricted to particular taxonomic groups. Additionally, *T. sessile* has repeatedly failed to show any substantial periodic components to behavior. Unfortunately, the comparative data are not sufficient to give a sense of the important characteristics of social organization that may be responsible for one species showing periodic activity and other species not.

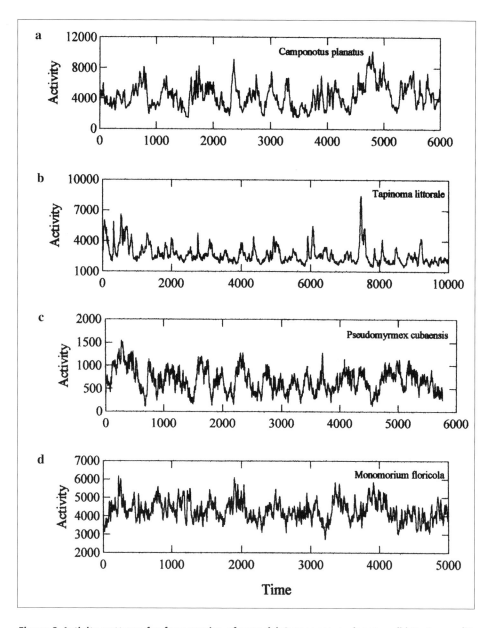

Figure 2 Activity patterns for four species of ants: (a) Camponotus planatus, *(b)* Tapinoma littorale, *(c)* Pseudomyrmex cubaensis, *and (d)* Monomorium floricola

The axes are as in Figure 1 with the time scale in five second intervals.

Most data collection runs have been for 4–8 h, but there have been longer observations runs of >24 h. During these longer runs, the period of the oscillation can occasionally change, rather abruptly. This blurs the periodicity, over the entire data record. If one looks at a sliding window, one can see the presence of a single period, the appearance of a second periodic component and the fading out of the first periodic component. These results suggest that a simple oscillator, governing the production of periodicity, is unlikely to be the cause of colony activity cycles. Rather, it is consistent with the interpretation that this is a decentralized, group phenomenon that emerges through the interaction of workers.

There are two further results from intact colonies that give information about the mechanism of spread of activity and the rate at which activity spreads. The activity record of an intact colony of *L. allardycei* was separately recorded in 16 separate grid squares (4 × 4). The activity of regions in two halves of the nest was synchronous (shown by the peak in the crosscorrelation function in Fig. 3a). A double screen partition was inserted into the nest into two portions that had air connections, but did not permit physical contact among ants. Separate regions on the same side of partition were still synchronized (Fig. 3b), whereas regions on opposite sides of the screen partition were not (Fig. 3c). Although they continued to oscillate, they did so with different periods. Finally, when the screen partition was removed, synchrony between the two halves of the nest was reestablished (Fig. 3d). Direct physical contact among workers seems to be required for synchronous activity. By comparing the activity records at two portion of the nest, and measuring the time required for activity at one location to spread to another location, it was established that activity traveled through the nest at the rate of about .8 mm/s. Both of these experiments suggest that it is the movement of workers themselves that transmits activity, not the diffusion of a pheromone.

Assembling activity patterns

In this section we shall describe the activity of workers and the interactions that lead to periodic activity and factors that influence activity cycles. We shall examine the activity patterns of isolated workers, small groups of workers, and the interactions among workers. This information exists only from *L. allardycei*. We shall also describe some of the complications of activity patterns including the effect of age, brood and feeding status.

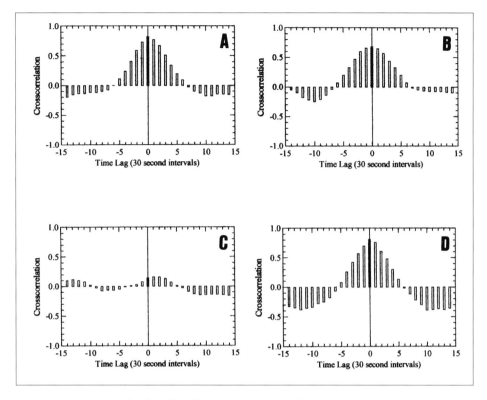

Figure 3 Cross-correlation functions for split colony experiments
(A) Cross-correlation in activity [2] between two portions of a nest before the experiment demonstrating synchrony. (B) Cross-correlation between two portions of the nest on the same side of a double-screened partition, demonstrating synchrony. (C) Cross-correlation between two portions of the nest on opposite sides of a double-screened partition (which allows air flow, but no physical contact among ants), demonstrating no synchrony. (D) Cross-correlation between these sectors after the double-screened partition has been removed, showing that synchrony is reestablished.

If colonies of *L. allardycei* characteristically show periodic activity, individual workers do not show any evidence of periodicity (Fig. 1b). An isolated worker becomes active spontaneously, remains active for a relatively short interval (typically about 5 min) and becomes spontaneously inactive. The interval between episodes of inactivity has a negative exponential frequency distribution (Fig. 4), suggesting that there is no periodic recurrence to the activity of an isolated individual. Cole [5, 6] has argued that the activity pattern of individual workers shows deterministic chaos. Although the interval between episodes of

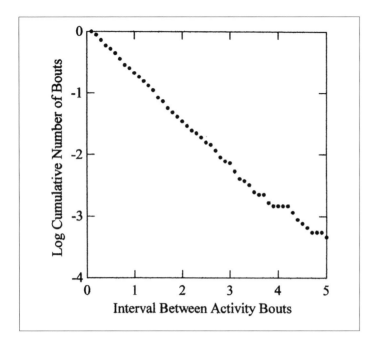

Figure 4 The cumulative distribution of intervals between activity bouts
The linear relationship on this log-linear plot indicates a negative exponential frequency distribution.

activity is not predictable over long intervals, the magnitude and timing of activity is predictable over a short interval. The activity pattern of a single ant has the characteristics of a chaotic switch [7]. Franks et al. [1] demonstrate that the distributions of the duration of activity bouts and interbout intervals for single workers within colonies of *L. acervorum* do not suggest periodic individuals either.

Age influences the amount of activity of individual ants. Cole [8] suggested that activity declines as a function of worker age, and the probability that a worker becomes spontaneously active decreases with age. Although the situation is somewhat more complicated, it is a function of age. The degree of activity of an isolated ant (measured as the fraction of time active) first increases for the first several months of life and then declines. There is a highly significant quadratic term (P < 0.001) in the regression of activity on age. It is only during the latter half of the life of the worker in the lab that activity declines monotonically.

The degree of periodicity increases with the number of workers in the aggregate. Cole and Cheshire [3] used aggregates of 2, 3, 5, 7, 11 and 15 workers.

Aggregates that had 15 workers had the same degree of periodicity as intact colonies, and there was a nearly linear trend to become more periodic with larger numbers of workers. There is not a threshold number of workers that results in periodicity, nor is there any evidence that there are special individuals that govern the oscillation; activity cycles are a property of the aggregate.

The presence of brood also influences the production of activity cycles [1, 3, 4]. Cole and Hoeg [4] showed that different types of brood are differentially effective in producing this change. They used four group sizes (1, 3, 7 and 15 workers) and 20 pieces of four types of brood, eggs, small larvae, large larvae or pupae. The results were that eggs, small larvae and large larvae all had identical effects, and they increased the level of periodicity, whereas in the presence of pupae workers were no more periodic than workers without brood. Although brood influence activity cycles, they are neither necessary—large aggregates without brood show cycles of activity, nor sufficient—small aggregates in the presence of brood do not show activity cycles.

Although this experiment controlled for the number of brood, the ratio of workers to brood was not controlled. Even in the largest aggregates, the number of brood per worker was greater than 1. Hemerik et al. [9] suggest that there is a threshold brood:worker ratio, above which brood can produce activity cycles, and below which activity cycles cannot be produced. An experiment by Tofts et al. [10] did look at the brood:worker ratio and found no difference between activity cycle characteristics and this ratio.

The feeding status of a colony influences the production of activity cycles in *L. acervorum* [1]. During starvation, a larger proportion of the workers become foragers and cycles seem less prominent.

Active ants affect the activity states of other ants. When an active ant encounters an inactive ant, the inactive ant will become active sooner than if it does not encounter an active ant. In *L. allardycei* there is a "phase advance" in the inactive ant whose amount depends on the timing of the interaction [11]. In *L. acervorum* active ants are effective in stimulating inactive ants to become active [1]. In *L. allardycei*, when two active ants encounter one another, they tend to lengthen the time that each will spend in the active state [3]. The effect is to double the amount of time that is spent active after an interaction.

Models for the production of activity cycles

There have been several attempts to model the production of colony activity cycles. The models take different approaches and make different assumptions. The purpose of this section is to categorize the different models and to point out the essential differences and similarities in the underlying organization, if not the approach. Then we shall use the results summarized above to see which models have empirical support and which seem less likely given the data available.

The autocatalytic model

The first model to explain the processes underlying the existence of activity cycles in insect colonies was the autocatalytic model developed by Goss and Deneubourg [12]. They demonstrated that activity cycles, such as those that had been observed in the ant *L. acervorum,* could be generated by a population of individuals without an exogenous driver.

The model consists of three core assumptions. At each time step there is a fixed probability (following an exponential distribution) that a currently active worker will become inactive. Once a worker becomes inactive, it experiences a refractory period of fixed duration during which it cannot be reactivated. Thereafter, the now susceptible ant can become active, either spontaneously at a low, fixed probability during each time step, or alternatively, through stimulation by the activity of any active worker that encounters it. The greater the current number of active workers in the colony, the greater the likelihood that an inactive worker will become active.

At each time step, individuals are randomly paired, and any inactive but susceptible ant that is paired with an active ant also becomes active. This process soon results in synchronous bouts of activity with a period between activity peaks that is approximately equal to the refractory period. However, the model does not produce synchronous activity in simulations where the refractory period is too short relative to the duration of activity, or when the ability of an active ant to stimulate an inactive ant is too low.

The fundamental outcome of the autocatalytic model is the demonstration that synchronous, periodic patterns similar to those produced by ant colonies can be generated without the need for an external pacemaker mechanism. The combination of a positive feedback mechanism, such as stimulation, and a post-activity latency period can be sufficient.

Cole [8] elaborated on these models to include several types of stimulation, including type 1 and type 0 phase resetting in the periodic activity of isolated workers. In type 0 resetting ants would become active after stimulation regardless of when, during the inactive phase, they are stimulated. In type 1 resetting, the phase change does not always result in activity, but is a function of the phase of stimulation. These models have less current validity since individual ants are known not to have periodic activity. All of these models are formally identical to epidemic models in which the "disease" being transmitted is activity and individuals move through infective, recovered but immune and susceptible stages [13].

Tofts et al. [10] present a different formalism of the Goss and Deneubourg model. The model has many of the same elements, although they are treated using a WSCCS (weighted synchronous calculus of communicating systems) description. In this model ants remain inactive for a fixed period, and then become wakeable. If they wake up, they wake up all other wakeable ants, then they immediately become inactive. The activity of colonies of such ants rapidly synchronize and remain synchronized. The length of the cycle is primarily a function of the length of the inactive period.

The energy model

The energy model developed by Hemerik et al. [9] is an analytical model that focuses more directly on how physiological factors might regulate activity cycles in ant nests. Goss and Deneubourg's autocatalytic model assumes that the phenomenon is driven by a combination of stimulation and a fixed post-activity latency period; the energy model is based on a threshold time for the digestion of food. The model demonstrates that the overall energy level in the colony coupled with the number of active ants is sufficient to produce synchronized, cyclic behavior in the nest. The energy model is mathematically similar to a number of models that have been proposed for the regulation of other physiological processes [14].

In order to model the interaction between colony nutrition and colony activity, they assume that the energy level in the nest is the total energy level of food both in the nest and within the adults and larvae. As in the autocatalytic model, individual ants are considered to be either active or inactive. The digestive process that is at the foundation of the energy model is expressed as exponentially decreasing with time, whereas the rate of increase of energy level in the colony is proportional to the total number of active ants within the nest. The assumption is thus that only active ants are able to provision the nest.

In this model, the rate at which active ants become inactive is a constant, such that the rate of change in the number of active ants is proportional to the number of active ants. The rate at which inactive ants become active is a decreasing function of colony energy, such that activity is stimulated by low energy level. This feature is built into the model because of the observation that starved colonies show more activity. Active ants are also able to activate inactive ants. Thus, the rate of increase in the number of active ants must also increase as the number of active ants increases. This is a point of similarity between the autocatalytic and energy models.

The energy model predicts that manipulation of colony energy levels will change the number of active ants. The validity of the model could be tested by examining the results of simulations under two conditions (starvation and increased brood:worker ratio). Franks et al. [1] found that after prolonged starvation, the proportion of active ants in a nest is at a higher and more constant level. The energy model also demonstrates that the proportion of active ants remains at a constant level that is greater than the maximum proportion that are active normally. When the brood:worker ratio is increased or decreased, there is a similar response in the rate of change of the colony energy level. When brood:worker ratio is high, the model predicts that fewer ants are synchronously active and that periods of activity are shorter, but occur more frequently.

The model assumes that there is a positive relation between the number of ants active and the number of ants that become active. It requires a quadratic term relating the rate of increase of active ants to the number of active ants in order to produce a rhythmic pattern of activity. Versions which only used the linear term resulted in a stable equilibrium. This is in keeping with the activation assumption of Goss and Deneubourg's [12] autocatalytic model, and the observation of Franks et al. [1] that individual *L. acervorum* workers become active significantly more often when contacted by other ants.

Mobile cellular automata models for activity

The spatially explicit models that are described in this section are direct descendants of the probabilistic model of Goss and Deneubourg [12]. The essential components of the probabilistic models, that ants are in different activity states and they interact with one another to change each other's activity are the same. The essential difference is that there is an explicit spatial component and interactions occur only locally. This is much more realistic and produces substantial differences in the behavior of the models.

These models are designed specifically to examine the behavior of *Leptothorax*, but are not limited to *Leptothorax* ants, ants in general or even social insects. We will first present the rationale for the modeling approach, then describe the model and define the conditions under which the model shows periodic behavior.

Mobile cellular automata (MCA) models were developed by Solé and co-workers [15]. Subsequently these models were elaborated and modified by Miramontes and co-workers [16], Solé and Miramontes [17] and Cole and Cheshire [3]. They are spatially explicit models in which the ants (the mobile automata) are allowed to move on a lattice representing the nest. An ant's level of activity can range from some maximum to some minimum. The behavioral state of an ant is active if its activity level exceeds some threshold, typically zero. An active ant may move one step on the lattice; an inactive ant remains in its current location. If an ant is inactive, there is a small probability that it can become active spontaneously. If an ant does not interact with another ant, its level of activity declines until it reaches zero.

An ant can interact with others if they are in close proximity, that is, the eight nearest lattice points. Rules for permissible interactions are based on the behavioral states of the ants involved. Since each ant is either active or inactive, and may have a neighbor that is either active or inactive, there are four types of pairwise interactions. If an interaction between two ants is allowed, it changes the focal individual's state; the activity of an ant in the next time step is a function of the activity of its neighbors and its current activity.

Solé and co-workers [15] showed that this MCA system exhibits oscillations that are similar to the activity cycles of *Leptothorax* ants. They examined several rule sets, but concentrated primarily on models where all possible interactions were allowed. Cole and Cheshire [3] looked systematically at the behavior that is produced by each possible rule set. By allowing or forbidding each of the four types of potential interactions, 16 rule sets were produced that enumerated all possibilities. An interaction matrix, J_{ij}, is defined to indicate the effect that a neighbor, j, has on ant i, where i and j are either active or inactive. For example, if $J_{11} = 1$, an active neighbor is allowed to influence the activity of the active focal ant; if $J_{11} = 0$, then an active neighbor is forbidden from influencing the behavior of the active focal ant.

The essential result of Cole and Cheshire is that a single rule was essential to produce cycles of activity in these models: J_{11}, the effect that an active ant has on another active ant. When $J_{11} = 1$, the models show cyclic activity patterns. The degree of periodicity increases with increasing numbers of ants. When $J_{11} = 0$, the models did not show periodicity with any number of ants. If an active ant causes another active ant to remain active longer, then cycles of activity will emerge as more ants are added to the aggregate. If active ants do not mutually reinforce each other's activity, then there will be no activity cycles. Active workers of *L. allardycei* reinforce one another's activity.

The relation of the theories

The autocatalytic model is the simplest model of those that have been proposed for the production of colony activity cycles. It shares a number of similarities both with the energy model and with the cellular automata models. The autocatalytic model and the energy model can both easily be written in the form of differential equations. While the solution to the equations cannot be as easily written, the models can be numerically integrated, and a periodic solution may exist. The existence of a periodic solution does not verify the models' accuracy, only that it is possible for such models to produce the appropriate behavior. A set of differential equations can be produced for a colony because both models assume the existence of global parameters, and global connections among the workers. By this we mean that the behavior of individuals is a function of the aggregate state of colony. This is most conspicuous in the energy model, because a state variable, the energy level of the colony, describes something that is a property of the colony as a whole. Furthermore, each worker responds to changes in this state variable: colony energy level dictates change in another state variable, the probability that ants activate. Each ant is connected to every other ant through this state variable. In the autocatalytic model the state variables are the proportion of workers that are active, refractory and inactive but susceptible. Since the probability that any ant becomes active is a function of the global fraction of ants that are in each of the three possible states, and each ant, in whatever activity state, is capable of interacting with ants in any other activity state, each ant is connected to each other ant. The necessity of writing a simple, differential form for the model, necessitates postulating global connection among ants, and global variables.

The MCA models are fundamentally different in this regard. There are no global variables. Instead, spatial structure is an integral part of the model. While this has some advantages in terms of realism and simplicity of construction, there are some disadvantages as well. The most prominent disadvantage is that it is no longer possible to write down a simple set of differential equations that can be solved by inserting them into widely available numerical integrators. Because MCA models do not have the history of differential equations, there are not widely available flexible techniques for model construction. As this sort of model becomes more widely used, the availability of simulation packages for their development can be expected to make such models easier to use for a larger number of investigators.

Although the spatial structure of a cellular automata model makes modeling more dependent on computer simulation, it has the advantage of being more realistic. The data for *L. allardycei* show that direct, physical interactions among ants are required for activity to spread and for synchrony of activity. Had there been a pheromone which produced activity within the colony (or alternatively, a metabolite that when produced inhibited activity), then the actions of one worker in one part of the nest could influence the actions of a worker in distant portions of the nest as easily as it influenced nearby ants. Such a globally connected colony could make use of global colony state variables. If the pattern for *L. allardycei* is general, this is not an accurate description of the phenomenon.

Because the phenomenon has spatial properties, it seems essential to use a model that assumes the spatial properties, whatever other disadvantages it may have. The MCA models are quite similar to the autocatalytic model in another regard; they assume that ants move through states of activity and inactivity (although refractory periods are not essential). In this sense the MCA models can be viewed as an individual-based version of the autocatalytic models. Here autocatalysis proceeds only at the level of a local neighborhood; the colony state is not directly relevant. A consequence of using a spatially explicit model, without global connections, and of considering the types of interactions that can occur between individuals leads to a nonintuitive conclusion: cyclic behavior of colonies is due solely to the influence that active ants have on other active ants. This social facilitation effect is necessary and sufficient for the production of cycles in sufficiently large aggregates. The stimulus of activity in inactive ants (the sort of interaction assumed by the autocatalytic and energy level models) is not necessary and in fact retards the production of activity cycles [3].

The significance of activity cycles

Although we know a certain amount about the phenomenon of activity cycles and there is also a considerable body of theoretical work that can be applied to the subject, there is little information on the significance of activity cycles. This is in stark contrast with the situation in the other chapters in this section in which the importance of recruitment, nest construction or swarming is transparent. In this section we will discuss hypotheses for the functional significance of activity cycles. There are two classes of hypotheses, those that ascribe a functional significance to activity cycles and those that suggest that cycles are an epiphenomenon produced by other selective pressures.

The existence of synchronized activity pulses in *Leptothorax* colonies may be adaptive if this organization improves the efficiency with which colonies perform tasks that either conform to a cyclic pattern or else coincide with rhythmic stimuli. Several authors have discussed circumstances under which synchrony may enable workers to coordinate their activities more efficiently, particularly when tasks require interactions between workers [10, 12, 18].

Goss and Deneubourg [12] illustrate how synchronous activity may be more efficient than uniform activity under such circumstances. If A represents a colony's total amount of worker activity, and this is uniformly distributed over a period of T minutes, then the average level of activity is A/T. Likewise, if A occurs during a single minute, the average activity level is also A/T. However, if workers interact with one another such that efficiency is related to the square of the number of active workers, then efficiency becomes A^2/T for uniform activity, but A^2 for synchronous activity. Thus, synchronous activity is T times more efficient. It seems perfectly plausible to suppose that interactions between synchronous workers could result in a nonlinear increase in productivity. In fact, the exponent that describes the nonlinear increase in efficiency only needs to be greater than 1. We are aware of no data that address this issue.

Synchrony may also help to ensure that worker activity is uniformly distributed throughout the nest. Hatcher et al. [18] argue that if individual brood care workers randomly choose which brood to tend, some may remain untended for a substantial period of time, with a negative effect on growth and survivorship. When activity is synchronized, a more uniform distribution of care results from information exchange via the mutual exclusion of workers from brood that are already being cared for, forcing those workers to find other, currently untended, brood [18]. This also has the advantage of preventing unnecessary task repetition. Simultaneous activity (rather that periodicity *per se*) is advantageous because it reduces the overlap in brood tending. The quiescent periods that occur between bouts of activity in *Leptothorax* colonies may be a consequence of synchrony, reducing energy expenditure during routine periods. Recent work with *Solenopsis invicta* by Cassill and Tschinkel ([19] and this volume) has shown that workers feed larvae in discrete time units; all larvae are fed for the same length of time, but hungry larvae are fed more frequently than sated larvae. This fact is consistent with the mutual exclusion hypothesis of Hatcher et al. [18]; however, Cole and Cheshire [3] demonstrate that although brood has an effect on the amount of periodicity, it is more dependent on the addition of more workers than on the addition of brood. It must be noted that if the brood:worker ratios are critical, there have been no experiments that directly manipulate this factor. It is the number of workers, not the number of larvae that regulates synchrony and periodicity in *L. allardycei*. This does not mean that there is no relationship between brood and activity cycles, only that brood is not necessarily the cause. However, the presence of brood does tend to enhance periodicity [3, 4]; thus brood may exert some direct influence.

Alternatively, synchrony and periodicity may not have any direct adaptive significance. No one has demonstrated that synchrony is functionally more efficient, so while it is possible that it enables more coordinated behavior in ant colonies, it is also possible that the simultaneous activity of numerous individuals may result in colonial inefficiency due to interference between workers. In fact, there is no experimental evidence that activity cycles have any adaptive value. It has been argued that the existence of activity cycles is an epiphenomenon that is not itself adaptive, but is the by-product of interactions among individuals [2, 8].

Periodicity may be an incidental consequence of selection acting on the interaction of individual ants with nest mates.

Selection acts on the colony phenotype by changing the actions of individuals. What sort of selection on colonies may result in the production of activity cycles as an indirect effect? Here it is useful to mention the result of one further set of simulations using the MCA model. Recall that one of the features of the MCA model was that certain types of interactions could be permitted or forbidden. The essential interaction that results in periodic activity is the effect that an active ant has on another active ant. The type of interaction that has the greatest effect on the total amount of activity in simulated colonies is the ability of an active ant to influence an inactive ant. However, model colonies with both types of interaction have the highest total activity level; the other interactions do not matter. Colonies in which active ants stimulate the activity of both active and inactive ants have the highest total activity. They will also have periodic activity.

The ability of active ants to stimulate the activity of inactive ants [1, 2, 11] may not be important to the establishment of periodicity, but it may be the single most important factor determining the overall activity level in colonies. Therefore, selection on colonies for higher activity levels should inevitably result in production of colony activity cycles.

Acknowledgments

We are grateful for the assistance of Diane Wiernasz in reading and criticizing this chapter. We acknowledge the support of NSF-BNS-9120965.

References

1   Franks NS, Bryant R, Griffiths R, Hemerik L (1990) Synchronization of the behavior within nests of the ant *Leptothorax acervorum* (Fabricius). I. Discovering the phenomenon and its relation to the level of starvation. *Bull Math Biol* 52: 597–612

2   Cole BJ (1991) Short-term activity cycles in ants: generation of periodicity by worker interaction. *Amer Naturalist* 137: 244–259

3   Cole BJ, Cheshire D (1996) Mobile cellular automata models of ant behavior: movement activity of *Leptothorax allardycei*. *Amer Naturalist* 148: 1–15

4   Cole BJ, Hoeg L (1996) The influence of brood type on the periodic activity of *Leptothorax allardycei*. *J Insect Behav* 9:

539–547

5 Cole BJ (1991) Is animal behavior chaotic? Evidence from the activity of ants. *Proc Rl Soc Lond B* 244: 253–259

6 Cole BJ (1994) Chaos and behavior: the perspective of nonlinear dynamics. *In*: L Real (ed): *Behavioral mechanisms in evolutionary ecology.* University of Chicago Press, Chicago, 423–443

7 Liebovitch L, Toth T (1991) A model of ion channel kinetics using deterministic chaotic rather than stochastic processes. *J Theor Biol* 148: 243–267

8 Cole BJ (1992) Short-term activity cycles in ants: age-related changes in tempo and colony synchrony. *Behav Ecol Sociobiol* 31: 181–188

9 Hemerik L, Britton NF, Franks NR (1990) Synchronization of the behavior within nests of the ant *Leptothorax acervorum* (Fabricius). I. Modeling the phenomenon and predictions from the model. *Bull Math Biol* 52: 613–628

10 Tofts C, Hatcher M, Franks NR (1992) The autosynchronization of the ant *Leptothorax acervorum* (Fabricius): theory, testability and experiment. *J Theor Biol* 157(1): 71–82

11 Cole BJ (1991) Short-term activity cycles in ants: a phase response curve and phase resetting in worker activity. *J Insect Behav* 4: 129–137

12 Goss S, Deneubourg JL (1988) Imitation as a source of autocatalytic synchronized rhythmical activity in social insects. *Insect Soc* 35: 310–315

13 Bailey NTJ (1964) *The elements of stochastic processes with applications to the natural sciences.* John Wiley, New York

14 Murray JD (1989) *Mathematical biology.* Springer, New York

15 Solé RV, Miramontes O, Goodwin BC (1993) Oscillations and chaos in ant societies. *J Theor Biol* 161: 343–357

16 Miramontes O, Solé RV, Goodwin BC (1993) Collective behavior of random-activated mobile cellular automata. *Physica D* 63: 145–160

17 Solé RV, Miramontes O (1995) Information at the edge of chaos in fluid neural networks. *Physica D* 80: 171–180

18 Hatcher MJ, Tofts C, Franks NR (1992) Mutual exclusion as a mechanism for information exchange within ant nests. *Naturwissenschaften* 79: 32–34

19 Cassill DL, Tschinkel WR (1995) Allocation of liquid food to larvae via trophallaxis in colonies of the fire ant, *Solenopsis invicta. Anim Behav* 50(3): 801–813

The mechanisms and rules of coordinated building in social insects

Guy Theraulaz, Eric Bonabeau and Jean-Louis Deneubourg

Summary

This chapter presents an overview of the mechanisms, usually intertwined, used by social insects to build their often elaborate nests: templates, stigmergy, self-organization and self-assembly. A few models based on these mechanisms are also discussed, but they are to a large extent speculative because experimental evidence is scarce. Our conclusion is that it is not necessary to invoke individual complexity to explain nest complexity. Recent work suggests that a social insect colony is a decentralized system comprised of cooperative, autonomous units that are distributed in the environment, exhibit simple probabilistic stimulus-response behavior and have access to local information. The complexity of a nest is likely to result from the unfolding of a morphogenetic process during which past construction provides both constraints and new stimuli. This form of indirect communication between insects through the environment is an important aspect of collective coordination, and has been coined "stigmergy" by Grassé. We show that stigmergy has to be supplemented with a mechanism that makes use of these interactions to coordinate and regulate collective building. We suggest that at least two such mechanisms play a role in social insects: self-organization and self-assembly. When they are combined with templates, they became the building blocks of a powerful construction game.

Unrivalled builders

Among the great variety of collective activities performed by social insects, building is certainly the most spectacular where the difference that exists between individual and collective levels is the strongest. This capacity to build

C. Detrain et al. (eds) Information Processing in Social Insects
© 1999, Birkhäuser Verlag Basel/Switzerland

nests that possess often very complex architectures remains a fascinating issue for naturalists. The evolution and function of such complexity still remains an unsolved mystery. The building activity of social insects has been the focus of many studies [1–4], bees [5–7], wasps [8–15] and termites [16–19], but one can notice that among the vast literature devoted to the study of collective phenomena in social insects, very few works exist that deal explicitly with the analysis of behavioural mechanisms and rules that are involved in the building activity and its regulation. Figure 1 shows some examples of nest architectures built by tropical wasps and African termites. Some species of wasps, such as *Chartergus chartarius* or *Epipona tatua*, build nests made up with several stacked combs, of cells. In these nests, there exists a central or peripheral communication opening that goes through the successive combs allowing the wasps to move from one floor to another. Some paper nests built by *Polybia* wasps can reach one meter in height. In termites, some nests are like cathedrals of clay. *Macrotermes* nests can reach 6 or 7 meters, that is 600 times the size of a single worker. Moreover, termites build particularly complex architectures as regards the diversity, delicacy and extreme regularity of the elements of which these nests are constituted. For instance, *Apicotermes* termites build subterranean oval nests about 20 cm high, in which stacked horizontal chambers are connected by helix-shaped vertical walkways that are used by termites as spiral staircases. Moreover, the outer surface of the nest is covered by a set of regularly spaced pores that open towards corridors circulating inside the internal wall of the nest (see e.g. [16, 20]). Finally, it appears that the nest structure of social insects is more than the simple repetition of the same basic module; even if some basic elements are repeatedly present, these are organized in superstructures and networks. For instance, common bees' nests are is not just a whole array of juxtaposed hexagonal cells; cells are organized in combs, and each comb is itself organized in three distinct concentric regions that create a very specific pattern at the surface of the comb; this pattern, which is consistent throughout the season, is made of a central area of compact brood surrounded by a ring of cells that only contain pollen, and finally a large peripheral region of cells where honey is stored [21, 22].

Figure 1 Some examples of nest architectures built by social insects
(a) Parachartergus *sp. wasp nest; (b)* Chartergus chartarius *wasp nest; (c)* Epipona tatua *wasp nest; (d)* Apicotermes lamani *termite nest (© Guy Theraulaz and Muséum National d'Histoire Naturelle, Paris).*

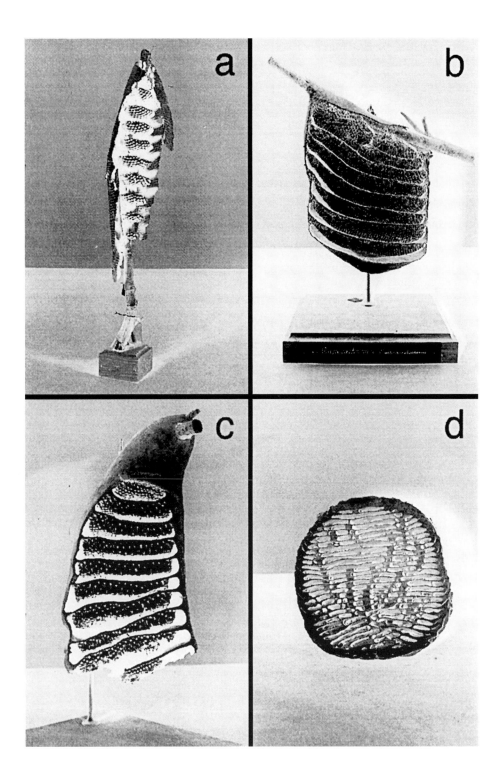

How to build a nest without blueprint?

It is generally believed that during the building of a nest, insects take part in a collective and cooperative enterprise of great complexity. This process leads to the production of a spatial structure, the characteristic scale of which is usually several times higher than the size of a single individual. Remarkably, such species-specific structures are able to develop at the scale of a colony, despite the fact that each insect's individual behaviour possesses a strong random component and remains basically very simple. Faced with such amazing architectures, one easily realizes the qualitative gap that exists between the behaviour of one individual insect and the shapes of the structures produced by a colony. Moreover, there is no evidence that would allow us to conclude that individual behaviours displayed by insects living in society are more sophisticated than those of solitary species. Then how is it possible for locally informed insects to coordinate their actions in order to build such complex architectures? Following Maeterlinck's thoughts, one could seek for the "brilliant architect" conducting all theses activities as well as "the spirit of the hive" [23].

The first hypotheses put forward to explain these feats were anthropomorphic and called for some individual intelligence or some knowledge of the global structure to be produced by individual insects. There was a tendency to hypothesize that the complexity of the overall architectures built by social insects relied in their ability to centralize and process information. Following this idea, an insect could decide the actions to be done thanks to the representation it builds of its own environment. But at present, no experimental data exist that would allow us to support the hypothesis that something like a map or a blueprint are used by a single individual insect during building.

Collective intelligence: there's no more need for individual complexity

Insects societies provide us with a functioning model that is rather different from the anthropomorphic one where a direct causal relationship was established between the complexity of the structures built and the complexity of the behavioural procedures implemented at the individual level to produce these structures. This new model relies on decentralized logic based on cooperative autonomous units (see [24]). These units distributed in the environment possess a relatively simple probabilistic stimulus-response-based behaviour. Insects have access only to local information that is without any reference to the global structure as a whole; they do not possess any representation or knowledge of the global structure they have to build. In other words, they do not use a global blueprint. On the contrary, these insects possess sensory-motor equipment with which they are able to respond to stimuli emitted either by their siblings (chemoreceptors, for instance) or coming from the environment (mecanoreceptors, thermoreceptors, hygroreceptors etc.). But these signals are not equivalent to signs that could have symbolic value. The way in which these signals that are simply attractive or repulsive, activating or inhibiting will affect insects' behaviour will depend on their intensity and the context in which they are released. In these societies, a global solution is not explicitly programmed at the individual level, but rather emerges from myriads of simple elementary interactions between individuals or between individuals and their environment; one could speak of a collective intelligence built on individual simplicity [24, 25].

There exists by now a growing literature in which numerous examples of this decentralized intelligence have been discussed, including the collective choice of a food source in ant and bee colonies [26–29], the formation of trail networks and foraging patterns in ant colonies [25, 30–35], the dynamical division of labour in ant and wasp colonies [36–38], the collective sorting of brood items in ant and bee colonies [21, 22, 39, 40] and some aspects of building behaviour in bee, wasp and ant colonies [2, 4, 6, 7, 12, 41]; for a review see [42]. So, in order to organize themselves and coordinate their behaviours, individual insects have to possess a set of very specific behavioural rules. We will present some of these rules in detail below.

The coordinating mechanisms of building activities

One of the most important problems faced by ethologists is to understand the structure and phylogeny of behavioural mechanisms that are used by the great variety of social insect species to build their nests. During the last 10 years, numerous studies have been undertaken, and some of them shed some light on the coordinating mechanisms of collective activities in social insects [42]. Even if much remains to be done in this field of investigation, several mechanisms are by now better understood. It is now possible to present a first overview and a tentative classification of the different mechanisms that are used by social insects to coordinate their building activities: templates, stigmergy, self-organization and self-assembly. Among these different potential mechanisms, the simplest is the template.

Templates

A first class of mechanisms widely used by social insects to organize and coordinate their building activity relies on templates. In this case, the blueprint of the future construction "already exists" in the environment under the form of physical or chemical heterogeneities. The consequence of the insects' activity will only be to reveal this "hidden" structure. There are many forms of templates that naturally exist in the environment such as temperature or humidity gradients. In particular, these kinds of gradients are used by ants to build their nests, to organize the spatial location of each caste of workers or simply to sort the different items of their brood (see e.g. [43, 44]).

Gradients may also be created by a member of the colony. This kind of mechanism figures when termites build the royal chamber. The physogastric queen of *Macrotermes subhyalinus* emits a pheromone that diffuses and creates a pheromonal template in the form of a decreasing gradient around her (see Fig. 2). Bruinsma [19] has shown that a concentration window exists, or a threshold, that controls the workers' building activities: a worker deposits a soil pellet if the concentration of pheromone is within this window or exceeds the threshold. Otherwise, they do not deposit any pellet or even destroy existing walls. If one places a freshly killed physogastric queen in various positions, walls are built at a more or less constant distance from the queen's body, following its contours, whereas a wax dummy of the queen does not stimulate construction [19]. With this very simple mechanism, the termite workers are able to produce at any moment an adjusted construction that fits the size of the queen. Indeed, when the queen gets fatter, the concentration thresholds move towards the periphery, and a new chamber is built instead of the old one (see Fig. 2).

The use of templates in social insects are nowadays very well known in numerous other examples ranging from dome construction in *Formica* ants studied in depth by Chauvin [2, 45, 46], to the formation of craters near the entrance hole of *Messor* ants' nests as was shown by Chrétien [47].

314

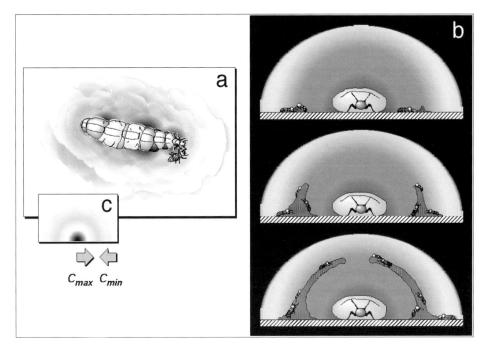

Figure 2 Pheromonal template created by a physogastric queen of Macrotermes subhyalinus
The queen's body emits a pheromone that diffuses and creates a decreasing gradient around her. A concentration window exists, or a threshold, that controls the workers' building activities: a worker deposits a soil pellet if the concentration of pheromone is within this window or exceeds the threshold. Otherwise, they do not deposit any pellet or even destroy existing walls. Walls and a chamber are progressively built around the queen's body following its contours, at a more or less constant distance. (a) top view; (b) cross section showing three construction steps of the royal chamber; (c) closeup view of the concentration thresholds.

Stigmergy

Stigmergy is a concept that was introduced at the end of the fifties by Pierre-Paul Grassé to explain task coordination and the regulation of building activity that appears in termites of the genus *Bellicositermes* [16]. Grassé formulated the hypothesis that a termite worker did not conduct its own work but rather was guided and directed by the result of its own previous building activity. Grassé gave this mechanism the name stigmergy, which came from the Greek *stigma*, meaning "sting" and *ergon*, meaning "work". As we will see, the blueprint is no longer lying in the environment as was the case in the template; it will progressively emerge through the sequence of numerous building activities performed in space and time by all the individuals of the colony.

The mechanism of stigmergy

How do things happen? Each time an insect encounters a stimulating structure, for instance a previous state of the nest, and performs a building activity, the construction is modified and results in the creation of a new material structure. This

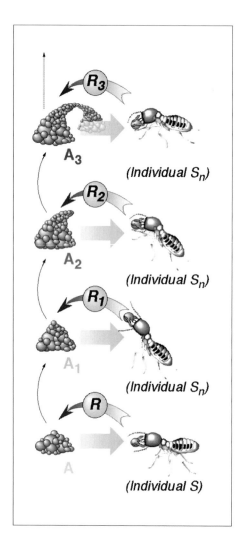

Figure 3 The mechanism of stigmergy as it was originally defined by P.P. Grassé

The successive steps in the initial building phases (A₁ to A₃) lead to the emergence of pillars and an arch joining two neighbouring pillars; (R₁ to R₃) are the corresponding building activities performed by termite workers (from Grassé [16, 17]). The diagram has to be read from the bottom, upwards.

structure will in turn guide and trigger a new building activity in another colony's insects if by chance one of them comes into contact with it and only if this insect carries building material (see Fig. 3). The main problem is then to determine what stimuli trigger building activity and especially how they are organized in space and time so as to lead to a perfect and coherent construction; this means that the insects belonging to a given species will build qualitatively the same architectural pattern. E.O. Wilson calls this process *"sematectonic communication"* since the only interactions taking place between individuals are those that occur through the environmental modifications the individuals make over time [48]. This process leads to an almost perfect coordination of work and gives the illusion that the colony as a whole is following some predefined blueprint. Grassé's introduction of the concept of stigmergy offered an elegant and stimulating framework to understand coordination mechanisms involved in building activities. But numerous unanswered questions still remained. In particular, what dynamics could such interactions produce, and how was it possible to achieve real on-line control of cooperativity using this dynamics? Trying to answer these questions, we realized that two types of mechanisms with very different dynamics were gathered together under the same concept of stigmergy. Each of these dynamics relies on a different way of controlling individual behaviour. Indeed, stigmergy is basically just a mechanism that mediates or implements indirect worker-worker interactions; thus, it has to be supplemented with a mechanism that makes use of these interactions to coordinate and regulate collective building. We discovered that there exist at least two such mechanisms that play this role in social insects: self-organization [31, 42] and self-assembly [14, 15]. These mechanisms will be discussed in the next sections.

Stigmergy and self-organization

Stigmergy can be thought of as a succession of stimulus-responses that are quantitatively different. One can observe this kind of mechanism in the initial phase of nest building in termites, where it leads to the emergence of regularly spaced pillars. A simple model was introduced by Deneubourg [18] and extended by Bonabeau et al. [49] to describe this phenomenon. This model shows how the different parameters characterizing the random walk of the termites, the attractivity of the cement pheromone, the diffusion of the pheromone and so on determine the regular distance between pillars. In order to build their nests, termites use mud pellets and stercoral cement. These pellets are soaked with building pheromone during buccal handlings and mixings. This pheromone diffuses in the environment, creating gradients the intensity of which decreases as we move away from the place where the pellet was dropped. The model that is consistent with the experimental data obtained by Bruinsma [19] and Grassé [16] assumes

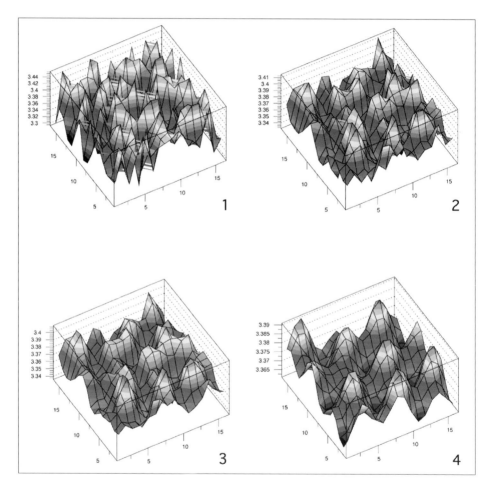

Figure 4 Four simulation steps showing the temporal evolution of the structure built by termites in a 2-D system

This simulation shows the evolution of the density of building material used by termites to build their nest. The simulation begins with a random distribution of building material in space (step 1), and the regularity of the interpillar spacing emerges progressively over time (steps 2, 3 and 4).

that the building behaviour and the movements of termite workers are locally controlled by pheromone concentration. The pathway of a termite results from the addition of two processes: random walk and response to the pheromone gradient (chemotactic behaviour). The termite's response is then proportional to the gradient: the greater the gradient, the more attracted termites are. Greater gradients attract termites towards peaks of concentration, which correspond to zones where much material has been deposited. When material is dropped, cement pheromone, is emitted and diffuses, thereby attracting more termites towards its area of origin, where more material is therefore dropped. This accumulation of material induces a stronger and stronger emission of pheromone, which attracts more termites. This positive feedback at different sites in the building area gives rise to competition between different pillars which are close to one another. This leads to an "inhibition" of pillar formation in the immediate neighbourhood of a pillar and also facilitates the emergence of another pillar further away. The byproduct of this amplification and competition is a regular distribution of the pillars in space, without any explicit coding of the interpillar distance (Fig. 4). In this way stigmergy appears as a particular instance of a self-organized process.

The self-organized regulation of building activity

Self-organizing processes are also involved in the growth and regulation of the size of the nest. The nest's growth is generally correlated with the growth of the insect population that lives inside. Several theoretical and experimental clues clearly indicate that different growth dynamics can be obtained when building behaviour obeys an amplification law as is the case in termites. In some cases, even a slight increase in population size leads to an immediate regulation of the nest's volume. By contrast, in other cases, regulation of the nest's volume is made by successive puffs. These global dynamics have been studied with a simple model [50]. Each time a place is dug by an insect, a pheromone marker is laid down, and before vanishing into the air this marker stimulates other insects to dig at the same place. The difference that exists between monotonous growth and growth that occurs with regular bursts depends on slight differences of numerical values assigned to environmental or behavioural variables (see Figs 5 and 6). The dynamics of this kind of growth can have important consequences for the shape of the nest. When the growth of the colony is characterized by regular

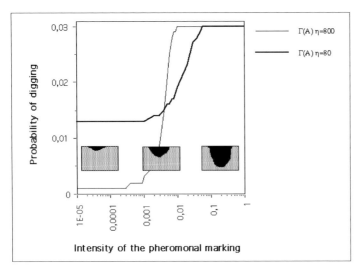

Figure 5 *Individual digging probability [Γ(A)] as a function of the pheromone density (A) located at the digging site*
This probability takes the following form Γ(A) = α(1 + e$^{-\eta(A-A_c)}$)$^{-1}$ *where* η *is a control parameter of the response. The higher* η, *the greater the likelyhood the response is all-or-nothing (modified from [50]).*

bursts, the corresponding nest structure will appear like a succession of similar modules stacked together, as in the nests of *Cubitermes* or *Noditermes* termites.

Stigmergy and self-assembly: the stigmergic processes of collective building in wasps

Stigmergy can give rise to another form of sequence where stimulus-responses differ from a qualitative point of view. The dynamics produced by such interactions are related to self-assembly phenomena. This is the kind of process that some wasp species used to build their nests. Wasp nest architecture can be very complex (see, for instance, Fig. 1 and [11]). The vast majority of wasp nests are made of wood fibres that are chewed by wasps and cemented with oral secretions. The resulting pulp is then used to build the different parts of the nest, the pedicel, the combs of cells or the external envelope. Modularity is another important feature of their nest architecture. Indeed, the repetition of the same basic structure is the simplest way to increase the size of a nest. In social wasps indi-

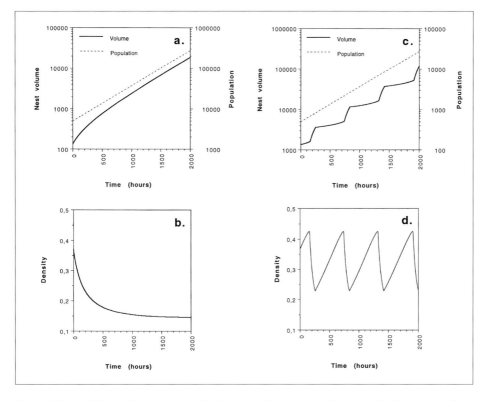

Figure 6 *Two different types of growth dynamics that can be obtained with the same behaviour: digging followed by the pheromonal marking of the excavated areas; individual digging probability defined in Figure 7 is the only difference existing between the examples. In both cases, the growth of the population in the nest is the same*
In the first case, with a low η value (a and b) there is a regular growth of the nest that matches the size of the population. In the second case, with a high η value (c and d) the growth of the nest occurs through successive pulses. As the population increases, long periods with very little digging activity can be observed; these periods are regularly interrupted by bursts of digging activity during which the nest's size nest increases dramatically (modified from [50]).

vidual building activities are mainly organized thanks to the patterns of cells and paper that are encountered by wasps moving on the nest. Moreover, several studies have shown that the individual behavioural program used by wasps in nest building relies on a series of decisions with conditional "if-then" loops [9, 10, 13, 41]. Building a nest starts with the construction of a pedicel that links the nest to the substrate and a first cell in the continuation. Then, as long as the building process is unfolding, the number of potential sites where a new cell can be added increases exponentially. Several parallel building activities can then be carried out. As a consequence, construction does not result from the unfolding of an *a priori* well-defined sequence. Indeed, swarm has the ability to build simultaneously in several different places on the same nest. This is certainly an important step towards the evolution and creation of complex architectures [8, 10]. But this fact has in turn introduced new constraints, since the local information (i.e. the patterns of matter that are sensed by wasps on the nest) have to be organized in space and time so as to ensure an almost coherent collective building. These constraints as well as the dynamics resulting from such interactions can be studied using a simple model, for example the use of lattice swarms, that describes the distributed building process in social wasps [14, 15].

The building behaviour of lattice swarms

Lattice swarms are systems of automata that use qualitative stigmergic-based rules. The swarm of automata moves randomly and independently on a three-dimensional cubic or hexagonal lattice [14, 15]. Using a limited number of bricks of different types with the same cubic or hexagonal shape, the swarm deposits these bricks according to a specified set of rules, embodied in a lookup table. The lookup table specifies what type of brick will be deposited in a particular site based upon the current configuration of bricks in the local neighbourhood. This neighbourhood consists of the 26 cells in a cubic lattice and 20 cells in a hexagonal lattice surrounding any particular central cell occupied by a given automaton. The bricks are the atomic building material, and all the agents using the same lookup table are able to put down the right brick whenever they meet a stimulating configuration. In this way the shape of the structure is what indirectly controls its own growth. If this model oversimplifies the processes used by wasps to build their nests, it is still a powerful tool for studying the behavioural constraints faced by wasps to coordinate their building activities, provided they use exclusively local patterns of matter that result from past construction. One important conclusion derives from the fact that when several individuals use a stigmergic building algorithm, the local configurations that are created at a given time and which will trigger new building actions have to be different from those created at a previous or a forthcoming building step so as to avoid disorganization of the building activity. The order in which stimulating configurations are produced must be strictly satisfied, and such configurations must not interfere. This ordering constraint further imposes a limit on the type of architecture that can be collectively generated.

To better understand what we mean, consider this simple fact: each single insect belonging to the same colony is able to build alone a complete nest architecture. From this vantage, building appears to be a purely individual behaviour. The problem comes when several individuals using the same behavioural building program must cooperate to build simultaneously the same architecture. In this case, the local patterns of matter that trigger building activity necessarily have to be organized in space and time so as to "coordinate" in turn the collective activity performed by the insects. This will indeed avoid disorganization of swarm global activity that could result from the combination of many building actions performed by insects without space-time coherence. So, the order in which the stimulating configurations appear on the nest as the construction progresses and as new cells are added must follow a well-defined sequence. This will ensure the coordinated building of a stable species-specific architecture. Figure 7 shows some of these stable architectures that have been obtained in simulations. The individual's reactions to these stimulating configurations of matter belongs to the genetic heritage of the species that is considered. The most important fact is that this constraint upon arrangement creates in turn strong constraints upon stable architectural shapes that can be built collectively. And one of the conclusions we came to is that the space of stable architectures that can be produced is strongly restricted by the very nature of the building coordination process [14, 15].

The role of air streams in shaping nest structure

The observation of termites' nests reveals a large range of building structures ranging from multiple pillars to galleries and walls. It is easy to show that a simple change in the propagation of the pheromone swept along by an air stream can greatly affect the shape of the resulting structures built by the termite workers. In particular, walls and galleries can be built instead of pillars without any modification of individuals' behaviour. This phenomena was originally experimentally shown by Bruinsma [19], who studied the effect of a laminar air flow of low velocity along the queen's body axis on the initiation of walls around the queen. Bruinsma, who measured the distance between the queen and deposition sites, observed a reduction in the mean distance at which soil pellets were deposited:

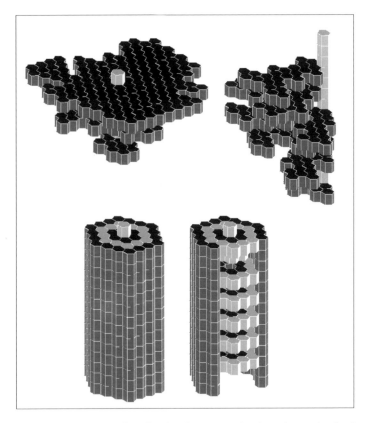

Figure 7 Three types of artificial architectures that have been obtained with 3-D hexagonal lattice swarms

Some of these architectural patterns possess a similar design to the structures built by wasps species such as Vespa *(top left),* Parachartergus *(top right) and* Chartergus *(bottom) where the external envelope has been partly removed to show the internal structure of the nest).*

whereas the average distance for the first 20 deposits was about 2 cm from the queen (in a direction orthogonal to the queen's body axis) in the absence of air flow, it was about 1.5 cm from the queen with the air flow. Other experiments indicate the primary role played by the pheromone produced and released by the queen, so that the experiment involving a slow air flow suggests that pheromone is convected along the air current away from the queen in the direction orthogonal to the air stream, leading to a smaller pheromonal template around the queen. As a corollary, it does not seem necessary to invoke a modification of the workers' behaviours to explain this modification in the size of the royal cell: the modified dynamics of the pheromone should be sufficient. A similar conclusion can be drawn to account for a transformation of pillars into walls as an air current is added in the model discussed above: although individual behaviours can in principle be modified by the air flow, and this is not absurd given the extreme sensitivity of termites to air movements, a single individual behaviour can lead to several types of structures in different environmental conditions [49] (see Fig. 8).

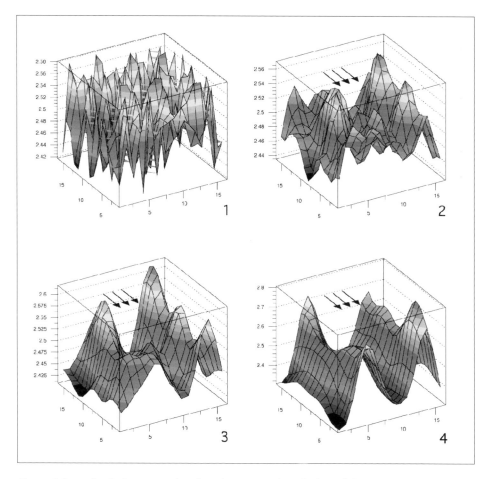

Figure 8 Four simulation steps showing the temporal evolution of the structure built by termites in a 2-D system with a monodirectional air stream along the x direction

This simulation was done with the same individual behavioural laws used without wind (see Fig. 4). Pillars observed in the absence of wind are replaced by walls, and the building face is moved in the direction of the wind.

The origin of nest complexity

Whenever behavioural biologists deal with issues related to collective construction, there is a rather general tendency to explain the production of different structures with different specific behavioural programs. In this way, different environmental conditions would elicit different specific building responses, and then the production of entirely new structures. In other words, it is generally believed that everything has to be "coded" at the individual level. Such hypotheses cannot be rejected, but we think that sometimes there is a tendency to confuse the individual behavioural rules that are neccessary to perform a collective work and the rules that only improve this performance. A whole set of simply quantitative behavioural alterations may occur when environmental heterogeneities are encountered or each time the physical properties of building materials are altered. In this way, as long as the nest's population increases, the space occupied by the colony increases. In fact, there is a strong probability that this growing space contains soil heterogeneities or simply temperature or humidity gradients. Such heterogeneities can greatly affect the individuals' building activity. Indeed, when humidity or temperature conditions are changing, insects display qualitatively different behaviours such as their speed of moving, and that leads to the production of rather different shapes and structures. Similar behavioural rules can then lead to the production of different structures as a result of a simple change of local conditions encountered by insects as the size of the nest increases.

So it seems that the size of a colony could be a first proximal driving element through which the increase of nest's complexity can be explained. Observation of *Lasius niger* ants shows that a change in humidity conditions results in a change in the shape of the craters built at the nest's surface, from a chimneylike shape to that of a flattened disc. If transporting and unloading behaviours remain unchanged whatever the humidity conditions, the cohesion of the building material strongly changes with humidity. When this variation is combined with the ants' unloading behaviour, it leads directly to the observed variety of craters' shapes. Another feature of *L. niger* ants is their hollow nests [51]. Chambers inside the nest have extremely different shapes depending on the environmental places where the nest has been dug. In the upper parts of the nest that are built with the soil particles coming from excavation, all the chambers have a sponge-like structure. On the other hand, in the subterranean part, one can distinguish very clearly a network of galleries and chambers. The spongiform structure of the upper parts can certainly be explained by a softer, less compact soil, where no particular places exist to focus digging activity. The diversity of shapes that arises is a direct consequence of the growth of the nest and of the past activity of the colony. Nest growth can lead as well to the formation of new structures when the distribution of the insects inside the nest is modified or when the signals used by insects are changed.

Thus nest complexity can be explained in part at least by the fact that different building mechanisms (template, self-organization and qualitative stigmergy) can be combined together like so many different elements of a very powerful construction game. In ants and termites, there exist a great number of nest architectures that result from a combination of templates and self-organization. In termites pillar building results from a self-organized process, but the arches joining the pillars are mainly produced using template mechanisms. Some ants species, such as *Leptothorax*, use their compact brood as a template to position and build walls. In this case, the brood functions in same way as the queen's body in termites when they build the royal chamber. However, in this case, the brood gathered in a particular place results form self-organization that precedes the building of walls [52]. One can also find a combination of qualitative stigmergy and templates in wasps. In this case, templates are chemical or gravitational gradients. Let us end with the fact that in all these examples, the solution of the building problem does not result from a preliminary analysis of the situation, followed by exchanges of informations between insects in order to solve the problem. We saw that a few simple behavioural rules lead to coordinated building and the production of amazingly complex architectures without needing to develop a strategy.

Acknowledgements

E.B. is supported by the Interval Research Fellowship at the Santa Fe Institute. G.T. is supported by a grant from the Conseil Régional Midi-Pyrénés. E.B. and G.T. acknowledge partial support from a grant of the Groupement d'Intérêt Scientifique (GIS) Sciences de la Cognition.

References

1 Hansell MH (1984) *Animal architecture and building behavior*. Longman, London

2 Gallais-Hamonno FG, Chauvin R (1972) Simulations sur ordinateur de la construction du dôme et du ramassage des brindilles chez une fourmi (*Formica polyctena*). *C R Acad Sci Paris* 275 D: 1275–1278

3 Sudd JA (1975) A model of digging behavior and tunnel production in ants. *Insect Soc* 22: 225

4 Franks NR, Wilby A, Silverman VW, Tofts C (1992) Self-organizing nest construction in ants: sophisticated building by blind buldozing. *Anim Behav* 44: 357–375

5 Darchen R (1959) Les techniques de la construction chez *Apis mellifica*. PhD thesis, Paris University

6 Belic MR, Skarka V, Deneubourg JL, Lax M (1986) Mathematical model of honeycomb construction. *J Math Biol* 24: 437–449

7 Skarka V, Deneubourg JL, Belic MR (1990) Mathematical model of building behavior of *Apis mellifera*. *J Theor Biol* 147: 1–16

8 Jeanne RL (1975) The adaptativeness of social wasp nest architecture. *Q Rev Biol* 50: 267–287

9 Downing HA, Jeanne RL (1988) Nest construction by the paperwasp *Polistes*: a test of stigmergy theory. *Anim Behav* 36: 1729–1739

10 Downing HA, Jeanne RL (1990) The regulation of complex building behavior in the paperwasp *Polistes fuscatus*. *Anim Behav* 39: 105–124

11 Wenzel JW (1991) Evolution of nest architecture. *In*: KG Ross, RW Matthews, (eds): *Social biology of wasps*. Cornell University Press, Ithaca, NY, 480–521

12 Karsai I, Penzes Z (1993) Comb building in social wasps: self-organization and stig-

mergic script. *J Theor Biol* 161: 505–525

13 Karsai I, Theraulaz G (1995) Nest building in a social wasp: postures and constraints. *Sociobiology* 26: 83–114

14 Theraulaz G, Bonabeau E (1995) Coordination in distributed building. *Science* 269: 686–688

15 Theraulaz G, Bonabeau E (1995) Modeling the collective building of complex architectures in social insects with lattice swarms. *J Theor Biol* 177: 381–400

16 Grassé PP (1959) La reconstruction du nid et les coordinations interindividuelles chez *Bellicositermes natalensis* et *Cubitermes* sp. La théorie de la stigmergie: essais d'interprètation du comportement des termites constructeurs. *Insect Soc* 6: 41–84

17 Grassé PP (1984) Réparation, reconstruction et remaniements internes du nid. Coordination des tâches individuelles et comportement stigmergique. Le déterminisme du comportement constructeur. *In*: PP Grassé (eds): *Termitologia, Tome* II. *Fondation des Sociétés – Construction*. Masson, Paris, 490–577

18 Deneubourg JL (1977) Application de l'ordre par fluctuations à la description de certaines étapes de la construction du nid chez les termites. *Insect Soc* 24: 117–130

19 Bruinsma OH (1979) *An analysis of building behaviour of the termite* Macrotemes subhyalinus. PhD thesis, Lanbouwhogeschool te Wageningen, Netherlands

20 Desneux J (1956) Structures "atypiques" dans les nidifications souterraines d'*Apicotermes lamani* Sj. (Isoptera, Termitidae) mises en évidence par la radiographie. *Insect Soc* 3: 277–281

21 Camazine S (1991) Self-organizing pattern formation on the combs of honey bee colonies. *Behav Ecol Sociobiol* 28: 61–76

22 Camazine S, Sneyd J, Jenkins MJ, Murray

JD (1990) A mathematical model of self-organizing pattern formation on the combs of honey bee colonies. *J Theor Biol* 147: 553–571

23 Maeterlinck M (1997) *The life of the white ant*. George Allen and Unwin, London

24 Bonabeau E, Theraulaz G (eds) (1994) *Intelligence collective*. Hermès, Paris

25 Franks NR (1989) Army ants: a collective intelligence. *Am Sci* March–April: 139–145

26 Pasteels JM, Deneubourg JL, Goss S (1987) Self-organization in ant societies (I): trail recruitment to newly discovered food sources. *In*: JM Pasteels, JL Deneubourg (eds): *From individual to collective behaviour in social insects*. Birkhäuser, Basel, 155–175

27 Beckers R, Deneubourg JL, Goss S, Pasteels JM (1990) Collective decision making through food recruitment. *Insect Soc* 37: 258–267

28 Seeley TD, Camazine S, Sneyd J (1991) Collective decision-making in honey bees: how colonies choose among nectar sources. *Behav Ecol Sociobiol* 28: 277–290

29 Camazine S, Sneyd J (1991) A model of collective nectar source selection by honey bees: self-organization through simple rules. *J Theor Biol* 149: 547–571

30 Aron S, Deneubourg JL, Goss S, Pasteels JM (1990) Functional self-organization illustrated by inter-nest traffic in ants: the case of the argentine ant. *In*: W Alt, G Hoffmann (eds): *Biological motion, lecture notes in biomathematics*, 89: 533–547

31 Deneubourg JL, Goss S (1989) Collective patterns and decision making. *Ethol Ecol Evol* 1: 295–311

32 Deneubourg JL, Aron S, Goss S, Pasteels JM (1989) The self-organizing exploratory pattern of the Argentine ant. *J Insect Behav* 2: 719–725

33 Goss S, Deneubourg JL, Aron S, Beckers R, Pasteels JM (1990) How trail laying and trail following can solve foraging problems for ant colonies. *In*: RN Hughes (eds): *Behavioural mechanisms of food selection*. NATO ASI Series vol G20, Springer, Berlin, 661–678

34 Franks NR, Gomez N, Goss S, Deneubourg JL (1991) The blind leading the blind in army ant raid patterns: testing a model of self-organization (Hymenoptera: Formicidae). *J Insect Behav* 4: 583–607

35 Deneubourg JL, Goss S, Franks NR, Pasteels JM (1989) The blind leading the blind: modeling chemically mediated army ant raid patterns. *J Insect Behav* 2: 719–725

36 Deneubourg JL, Goss S, Pasteels JM, Fresneau D, Lachaud JP (1987) Self-organization mechanisms in ant societies (II): learning in foraging and division of labour. In JM Pasteels, JL Deneubourg (eds): *From individual to collective behaviour in social insects*. Birkhäuser, Basel, 177–196

37 Theraulaz G, Gervet J, Semenoff S (1991) Social regulation of foraging activities in *Polistes dominulus* Christ: a systemic approach to behavioural organization. *Behaviour* 116: 292–320

38 Theraulaz G, Goss S, Gervet J, Deneubourg JL (1991) Task differentiation in *Polistes* wasp colonies: a model for self-organizing groups of robots. *In*: JA Meyer, SW Wilson (eds): *Simulation of adaptive behavior: from animals to animats*. MIT Press/Bradford Books, Cambridge, MA, 346–355

39 Deneubourg JL, Goss S, Franks NR, Sendova-Franks A, Detrain C, Chretien L (1991) The dynamics of collective sorting: robot-like ant and ant-like robot. *In*: JA Meyer, SW Wilson (eds): *Simulation of adaptive behavior: from animals to animats*. MIT Press, Cambridge, MA, 356–365

40 Franks NR, Sendova-Franks AB (1992) Brood sorting by ants: distributing the

workload over the work-surface. *Behav Ecol Sociobiol* 30: 109–123

41 Deneubourg JL, Theraulaz G, Beckers R (1992) Swarm-made architectures. In FJ Varela, P Bourgine (eds): *Toward a practice of autonomous systems, proceedings of the first european conference on artificial life.* The MIT Press, Cambridge, MA, 123–133

42 Bonabeau E, Theraulaz G, Deneubourg JL, Aron S, Camazine S (1996) Self-organization in social insects. *Trends Ecol Evol* 12: 188–193

43 Brian MV (1983) *Social insects: ecology and behavioural biology.* Chapman and Hall, New York

44 Thomé G (1972) Le nid et le comportement de construction de la fourmi *Messor ebenius*, Forel (Hymenoptera, Formicoïdea). *Insect Soc* 19: 95–103

45 Chauvin R (1958) Le comportement de construction chez *Formica rufa. Insect Soc* 5: 273–286

46 Chauvin R (1959) La construction du dôme chez *Formica rufa. Insect Soc* 6: 307–311

47 Chrétien L (1996) Organisation spatiale du matériel provenant de l'excavation du nid chez *Messor barbarus* et des cadavres d'ouvrières chez *Lasius niger* (Hymenopterae: Formicidae). PhD thesis, Brussels University

48 Wilson EO (1975) *Sociobiology.* The Belknap Press of Harvard University Press, Cambridge, MA

49 Bonabeau E, Theraulaz G, Deneubourg JL, Franks NR, Rafelsberger O, Joly JL, Blanco S (1997) A model for the emergence of pillars, walls and royal chambers in termite nests. *Phil Trans R Soc Lond B 353: 1561–1576*

50 Deneubourg J-L, Franks NR (1995) Collective control without explicit coding: the case of communal nest excavation. *J Insect Behav* 8: 417–432

51 Frisch K von (1975) *Animal architecture.* Hutchinson, London

52 Franks NR, Deneubourg JL (1997) Self-organizing nest construction in ants: individual workers behaviour and the nest's dynamics. *Anim Behav* 54: 779–796

Decision-making in foraging by social insects

Claire Detrain, Jean-Louis Deneubourg and Jacques M. Pasteels

Summary

How are foraging decisions determined in social insects? Investigations implemented within the framework of the optimal foraging theory bring evolutionary and functional answers. In this respect, decisions of solitary foragers like bumblebees seem to be ruled by an optimization of the energy (and time) invested among different feeding sites. Similarly, in insects which can forage collectively, like ants or honeybees, decisions have been interpreted in terms of energetic reward assigned to single workers without any reference to recruitment. Evidence, however, supports the idea that (time and energy) investments in recruitment of nestmates can also alter foraging decisions of the individual. Additional questions arise as to how an insect processes information about food resources and environmental constraints and decides whether or not to recruit nestmates. In ants, adaptive collective decisions emerge from numerous interactions among individuals which use local information and follow simple decisional algorithms to modulate their recruiting behavior. The environment itself contributes to the emergence of foraging decisions by altering the dynamics of recruitment and trail reinforcement. Several experimental and theoretical findings will lead us to re-consider the level of complexity of information processing and coding needed for the emergence of adaptive foraging patterns.

Introduction

As argued by Tinbergen [1], behavioral studies should aim to answer four major questions about the causation, current function, development and phylogenetic history of a particular behavior. Most research effort on foraging behavior in

C. Detrain et al. (eds) Information Processing in Social Insects

social insects has concentrated (i) on the functional significance and (ii) on the mechanisms which underlie decision-making processes.

In the present chapter, we aim to synthesize major issues brought by these functional and mechanistic approaches to foraging decisions. The first section will provide an overview of foraging decisions evidenced in social insects exploiting food solitarily and interpreted in the light of the optimal foraging theory. Decision making is here analyzed through cognitive models which often assume, either implicitly or explicitly, that the insect possesses a relatively great capacity to process information. Such individual complexity remains, however, speculative without direct ethological and neurophysiological evidence.

The following sections will refer to insects which can forage collectively. As cooperation and recruitment occurs, additional questions arise as to how foragers decide (or not) to communicate information to nestmates. At first, we will summarize the contribution of the optimization concept to the understanding of recruitment decision in ant and honeybee societies (e.g. how the energy and time costs of recruitment alter the behavior of foragers). Thereafter, we will focus on the proximate mechanisms of decision making evidenced in ants. Indeed, these societies provide us with a system open to analytic studies of the mechanisms involved in foraging. By altering colony components or environmental constraints and by quantifying the individual's response, it is possible (i) to identify cues used by the foragers to take their decisions whether or not to recruit nestmates and (ii) to follow how information is transferred. Bottom-up investigations and simulations on foraging patterns will be reviewed since they enable to link individual decisions to collective ones. We will identify which information on food and environmental constraints are "measured" by scouts and are coded to nestmates by scent trails. In addition, we will ask whether the environment itself might alter the dynamic properties of recruitment and therefore might contribute to the emergence of collective decisions. We will demonstrate that simple decision-making algorithms and local processing of information may account for most foraging strategies of the insect society. The level of complexity in information processing and coding needed for optimal foraging will be discussed.

Decision making in social insects foraging solitarily

When dealing with foraging by solitary insects, investigators questioned how the individual invests energy (and time) among different feeding sites in order to harvest food in the most efficient way. In the ant species known to forage solitarily (e.g. *Cataglyphis* spp. [2, 3] or *Pachycondyla apicalis* [4]), workers leave the nest in the sectors pointing towards food sites and concentrate their effort on rewarding areas. While deciding to exit the nest in a particular direction, each forager relies on reward expectation, which is set by the food availability of the environment.

Similar questions on foraging decisions have arisen in bumblebees. Each insect may visit thousands of flowers, and after each flower visit it has to decide whether to depart from the plant or to stay and visit another flower on the plant. These decisions were often analyzed in the light of optimal foraging theory. This concept was initially developed from studies on vertebrates [5–7] and was extended to encompass social insects as bumblebees which carry food items back to a central place where food is eaten or stored [8]. As regards bumblebees [9, 18], three food parameters susceptible to altering foraging decisions attracted the attention of researchers: the amount of food available, the variability in availability (risk) and the spatial distribution of food resources.

Concerning the amount of available food, bumblebees capitalize on the most rewarding flowers by altering their foraging behavior. Forager's responses to increased amount of pollen [9] or nectar [10] include a decreased likelihood of departure from plants, reduced distances of movements within and between plants, and longer-lasting or more frequent visits to flowers.

As regards the variability in food supply, there is evidence that some bumblebee species are sensitive to risk (variability) in the amount of nectar rewarded [11–13], while others are risk-indifferent [14]. When risk-sensitive bumblebees are faced with two patches of flowers, one of which provides a constant nectar volume and the other a variable volume (with a same average amount), foragers often show preference for the constant type (risk-averse). However, they can become "risk-prone" as the amount of food stored within the nest decreases [15].

The spatial distribution of food resources also influences the foraging behavior of bumblebees. On plants that are characterized by a gradient of nectar amount decreasing from bottom towards the top of inflorescence, the insect moves systematically from one flower to the adjacent upper flower. Hence, the nectar received from the current flower is an accurate predictor of the reward in as-yet unvisited flowers and affects the forager's decision to leave the plant [16–18]. When neither flower arrangement nor nectar distribution show any consistent pattern, bees seem to integrate information from more than a single flower in making departure decisions [19]. Foraging decision rules in bumblebees appear to be straightforward adaptations to patterns of nectar availabilities.

The energetic optimization concept proved to be a stimulating framework to analyze decisions of solitary foragers within an evolutionary perspective. This ultimate approach, however, provides little insight into the mechanisms by which adaptive foraging behavior is achieved. In this respect, one can assume that, in order to forage optimally, each bumblebee has to monitor pollen and nectar availabilities, to attend to the colony needs and possibly to assess the presence of competitors, predators or disturbances. Only theoretical descriptions of information processing can be found in literature which all suggest that bumblebees use complex departure rules and computational algorithms [18, 20, 21]. Experimental evidence is, however, lacking on the capabilities and limitations of insect cognitive machinery.

Decision making in insects foraging collectively: an optimization approach

The impact of individual energetic balance on foraging decisions

When applied to insects which can forage collectively, the optimization approach often dealt with the individual, outside any context of recruitment. In this respect, honeybee foraging behavior (e.g. number of flower visits, flight patterns, handling time on the flower or crop loading) was analyzed exclusively in terms of individual energetic efficiency [22–24] without any reference to cooperation and information sharing between nestmates. Similarly in ants, optimality in prey choices under different food availabilities and external conditions (e.g. temperature) was interpreted in terms of cost and benefit, only for the individual forager [25, 26].

Since ants and honeybees are central place foragers, the impact of food distance on foraging optimization deserved special attention from the investigators. When single honeybees are faced with food patches at different distances from the hive, they seem to fill their crop with a nectar load which maximizes the energetic efficiency [(benefit – cost)/cost] in an average foraging excursion [22, 23]. The use of energetic efficiency as the food profitability criterion might be appropriate for those flying animals in which the foraging success seems more limited by energy expenditure than by time availability [27]. Constraints on foraging decisions are different for terrestrial insects in which the cost of walking per unit of time is much lower than the cost of flight and in which foraging decisions seem less influenced by energetic expenses. Hence, the energetic gain rate [(benefit – cost)/time] appears to be a more important currency than energetic efficiency for foraging ants. This was demonstrated for harvester ants [28], but seems less evident for nectar-feeding species with lower benefit/cost ratios (*Paraponera clavata*, [29]). Besides, the optimal foraging theory predicts that feeding selectivity increases as a function of the distance to a forage site. Though this trend was confirmed in some studies on seed harvesting [30] and leaf-cutting ants [31], similar tests showed no significant change in the range size of seeds collected at different foraging distances [32, 33].

The next section will raise questions about how recruitment and associated energetic/time constraints might influence foraging decisions at both the individual and collective level.

The impact of recruitment costs on foraging decisions

In the case of insects foraging solitarily, one assumes that their decisions take origin in the energetic balance of the forager itself. However, as sociality proceeds in ants or honeybees, foragers become a part of a large, integrated unit composed of related workers: the decision to recruit taken at the individual level ultimately leads the colony to a mean feeding rate higher than in foraging without any communication. One would expect that the individual energetic gain of those foragers that decide to recruit nestmates would be reduced owing to food sharing with nestmates, to the metabolic cost invested in recruitment signals (e.g. bee dances or trail deposits) and to the time spent in recruitment activities. Hence, foraging decisions appear as a trade-off between immediate maximization of the individual energetic reward and prospective higher returns due to collective food exploitation. Evidence for such a trade-off might be found in bees where foragers return from flowers with partially filled crops. Similarly, leaf-cutting ants sacrifice individual food delivery rate to recruit nestmates earlier to higher quality leaves. These seemingly suboptimal behavior might be explained by the long-term advantages of maximizing social efficiency through the exchange of information with nestmates [34–36].

Studies are lacking on recruitment costs, but several observations support the idea that time spent in recruitment of nestmates matters in decision making since it represents a substantial portion of each foraging cycle. In honeybees, large time investments in recruitment, through long-lasting dances, are restricted to food sources with a high profitability [37]. In ants, less frequent and weaker recruitments lead to food sources requiring higher energetic and time investments (e.g. to food further from the nest [38] or connected to the nest by routes with high vegetational cover [28]). Several foraging patterns also appear as a means to reduce the time costs of recruitment. For instance, temporary recruitments are directed towards the permanent network of trails in aphid-tending ants [39] or towards the raid column in army ants [40], which allows a quicker recruitment of extranidal foragers to newly discovered food. Similarly, *P. clavata* foragers match recruiting effort to increase in distances via two mechanisms (i) a reduction in the relative number of foragers that recruit to long-distance resources, and (ii) an extranidal recruitment (and hence reduced recruitment times) of specialized workers that concentrate their foraging on remote resources [41].

All these studies devoted a great deal of attention to the economics and to the functional significance of decision making. Within this perspective, the insect is assumed to balance global properties of food resources and to "measure" the pay-off of a foraging behavior by using the adequate energetic currency. Such an assumption raises questions as to how such a complex and flexible processing of information operates within insects whose cognitive capabilities may be limited. The aim of the next section is to review individual decision rules during food recruitment as well as their consequences on collective foraging behavior.

Decision making in ants' recruitment: a proximate approach

Collective decisions in ants

Foraging decisions in ants are intimately linked to recruitment mechanisms which enable the colony to focus foragers on top-quality sources and to readjust the work allocation when the environment and food profitability change [42, 43]. Key factors which may alter foraging patterns are listed in Tables 1 and 2, while collective decisions and recruitment behaviors reported in literature are briefly summarized. Special emphasis is put on bottom-up studies which try to link individual decision rules and recruitment processes to collective decisions.

Most quoted studies refer to a subset of food and environmental factors (e.g. food quality), whereas data are crucially lacking for other parameters (e.g. for the dynamics of food availability). Collective decisions reported in Tables 1 and 2 always appear as efficient choices which tend either to maximize the energetic return (e.g. selection of the richest source) or to minimize the costs due to foragers' movements (e.g. selection of the shortest route) or due to competitors' pressure (selection of safe sources). However, such conclusions about the adaptive value of foraging decisions are more frequently supported by common sense and intuition than by actual energetic data.

While most studies discuss the ultimate issues of collective decisions, fewer investigate recruitment mechanisms which lead to their emergence. Two main recruiting behaviors which differ in whether they provide directional information deserve attention from investigators (i) invitation displays (body jerking or antennation) to nestmates which stimulate them to leave the nest without directional information, and (ii) behaviors such as trail recruitment which stimulate workers to leave the nest and lead them to food sources.

Recruitment dynamics are then shown to closely depend on the percentages of ants laying a trail (TLA) or inviting congeners in the nest (IA), and on the intensities of invitation behavior (IIB) or chemical trail marking (ICM) performed per recruiting ant.

Most frequently, experiments show that higher rates of foragers mobilized (RMF) to a food source are essentially due to higher percentages of inviting and/or trail-laying ants (Tab. 1). Additional modulation may occur at the individual level (e.g. in the amount of pheromone laid per recruiter), but seems restricted to the coding of some parameters such as the quality of food.

The two following sections will identify parameters measured by the foragers and criteria used to modulate their recruiting behavior. This review of recruitment decision rules should not, however, shade the influence of other factors on the emergence of collective decisions such as the evaporation rate of scent trails. In this respect, assumptions on how the environment itself might contribute to the emergence of adaptive foraging patterns will be presented.

Table 1 *Review of ants' foraging responses to different food variables*

Food variables	Collective decision	Individual recruitement behavior			Species [reference]
		Trail-laying	Invitation	Decision criteria	
Food quality Concentration of sucrose solution	higher RMF to rich food sources	higher percentage of TLA / higher intensity of ICM to rich food sources	higher percentage of IA to rich food sources	sucrose concentration	*Acanthomyops interjectus* [44] *Formica oreas* [45] *Lasius niger* [46] *Monomorium* spp, *Tapinoma* spp. [47] *Myrmica sabuleti* [48] *Paraponera clavata* [49] *Solenopsis geminata S. saevissima* [38, 50, 51] *Tetramorium impurum* [52]
Food weight	higher RMF to heavy prey	higher percentage of TLA to heavy prey	higher percentage of IA and intensity of IIB to heavy prey	resistance to retrieval	*Lasius neoniger* [53] *Myrmica* spp [54, 55, 56] *Paraponera clavata* [29] *Pheidole pallidula* [57, 58, 59]
Food Volume Droplet below the volume of the ant crop	higher RMF to large food droplets	higher percentage of TLA but same intensity of ICM to large droplets	higher percentage of IA but same intensity of IIB to large droplets	ability to fill the crop to a desired volume	*Lasius niger* [60] *Myrmica sabuleti* [56] *Paraponera clavata* [49] *Solenopsis geminata* [38]
Droplet above the volume of the crop	higher RMF to large food droplets	same percentage of TLA and intensity of ICM (excepted for *Paraponera clavata*)	same percentage of IA and same intensity of IIB	ability to fill the crop to a desired volume	*Lasius niger* [60] *Paraponera clavata* [29] *Solenopsis geminata* [51]

Table 1 (continued)

| Food variables | Collective decision | Individual recruitement behavior | | | Species (reference) |
		Trail-laying	Invitation	Decision criteria	
Number of food items	higher RMF to numerous food items	–	same percent-age of IA and same intensity of IIB	–	*Myrmica sabuleti* [56]
Spatial and temporal distri-bution of food	higher RMF to dense and/or persistent food patches	–	–	–	*Messor rufitarsis* [61] *Pogonomyrmex occidentalis* [38]

Collective decisions are related to the recruitment behavior of individuals. Trail-laying behavior, invitation displays within the nest and decision criteria actually used by the individual forager are summarized. Recruitment mechanisms and decision criteria are mentioned if they were evidenced in at least one (but not necessarily all) quoted ant species. Blank cells mean that no data are available on the topic. RMF, rate of mobilized foragers; TLA, trail-laying ants; ICM, individual chemical marking; IA, inviting ants; IIB, individual invitation behavior.

Table 2 Review of ants foraging responses to environmental constraints

Environmental constraints	Collective decision	Individual recruitement behavior			Species (reference)
		Trail-laying	Invitation	Decision criteria	
Number of predators/competitors					
	higher RMF at the safe food source	–	–	–	*Lasius pallitarsis* [62, 63]
Deviation from the nest-food axis					
	higher RMF on routes leading straightfully to food	–	–	–	*Lasius niger* [64]
Light exposure	higher RMF on routes with low trail degradation by UV	–	–	–	*Acromyrmex octospinosus* [65]
Nature of the substrate					
adsorption/evaporation rate of the trail pheromone	higher RMF on substrate with low evaporation rate	same percentage of TLA and same intensity of ICM	–	–	*Lasius niger* [66]
Vegetational cover	higher RMF on routes with low cover	–	–	–	*Pogonomyrmex occidentalis* [66]

For meaning of abbreviations se Table 1 captions.

Individual assessments of food and recruitment decision rules

Though one might suspect that the "correct" assessment of both the environment and available resources would be a matter of importance for colony fitness, the pertinent criteria used by the insect in mediating an adaptive behavior remain largely uninvestigated. Since chemical trails in ants act as stimulating signals increasing the exit of nestmates as well as orientation cues channeling them to feeding areas, each scout can influence the global foraging strategy by deciding whether or not to lay a trail and by modulating the rate of pheromone emission. The present section will review how decision rules and coding of information through chemical trails are implemented in ants.

Coding of food quality

For food droplets differing in sucrose molarity, the intensity of individual recruitment behavior (and hence the amount of trail pheromone laid by each recruiter) increases as a function of the concentration perceived by the forager [38, 44–52]. Though this recruitment decision rule seems trivial, not in terms of physiological abilities of ants but in logical terms, it appears to be an efficient means for the colony to select energetically valuable food sources.

Coding of prey weight

In *Pheidole pallidula* [57–59], for a pile of small prey (e.g. fruit flies), the mobilization of foragers is very slow, whereas a large prey item (e.g. a cockroach) induces a strong recruitment. These foraging patterns can be generated using the following individual rules of thumb. Success in prey-carrying encourages the forager to move on and to lay a weak trail on its way back to the nest. This can possibly lead to slow monopolization of the source when small items are numerous. A failure by the forager to retrieve the prey results in a shortened stay at the food source followed by intense trail recruitment. Recruited ants then gather around the prey, suck its hemolymph or dissect it on the spot into smaller pieces which are retrieved by individual or collective transport. It was demonstrated that the individual decision to recruit depends on a key parameter used by the forager to estimate prey size: prey resistance to retrieval. Though crude in appearance, this tractive resistance is a decision criterion of higher functional value than any sophisticated measure of prey size or weight. Indeed, it provides indirect information at every moment about not only prey size but also prey vitality, current force of cooperating carriers or any factors affecting resistance to retrieval such as microtopography.

Coding of food volume

Several ant species are known to increase their foraging activity to large food sources by increasing the proportion of scouts laying a trail [49, 60] and/or recruiting workers within the nest [49, 56]. Experiments on *Lasius niger* showed that scouts make no absolute measure of food volume. Indeed, when they are faced with sugar droplets of different sizes, their decision to recruit depends simply on their ability to fill their crop to a desired volume [60]. When the source is too small to provide such a volume at once, the ant returns to the nest without laying a trail or goes on exploring for additional droplets. The well-known statement that an unfed ant does not recruit [50] thus appears as one extreme case of a more general decision rule. This simple individual decision rule generates many adaptive responses at the collective level: (i) the dynamics of food recruitment (e.g. the rate of trail reinforcement) is expected to fit to the patch size of droplets (e.g. to the number of aphids) and to be synchronized to their renewal rate (e.g. rate of honeydew production). (ii) As first highlighted by Wilson [50], no overcrowding of foragers will occur at one large food source. As recruitment proceeds, the increasing difficulty (or even inability) of newcoming foragers to reach food and to ingest a desired volume will lead them to delay (or even to suppress) their recruiting activity.

These experimental findings on decision rules lead us to reconsider the idea that collective foraging patterns find their roots in behavioral complexity and extensive cognitive abilities of the individual. Adaptive strategies can emerge from numerous interactions among individuals which use local information and follow simple rules of thumb to modulate their recruiting behavior. The next section will show that the coding itself of information may be unnecessary and that the environment can directly influence recruitment dynamics.

Environment as an agent of the decision making process

The following models will demonstrate that environmental constraints which alter the dynamics of trail reinforcement may contribute to the emergence of adaptive foraging decisions without any explicit measure or coding of information by the foragers.

The distance effect

In theoretical simulations, ants are given the choice between two paths of different lengths leading to two food sources of equal quality. A major assumption of the model is that no modulation of trail laying nor of any other recruiting displays occurs according to distance to the food source (for more details about the model, see legend, Fig. 1). Simulations show that, for a low flow of 0.03 ant/s and for one source placed at a distance twice as long as the nearest food (e.g. dr = 30 cm and dn = 15 cm), this nearest source is preferred in less than 60% of simulations. For a same distance ratio but longer absolute distances (e.g. dr = 120 cm, dn = 60 cm), higher percentages of simulations end with the choice of the nearest source. The choice of the nearest source is also facilitated by high flows of foragers (0.2 ant/s). In this case, even very small differences in distances between sources (e.g. 10%, 17 versus 15 cm) lead to a marked selection of the nearest source (in around 80% of the simulations).

As demonstrated by these simulations, the distance by itself can determine collective choices as far as this parameter significantly alters traveling times and hence rates of trail reinforcement. Evidence for a decrease in the percentage of trail-laying behavior with increasing distances of food from the nest [29, 51] appears more as an improvement rather than a prerequisite for the selection of near food sources by the colony. One could suspect that this trail modulation would be most useful to discriminate between food sources with small or medium distance differences.

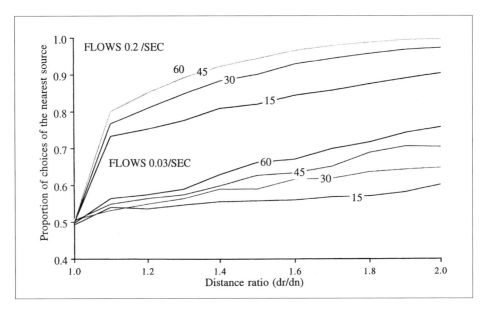

Figure 1 Results of 1000 simulations describing choices of ant colonies between two food sources of equal quality but offered at different distances from the nest

The x-axis gives the distance ratio between the remote (dr) and the nearest food source (dn). A value (15, 30, 45 or 60 cm) is associated to each curve and gives the absolute distance from the nest to the nearest source (dn). The y-axis gives the proportion of simulations where the nearest source is chosen. For each simulation, a food source is considered as chosen when, after 2 h, the corresponding branch bears more than 50% of the ant traffic. These simulations rely on the following model. The probability that an ant will choose the path leading to the nearest source (Pn) is ruled by the equation $Pn = (k + Cn)^n / (k + Cn)^n + (k + Cr)^n$ where Cn and Cr are the trail concentration on the shorter and longer paths respectively. k is a constant standing for all factors other than the trail pheromone (e.g. thigmotactism), which might influence the choice made by foragers. Except for the distance to food, the following parameters were kept fixed with values drawn out of empirical data on Lasius niger: $k = 6$, $n = 2$, the time spent at the food source (100 s), ant velocity (1 cm/s), the lifetime of trail pheromone (1 h). No recruitment of foragers occurs but only a constant input flow of 0.2 or 0.03 forager/s.

The effect of risk and competition

As described for *L. pallitarsis*, food associated with high risk is neglected, and foraging decisions are affected by encounters with a competitor or a predator [62, 63]. Similarly, in *L. neoniger* [67] and *Pogonomyrmex* spp. [68], confrontations between workers of adjacent colonies in a common foraging area will result in their segregation on two newly established trails and in their partitioning on distinct areas.

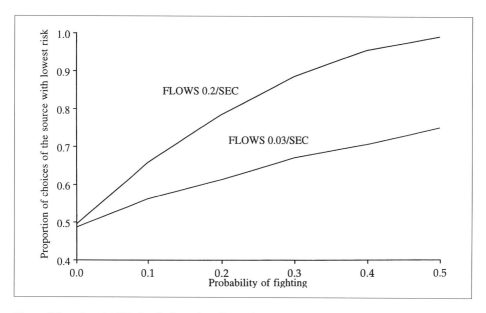

Figure 2 Results of 1000 simulations describing the choices of ant colonies between two food sources of equal quality but exposed to different competition risks
The x-axis gives the probability for each ant to fight with an enemy during food exploitation. The y-axis gives the percentage of simulations in which the safest source was chosen. The model equation is the same as in Figure 1. The probability that an ant will choose the safe source is $Ps = (k + Cs)^n/(k + Cs)^n + (k + Cd)^n$ where Cs and Cd are the trail concentration on paths leading to the safe and dangerous sources, respectively.

Simulations (Fig. 2) demonstrate that this adaptive selection of the safest food might occur without any coding of risk. In this model, ants choose between food sources of equal quality exposed to different probabilities of enemy encounters. Each hostile confrontation is assumed to result in physical blocking: each fighting forager will stay 300 s longer around the food than a nonfighting one before returning to the nest. For a low flow of foragers (0.03 ant/s), the safest side is preferred as the risk associated to the other food increases. A higher flow (0.2 ant/s) induces a more marked selection of the safest food: all simulations end up with a unanimous choice of this source, when the probability of enemy encounters reaches 0.5.

Although the desertion of dangerous food sources occurs simply due to delays in the return of fighting foragers to the nest, the ant society may refine this basic response both on short and long time scales, by achieving defensive recruitments, by mobilizing specialized castes of soldiers [69] or by adapting demography (e.g. by increasing the colony production of soldiers [70]).

Integration of multiple information and complex decision making

In nature, ant foragers encounter complex situations where it may be advantageous to assess a wide array of variables. For instance, as the quality of a food source improves, its attractiveness and hence the occurrence of competitive interactions increases. In these conflict situations, foraging choices may demand the sacrifice of some amount of foraging success in order to achieve other goals as well. In *L. pallitarsis* and *Myrmica incompleta* colonies [71, 72], the use of the higher-quality patch is depressed as an associated mortality risk increases. The authors assume that this trade-off is related to an integrated balance between nutritive returns and losses of workers due to predation. However, they give no idea how sensory inputs about quality and risk may be processed by the forager and how collective decisions may emerge. On the one hand, one can speculate about a recruitment trail which might be a complex blend of attractive and repulsive signals (see contribution of Stickland et al., this volume) and which reflect both the quality and the danger of feeding in the patch perceived by the forager. An alternative way to explore trade-offs in ants is through models based on the minimalist assumption that no sophisticated coding of information occurs at the individual level (Fig. 3, J.C. Romond et al., unpublished data). Simulated ant colonies are faced with two food sources which differ in quality (sucrose concentration of S1 > concentration of S2) as well as in the risk of fatal encounters at the richest one (Pm: probability of mortality at S1). The model assumes that the amount of pheromone laid per recruiter (Q1,Q2) is modulated according to sucrose concentration only and that no additional coding of danger occurs. For a given ratio of pheromone quantities laid from the poorest and the richest food (e.g. Q2/Q1 = 0.6), the most profitable (in terms of energetic return) source S1 remains preferentially exploited as long as predation pressure is low (Pm < 0.4, Fig. 3a). As the mortality risk increases, less than half of colonies forage at the source S1 and even completely abandon this rich but dangerous source (Pm > 0.7). It should be stressed that the desertion of the most profitable food S1 shifts to higher predation pressure as the difference increases between the energetic returns expected from the rich and poor sources (to Pm values >0.9 for

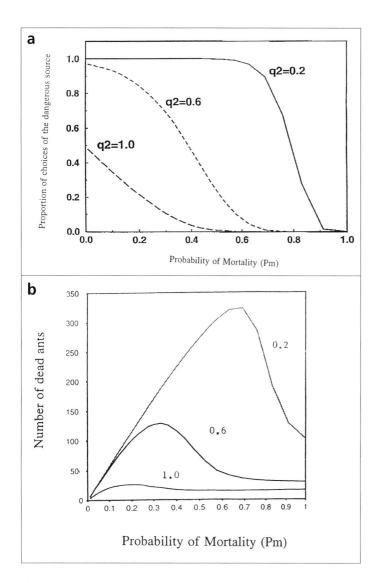

Figure 3

(a) Results of 1000 simulations describing the choices of ant colonies between two food sources (S1 and S2) which differ in both quality (sucrose concentration of S1 > concentration of S2) and associated predation risks (no risk at source 2, probability of mortality Pm at source 1 between 0 and 1). The pheromone quantities laid from the poorest and the richest food source are, respectively, Q2 and Q1. The x-axis stands for the mortality risks (Pm) at the richest food S1. The y-axis gives the proportion of simulations in which the safest source was chosen. Values (1, 0.6 or 0.2) associated with each curve give the ratio Q2/Q1, Q1 being kept constant (= 1). The model underlying these simulations is basically the same as in Figure 2. (b) Same as in (a). The y-axis gives the mean number of dead ants found at the end of simulations.

Q2/Q1 = 0.2). As regards the number of dead foragers, it does not simply increase with risk but shows a maximum value followed by a decrease in mortality when ants undergo severe predation (Fig. 3B). This counterintuitive result can be explained as follows: when the higher trail-laying intensity to the rich but dangerous food source no longer makes up for the loss of recruiters, it becomes deserted, resulting in a decrease in global mortality. Peak values of foragers' death are observed when despite a high predation pressure, the colony is "trapped" in its first choice of the most profitable source and is unable to shift its activity to the safest one. As the difference in profitabilities (and hence in the amounts of trail pheromone laid per recruiter) between the two sources increases, mortality peak value shifts to higher predation pressures, resulting in higher loss of foragers (e.g. for Q2/Q1 = 0.2, maximum number of dead foragers = 300 versus 100 for Q2/Q1 = 0.6, Fig. 3B).

In the foregoing sections, foraging decisions were studied in the laboratory on a limited subset of food parameters while trade-offs were investigated by theoretical simulations dealing with two parameters only. However, under natural conditions, a colony frequently has to choose between multiple (more than two) food sources, whereas recruitments may be altered by several factors such as by food reserves stored within the nest or by the network of preexisting trails. Hence, we might wonder whether the simplicity in behavioral mechanisms evidenced so far is compatible with the complexity and flexibility of foraging patterns shown by ants in nature. We believe this to be the case due to the property of chemical trails which is to convey information, dynamic in both time and space. Spatiotemporal plasticity in foraging trails is due to the fact that increased amounts of pheromone induce not only higher rates of nest exits but also a higher directionality of recruited nestmates. As a result, either the succession of trail reinforcements to long-lasting food sources or the spatial proximity of individual trails to clumped food items will elicit an adaptive shift from nondirectional foraging to oriented recruitment and will allocate more foragers to areas where the expectancy for food is higher. Even more complex spatial patterns like the highly coordinated swarms of army ants or the rotation of trails in harvester ants may theoretically arise without any centralized control of foraging from simple decision rules and autocatalytic trail reinforcements [42, 73, 74].

Conclusion: is foraging optimization dependent on complexity of decision-making?

One can examine foraging in social insects without investigating mechanisms by which optimal decisions are achieved. However, the advantages of a proximate approach is that it helps: to clarify the pertinent variables used by the insect to modulate its behavior, and to determine what level of physiological and cognitive complexity is required at the individual level to achieve adaptive strategies. Researchers brought, either implicitly or explicitly, several answers which vary greatly in the level of complexity assumed.

At first sight, the integration of all pertinent variables by the animal might be seen as a prerequisite for optimal foraging strategy. In the case of social insects foraging collectively, an exhaustive and precise coding of information might be favored by natural selection, since it increases colony efficiency in recruitment and task performance. However, several constraints might limit this behavioral complexity as for example:

1) The physical, physiological and cognitive abilities of the insect. The individual has an experience limited in time and space which cannot follow all fluctuations of an unpredictable environment. Each forager is biologically limited in its perception of environmental cues as well as in its integrative power of multiple signals at high neuronal networks.
2) The time costs spent in assessing food or environmental features. One might suspect that the more precise the assessment of food, the more careful its inspection carried out by foragers. Hence, complex individual assessments which are time-consuming might expose foragers to higher predation risks, delay nestmates' recruitment and prevent food monopolization against competing species.
3) The risk for erroneous transmission of information to nestmates, which potentially increases with the complexity of individual assessment and the multiplicity of coding vectors.

An alternative scheme to behavioral complexity is that individual decisions rely on a small set of simple and functional criteria. Instead of assessing multiple parameters, foragers measure and hence communicate to nestmates only a few cues (e.g. prey resistance to traction) which most pertinently summarize food properties (e.g. weight) and environmental constraints (e.g. microtopography). All cues evidenced so far were indirect measures of energetic benefit and/or expenses: they allow the individual to make optimal decisions and to save time during food exploitation due to limited processing of information. Moreover, any parameter (e.g. food distance from the nest) which influences the dynamic properties of recruitment can be an intimate part of the decision-making process and hence does not need to be specifically measured or coded by foragers, at least in theory.

Though we do not deny the possible existence of more sophisticated decision making, we point out that adaptive foraging strategies can emerge out of simple decision rules through the iterative processes of recruitment and trail reinforcement. As they have been determined by their particular evolutionary history, the decision rules evidenced in ants do not apply in their details to any other group-living animals, but analogous mechanisms underlie the functioning of societies like honeybees [37, 75], tent caterpillars (see Fitzgerald and Costa, this volume) or even mole and Norway rats [76, 77]. From the examples described above, we believe it will be possible to conciliate some "panglossian" view of foraging optimization with simplicity in the decision-making processes of social insects.

Acknowledgments

We would like to thank S. Camazine for his critical reading of the manuscript. This work was supported by the Belgian Fund for Joint Basic Research (grant no. 2451393F). C. Detrain and J. L. Deneubourg are research associates from the Belgian National Fund for Scientific Research.

References

1 Tinbergen N (1951) *The study of instinct.* Oxford University Press, Oxford

2 Wehner R, Harkness RD, Schmid-Hempel P (1983) Foraging strategies in individually searching ants *Cataglyphis bicolor* (Hymenoptera: formicidae). *In*: M Lindauer (ed): *Information processing in animals.* G. Fischer, Stuttgart, 1–79

3 Schmid-Hempel P (1987) Foraging characteristics of the desert ant *Cataglyphis. In*: J Pasteels and JL Deneubourg (Eds): *From individual to collective behavior in social insects.* Birkhäuser, Basel, 43–63

4 Fresneau D (1985) Individual foraging and patch fidelity in a ponerine ant. *Insect Soc*

32: 109–116

5 Maynard Smith J (1978) Optimization theory in evolution? *Annu Rev Ecol Syst* 9: 31–56

6 Stephen DW, Krebs JR (1986) *Foraging theory*. Princeton University Press, Princeton

7 Krebs JR, Davies NB (1992) *Behavioural ecology: an evolutionary approach*. Blackwell Scientific Publications, Oxford

8 Orians GH, Pearson NE (1979) On the theory of central place foraging. *In*: DJ Horn, GR Stairs, RD Mitchell (eds): *Analysis of ecological systems*. Ohio State University Press, Columbus

9 Harder LD (1990) Behavioral responses by bumble bees to variation in pollen availability. *Oecologia* 85: 41–47

10 Chittka L, Gumbert A, Kunze J (1996) Foraging dynamics of bumble bees: correlates of movements within and betweeen species. *Behav Ecol Sociobiol* 8(3): 239–249

11 Waddington KD, Allen T, Heinrich B (1981) Floral preferences of bumblebees (*Bombus edwardsii*) in relation to intermittent versus continuous rewards. *Anim Behav* 29: 779–784

12 Harder LD, Real LA (1987) Why are bumble bees risk averse. *Ecology* 68(4): 1104–1108

13 Kacelnik A, Bateson M (1996) Risky-theories: the effects of variance on foraging decisions. *Amer Zool* 36: 402–434

14 Waddington (1995) Bumble bees do not respond to variance in nectar concentration. *Ethology* 101: 33–38

15 Cartar RV, Dill LM (1990) Why are bumble bees risk sensitive foragers? *Behav Ecol Sociobiol* 26: 121–127

16 Pyke GH (1978) Optimal foraging: movement patterns of bumble bees between inflorescences. *Theor Pop Biol* 13: 72–98

17 Hodges CM (1985) Bumblebees foraging: the threshold departure rule. *Ecology* 66: 179–187

18 Pleasants JM (1989) Optimal foraging by nectarivores: a test of marginal value theorem. *Amer Naturalist* 134: 51–71

19 Waddington KD (1980) Flight patterns of foraging bees relative to density of artificial flowers and distribution of nectar. *Oecologia* 44: 199–204

20 Real LA (1992) Information processing and the evolutionary ecology of cognitive architecture. *Amer Naturalist* 36: 518–529

21 Real LA (1996) Paradox, performance and the architecture of decision-making in animals. *Amer Zool* 36: 518–529

22 Schmid-Hempel P, Kacelnik A, Houston AI (1985) Honeybees maximize efficiency by not filling their crop. *Behav Ecol Sociobiol* 17: 61–66

23 Schmid-Hempel P (1987) Efficient nectar-collecting by honeybees. I. Economic models. *J Anim Ecol* 56: 209–218

24 Schmid-Hempel P, Schmid-Hempel R (1987) Efficient nectar-collecting by honeybees. II. Responses to factors determining nectar availability. *J Anim Ecol* 219–227

25 Traniello JF, Fujita MS, Bowen RV (1984) Ant foraging behavior: ambient temperature affects prey selection. *Behav Ecol Sociobiol* 15: 65–68

26 Traniello JF (1987) Social and individual responses to environmental factors in ants *In*: J Pasteels and JL Deneubourg (eds): *From individual to collective behavior in social insects*. Birkhäuser, Basel, 63–80

27 Heinrich B (1975) Energetics of pollination. *Annu Rev Ecol Syst* 6: 139–170

28 Fewell JH (1988) Energetic and time costs of foraging in harvester ants *Pogonomyrmex occidentalis*. *Behav Ecol Sociobiol* 22(6): 401–408

29 Fewell JH, Harrison JF, Lighton JR, Breed M (1996) Foraging energetics of the ant *Paraponera clavata*. *Oecologia* 105: 419–427

30 Davidson DW (1979) Experimental tests of the optimal diet in two social insects. *Behav Ecol Sociobiol* 4(1): 35–41

31 Roces E (1990) Leaf-cutting ants cut fragment sizes in relation to distance from the nest. *Anim Behav* 40: 1181–1183

32 Baroni-Urbani C, Nielsen MG (1990) Energetics and foraging behaviour of the European harvesting ant, *Messor capitatus* (Latreille). Do ants really optimize their harvesting? *Physiol Entomol* 15: 449–461

33 Wetterer JK (1992) Source distance has no effect on load size in the leaf-cutting ant. *Atta cephalotes. Psyche* 98: 355–359

34 Nunez JA (1982) Honeybee foraging strategies at a food source in relation to its distance from the hive and the rate of sugar flow. *J Apicult Res* 21: 139–150

35 Nunez JA, Giurfa M (1996) Motivation and regulation of honey bee foraging. *Bee World* 77 (4): 182–196

36 Roces F, Nunez JA (1993) Information about food quality influences brood size selection in recruited leaf cutting ants. *Anim Behav* 45: 135–143

37 Seeley TD (1995) *The wisdom of the hive.* Harvard University Press, Cambridge, MA

38 Taylor F (1977) Foraging behavior of ants: experiments with two species of myrmicine ants. *Behav Ecol Sociobiol* 2: 147–167

39 Quinet Y, Pasteels JM (1991) Spatio-temporal evolution of the trail network in *Lasius fuliginosus* (Hymenoptera, Formicidae). *Belg J Zool* 121: 55–72

40 Chadab R, Rettenmeyer CW (1975) Mass recruitment by army ants. *Science* 188: 1124–1125

41 Fewell JH, Harrison JF, Stiller TM, Breed MD (1992) Distance effects on resource profitability and recruitment in the giant tropical ant *Paraponera clavata Oecologia* 92: 542–547

42 Deneubourg JL, Goss S (1989) Collective patterns and decision-making. *Ethol Ecol Evol* 1: 295–311

43 Beckers R, Deneubourg JL, Pasteels JM (1990) Collective decision-making through food recruitment. *Insect Soc* 37(3), 258–267

44 Hantgartner W (1970) Control of pheromone quantity in odor trails of the ant *Acanthomyops interjectus. Experientia* 26: 664–665

45 Crawford DL, Rissing SW (1983) Regulation of recruitment by individual scouts in *Formica oreas* Wheeler (Hymenoptera, Formicidae). *Insect Soc* 30(2): 177–183

46 Beckers R, Deneubourg JL, Goss S (1993) Modulation of trail-laying in the ant *Lasius niger* (Hymenoptera: Formicidae) and its role in the collective selection of a food source. *J Insect Behav* 6 (6): 751–759

47 Szlep R, Jacobi T (1967) The mechanism of recruitment to mass foraging in colonies of *Monomoriuum venustum, M. subopacum* ssp. *phoenicum, Tapinoma israelis* and *T. simothi* v. *phoenicum. Insect Soc* 1: 25–40

48 de Biseau JC, Deneubourg JL, Pasteels JM (1991) Collective flexibility during mass recruitment in the ant *Myrmica sabuleti* (Hymenoptera: Formicidae). *Psyche* 98(4): 323–336

49 Breed MD, Fewell JH, Moore AJ, Williams KR (1987) Graded recruitment in a ponerine ant. *Behav Ecol Sociobiol* 20: 407–411

50 Wilson EO (1962) Chemical communication among workers of the fire ant *Solenopsis saevissima* (Fr Smith): the organization of mass foraging. *Anim Behav* 10: 134–147

51 Hantgartner W (1969) Structure and variability of the individual odor trail in *Solenopsis geminata* (Hymenoptera; Formicidae). *Z Vergl Physiol* 62: 111–120

52 Verhaeghe JC (1982) Food recruitment in *Tetramorium impurum. Insect Soc* 29: 67–85

53 Traniello JFA (1983) Social organization and foraging success in *Lasius neoniger*

(Hymenoptera:Formicidae): behavioral and ecological aspects of recruitment communication. *Oecologia* 59: 94–100

54 Cammaerts MC (1980) Systèmes d'approvisionnement chez *Myrmica scabrinodis* (Formicidae). *Insect Soc* 27(4): 328–242

55 Cammaerts MC, Cammaerts R (1980) Food recruitment strategies of the ants *Myrmica sabuleti* and *Myrmica ruginodis*. *Behav Process* 5: 251–270

56 de Biseau JC, Pasteels JM (1994) Regulated food recruitment through individual behavior of scouts in the ant *Myrmica sabuleti* (Hymenoptera: Formicidae). *J Insect Behav* 7(6): 767–777

57 Detrain C, Deneubourg JL (1997) Scavenging by *Pheidole pallidula*: a key for understanding decision-making systems in ants. *Anim Behav* 53: 537–547

58 Detrain C, Pasteels JM, Deneubourg JL, Goss S (1990) Prey foraging by the ant *Pheidole pallidula*: decision-making systems in food recruitment. *In*: GK Veeresh, B Mallik, CA Viraktamah (eds): *Social insects and the environment*. Oxford and IBH Publishing, New Delhi, 500–501

59 Detrain C, Pasteels JM (1991) Caste differences in behavioural thresholds as a basis for polyethism during food recruitment in the ant *Pheidole pallidula* (Nyl.). *J Insect Behav* 4(2): 157–176

60 Detrain C, Mailleux AC, Deneubourg JL (1997) Coding of food volume in the ant *Lasius niger*. *Ethology* 32: 183

61 Hahn M, Maaschwitz U (1985) Foraging strategies and recruitment behaviour in the European harvester ant *Messor rufitarsis*. *Oecologia* 68: 45–51

62 Nonacs P (1990) Death in the distance: mortality risk as information for foraging ants. *Behaviour* 112: 23–35

63 Nonacs P, Dill LM (1988) Foraging response of the ant *Lasius pallitarsis* to food sources with associated mortality risk.

Insect Soc 35: 293–303

64 Beckers R, Deneubourg JL, Goss S (1992) Trails and U-turns in the selection of a path by the ant *Lasius niger*. *J Theor Biol* 159: 397–415

65 Therrien P, Mc Neil JN, Wellington WG, Febvay G (1987) Ecological studies of the leaf-cutting ant, *Acromyrmex octospinosus* in Guadeloupe. *In*: CS Logren, FRK Vander Meer (eds): *Fire ants and leaf-cutting ants: biology and management*. 172–183

66 Natan C (1997) Rôle de l'exploration et du substrat dans l'organisation du réseau de pistes chez *Lasius niger*. Mémoire de licence-Université Libre de Bruxelles

67 Traniello JFA (1987) Chemical trail systems, orientation, and territorial interactions in the ant *Lasius neoniger*. *J Insect Behav* 2: 339–354

68 Hölldobler B (1974) Home range orientation and territoriality in harvesting ants. *Proc Natl Acad Sci USA* 71: 3274–3277

69 Detrain C, Pasteels JM (1992) Caste polyethism and collective defense in the ant *Pheidole pallidula*: the outcome of quantitative differences in recruitment. *Behav Ecol Sociobiol* 29: 405–412

70 Passera L, Roncin E, Kaufmann B, Keller L (1996) Increased soldier production in ant colonies exposed to intraspecific competition. *Nature* 379: 630–631

71 Nonacs P, Dill LM (1990) Mortality risk versus food quality tradeoffs in a common currency: ant patch preferences. *Ecology* 71(5): 1886–1892

72 Nonacs P, Dill LM (1991) Mortality risk versus food quality trade-offs in ants: patch use over time. *Ecol Entomol* 16: 73–80

73 Deneubourg JL, Goss S, Franks N, Pasteels JM (1989) The blind leading the blind: modeling chemically mediated army ant raid patterns. *J Insect Behav* 2: 719–725

74 Goss S, Deneubourg JL (1989) The self-

organizing clock pattern of *Messor pergandei. Insect Soc* 36(4): 339–346

75 Camazine S, Sneyd J (1991) A mathematical model of colony level nectar source selection by honey bees: self-organization through simple rules. *J Theor Biol* 149: 547–571

76 Judd TM, Sherman PW (1996) Naked mole-rats recruit colony mates to food sources.*Anim Behav* 52: 957–969

77 Galef B, Buckley L (1996) Use of foraging trails by Norway rats. *Anim Behav* 51: 765–771

The mystery of swarming honeybees: from individual behaviors to collective decisions

P. Kirk Visscher and Scott Camazine

Summary

Thousands of individuals in a honeybee swarm make a collective decision for one among many nest sites discovered. We recorded the waggle dances on swarms in a forested area, where one swarm's search encompassed about 150 km^2 and discovered about 50 different sites. We also analyzed swarms in a more controlled situation, with only nest sites which we provided and monitored. Most bees did not visit any site; very few visited more than one. Apparently choices were made with little or no direct comparison, through the interaction of two mechanisms: positive feedback, through recruitment leading to growth in the number of scouts visiting good nest sites, and attrition-reducing activity and recruitment for nonchosen sites. Individual differences between bees substantially affected these dynamics. Scouts varied considerably in the amount of dancing and persistence, but most that danced did so vigorously for a site after their first few visits, then ceased, though continuing to visit. Scouts followed dances of others, and occasionally visited alternative sites, but rarely switched their dancing. Our results suggest that the choice among nest sites relies less on direct comparison of nest sites, and more on inherent processes of feedback and attrition.

Introduction

Periodically, a colony of honeybees, having outgrown the confines of its nest cavity, will swarm, and the old mother queen and approximately half of the colony's inhabitants depart to find a new home. This process of reproductive swarming is a critical moment in the colony's life cycle. After leaving the old nest, the swarm generally sets down for several days, usually on a tree branch, and begins a deci-

C. Detrain et al. (eds) Information Processing in Social Insects
© 1999, Birkhäuser Verlag Basel/Switzerland

sion-making process that will have enormous consequences in terms of the survival and fitness of the swarm. Some of the workers act as scouts, flying throughout the surrounding countryside in search of appropriate cavities for the new bee colony. A scout returning to the swarm after inspecting a high-quality site communicates that information to other bees by means of waggle dances performed on the surface of the cluster. These dances encode the distance and direction to the site.

The most distinctive and mysterious feature of this process is that it is a collective decision. Since a swarm has many scouts which discover an array of potential nest sites, dances for alternative sites will be performed simultaneously on the swarm. However, in time, all the scouts reach a consensus, and only dances for a single site are advertised. Shortly thereafter, the cluster abruptly breaks up, and a cloud of bees takes off for its new home.

Although swarming has been the subject of several studies, we know very little about how a swarm reaches its collective decisions. Few, if any, of the bees will have the opportunity to acquire information from multiple nest sites on the many variables affecting site selection. Nonetheless, as a group the bees of the swarm are able to consistently evaluate many characteristics of the site which affect its suitability as a shelter for their future nest, and use that information to come to a unanimous agreement on a single site.

Many aspects of swarming involve the type of collective decision-making processes which form a focus of this book. The key feature of all these collective processes is the acquisition and processing of information by the individual social insects. To fully understand swarming will require a detailed examination of how the bees process and act upon information acquired during scouting. Many questions remain unanswered: (i) How does the colony decide when to swarm? (ii) What triggers the abrupt exodus of approximately half the colony's inhabitants? (iii) What determines which individuals remain with the new queen and which leave with the old queen? (iv) How do scouts on the swarm find and evaluate potential nest sites? (v) How does the swarm choose and reach agreement among alternative nest sites? (vi) How does the group know when a collective decision has been reached? (vii) What triggers the swarm to take off for the selected site? (viii) How do the bees orient as a group while flying to the chosen nest site whose location is unknown to a majority of the bees?

In this chapter, we focus on question (v): How does the swarm choose and reach agreement among alternative nest sites?

Previous studies

The first studies to address how a swarm finds and evaluates nest sites were conducted by Lindauer [1–3] and these were followed up by the studies of Seeley and Morse [4] and Seeley [5], who examined which properties of nest sites scout bees evaluate, and how they measure the volume of the nest cavity. These studies showed that scouts evaluate many properties of nest sites, including volume, exposure, entrance size, and height from the ground, and show consistent preferences in these properties.

Lindauer [3] also investigated which bees participate in househunting as scouts, and reported that they were often bees which had been nectar foragers in the preswarming bee colony. Seeley et al. [6] estimated that about 5% of the bees in the swarm participate in scouting nest sites, with only about one-third of the scouts seen to visit the nest box participating in dancing.

Lindauer was particularly interested in how the swarm reaches a unanimous collective decision for a single nest site. In his classic 1955 study, he presented observations on the decision-making processes of 19 swarms. He obtained detailed records of the recruitment activity of the scouts by marking them and observing their dances on the swarm cluster. Figure 1 represents a typical record of the "debate" leading to consensus for a single site. During the first 3 days, no sign of agreement could be seen. On the 4th day, interest picked up for a site 350 m to the southeast. Over the course of the day, activity became increasingly focused on this site, whereas interest for the other sites waned. The swarm did not take off, however, until the 5th day when "all competitors had been silenced and an agreement had been reached" [7].

Based upon observations such as these, Lindauer [7] summarized his findings in the following hypothesis: "the better the qualities a nesting place exhibits, the livelier and longer will be the messengers' dance after the inspection. In this way new messengers are recruited in the cluster for this place, and then they too solicit by means of the same lively dances. If those scouting bees which at first had only inferior or average dwellings to announce are persuaded by the livelier dances of their colleagues to inspect the other nesting place, then nothing more stands in the way of an agreement. They can now make a comparison between their own and the new nesting place, and they will solicit in the cluster for the better of the two."

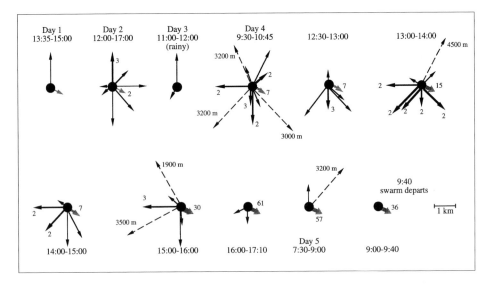

Figure 1 Record of the decision-making activities of one of Lindauer's [3] swarms (Eckschwarm) North is up, the length of each arrow indicates the distance to the site being advertised, and the number next to the arrow corresponds to the number of dancers for that site marked during the period of observation. (Arrows without numbers indicate a single dancer.) The site that was chosen by the swarm is shown with grey arrows.

Lindauer's hypothesis has two components. First, there is positive feedback for good sites, in which scouts recruit more scouts who in turn recruit more, potentially leading to exponential increases in scout numbers. Second, there is comparison of alternative sites by individual scouts, in which scouts that have evaluated nest sites follow dances for others, visit them, and recruit for the one they find superior.

How is the consensus for a single site perceived? To a human observer, the emergence of a consensus is apparent as the number of different sites advertised by dances on the swarm surface diminishes, until all of the dances indicate the same site. Some time after this occurs, the bees seem to sense that a decision has been made, and many scouts begin to perform "schwirrlaufen" [3] or buzz-running on the nest sites, which apparently signals the bees at the nest site to return to the swarm [3, 6], and also on the swarm, which appears to cause the cluster to break apart and take to the air. The mechanism which informs scouts that a decision has been reached remains unknown; it may involve perceiving unanimity in the dances, or buildup of scouts at the nest box to critical densities, or other cues on either the swarm cluster or nest boxes.

Two additional aspects of the househunting process involve coordinated collective behavior. First, how is the takeoff triggered? Probably relatively few bees participate in the decision-making process. These bees then need to activate the entire colony from a quiescent cluster to an airborne swarm. Lindauer [3] described bees performing buzz runs, burrowing through the swarm cluster and butting into other bees as the critical signal, but there are also components as well. Near the time of takeoff, bees in the swarm emit a piping noise [8] that may also activate the bees.

Second, how do swarms orient during their flight to the selected nest cavity? The flying swarm consists largely of bees which have never been to the nest site, and most have probably not followed dances indicating it, so the mechanism by which the swarm is steered remains a challenging, and unanswered question. Lindauer [3] hypothesized that once airborne, the swarm is directed toward the chosen nest site by scouts who streak rapidly through the airborne swarm cloud in the appropriate direction. Seeley et al. [6] also observed some bees flying through the swarm cluster which may have been "streakers," and Avitabile et al. [9] suggested that pheromones might also be used by scout bees in leading the swarm. However, no experimental work has been done to test these hypotheses.

Experimental methods

For these studies, one needs a readily available supply of honeybee swarms which are in the househunting mode of behavior. At certain times of the year, natural swarms can be obtained as they emerge from strong colonies. However, this source is unpredictable—at the whim of weather conditions, location, season, and nectar availability. Years ago, it was shown that artificial swarms can be prepared from established colonies by shaking bees from their combs into a screened cage [3, 9, 10]. The colony's queen is then placed in a small cage with several worker bees, and suspended this inside the larger cage, where the bees cluster around it. The swarm is fed a 50% sugar solution over the next 3 to 4 days. Within several days, almost all the bees in the swarm have full crops, and wax scales are seen on the abdomens of many bees. Artificial swarms prepared by this method can be placed on a suitable stand and will behave like normal swarms (Fig. 2). Scout bees seek new nest sites, and advertise the sites with waggle dances on the swarm. Such swarms take off for the selected nest site, but unless

the queen has been released, the swarm returns several minutes later and reclusters around her.

Under natural conditions, a swarm of bees will discover many potential nest sites over a wide area. Under these conditions, it would be impossible to make direct observations of scout behavior at the potential nest sites. Therefore, to study the house-hunting process at both ends—at the swarm and at the nest sites—and to manipulate the system experimentally, we located an area nearly devoid of good nest cavities in an uninhabited desert approximately 20 km east of Indio, California. Trees in the area are too small to provide nesting cavities, and there are almost no human structures. Around our study site, the soil is too sandy to contain large underground cavities. Most potential nest sites are several kilometers from the study site in rock outcroppings at the base of the Orocopia mountains. As a result, bees will reliably scout and move to nearby sites we provide.

As nesting sites, we offered the bees empty commercial beehive bodies, with plywood tops and bottoms screwed in place. These have an internal volume of approximately 40 liters, and a single round entrance hole about 4 cm in diameter. These boxes had previously been occupied by honeybee colonies and contained residues of wax and propolis. Prior work [11] has shown that such previously occupied cavities are especially attractive, and these were readily accepted as nesting sites when placed in the shaded crotches of palo verde trees within several hundred meters from the swarm. We conducted many such experiments, with swarms whose populations ranged from approximately 2500 to 7500 bees. We named each swarm for relevant events and surroundings.

To obtain detailed records of the activities of individual scout bees, it is essential to be able to individually identify particular bees. For small numbers of bees, this is easily accomplished by marking the bees with spots of colored paint [12] dabbed on the thorax or abdomen. When marking large numbers of bees, we used tiny plastic tags glued to the thorax. The tags come in five colors, numbered 0–99, and when used in combination with colored paint spots on the abdomen, we can keep track of thousands of different bees in a swarm.

Figure 2 Artificial swarm on swarm stand
By placing the swarm on a flat, wooden surface, the swarm assumes a flattened structure making it easier to "read" the recruitment dances. In addition, translucent side and upper panels are placed around the swarm to prevent the dancing bees from directly viewing the sun. This assures that the dances will be oriented with respect to gravity, making it easier for human observers to interpret the direction to the nest site.

As a permanent record, we made VHS videotape recordings of all the activity on the swarms. To keep track of time to the nearest minute, we mounted a clock on the face of the swarm stand, next to the bees, within the field of view of the video camera (as shown in Fig. 2). Even though the bees of interest were individually marked, the resolution of the video camera was not sufficient to distinguish among the marked bees. Therefore, at intervals of 1 min or less, an observer at the swarm pointed to each marked bee and announced its name for the audio portion of the videotape.

Studies of collective decision making by swarms

Collective behaviors

Nest site selection under natural conditions

To carefully observe and document the group-level process of decision making, we set up an artificial swarm of approximately 7240 bees in a forested area of upstate New York, and observed the activity of bees on the cluster. The overall pattern of househunting is illustrated in Figure 3. In this forested area the bees found and recruited to a large number of nest sites. Lumping fairly broadly the scattered sites in Figure 3, we estimate that the scout bees danced for at least 25 sites over the course of 2 days, ranging in distance from 554 to 9520 m from the swarm cluster. The distances over which these swarms found and advertised nest sites were larger than those reported in Lindauer's 1955 studies. Figure 4 shows the distribution of distances in the dance information during the first part of nest site selection, until the time at which there was nearly a consensus on one site. For comparison, this same information is presented for the five swarms househunting in an urban area for which Lindauer [3] reported dance information.

The swarm as an information center

Like a honeybee colony in an established nest, a swarm of bees serves as an information center in which the findings of many scouts are disseminated at a central location. Figure 5 shows data from an experiment in which we set up an

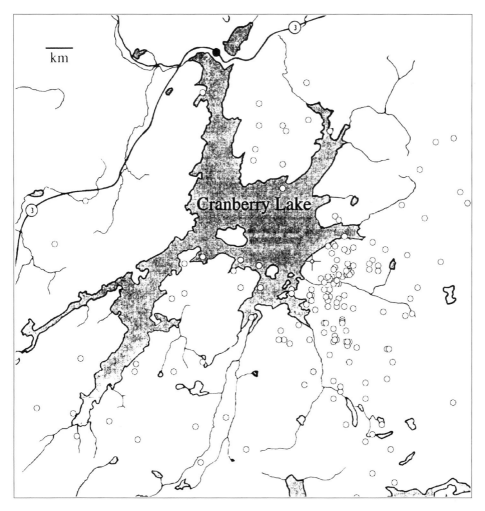

Figure 3 The overall (collective) pattern of house-hunting by a swarm of over 7000 bees set up in a forested area of upstate New York (Cranberry Lake)
The cross at the southeast edge of the lake marks the location of the swarm cluster.

artificial swarm (the Cholla swarm) in the desert and offered it a choice between two similar nest boxes, each approximately 160 m from the swarm, one to the east and the other to the west. The figure shows the location of dances on the surface of a swarm during three successive periods of the nest site selection process. On the swarm cluster the dances are initially concentrated in a restricted "dance floor" (the lower left quadrant of Fig. 5A) [13–15], but this concentration reduces later in the househunting process, when dances spread out over much of

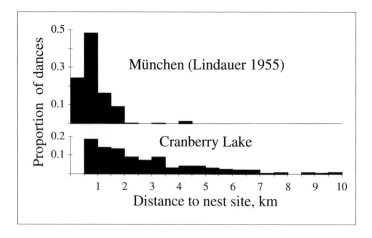

**Figure 4 The distribution of distances (n = 176) in the dance informa-
tion during the first part of nest site selection, until the time at which
there was nearly a consensus on one site**
For comparison, this same information is presented for the FIVE
swarms house-hunting in an urban area for which Lindauer [3] reported dance information (n = 766).

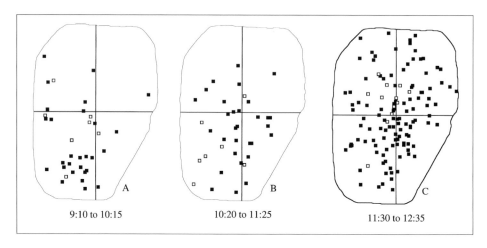

**Figure 5 The location of dances on the surface of the Cholla swarm during three successive por-
tions of the nest site selection process**
The swarm began househunting on December 13 at 08:42 and departed at 12:38.

the swarm cluster (Fig. 5C). One might hypothesize that the dances would not
only be concentrated on a particular portion of the swarm, but also that dances
for each site would be spatially segregated on the dance floor. However, this

appears not to be the case. One can see a broad overlap in the location of dances for the two competing nest sites.

Patterns of recruitment on the swarm and at the nest box

To obtain an accurate picture of the collective process both at the swarm, and at the nest sites, we made several repetitions of the experiment in which a swarm was offered a choice between two similar nest sites. In some cases, we marked as they appeared at the nest boxes. In other cases, we marked every dancer when we first observed her dancing on the swarm. In both of types of experiment, we observed events at the swarm as well as at the two nest boxes throughout the discovery and decision-making process.

Figure 6 shows the dynamics of the buildup of individual scouts at two nest boxes from the Ocotillo swarm. Since scouts visited the sites repeatedly after they were marked, the buildup in visitation is even greater than the number of scouting individuals shown here. In this experiment, the North site was found first, but recruitment of new scouts at the nest site was somewhat more intense for the South site the first day, with 38 bees at the South and 26 at North box marked by 15:00. Note, however, that the rates of recruitment for these first cohorts of recruits was similar for the two sites. On the 2nd day recruitment

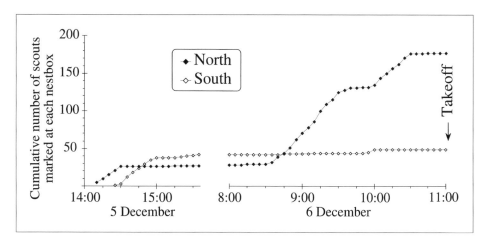

Figure 6 The dynamics of the buildup of individual scouts at the two nest boxes from the Ocotillo swarm

began first at the North site, and then increased (in two bursts) until swarm take-off. There were flat areas on the curve of the North site, and throughout most of the 2nd day on the South site, corresponding to periods in which no new recruits arrived. Note that one of these occurred at the North site just before takeoff.

For the Cholla swarm, we obtained simultaneous records of recruitment on the swarm cluster, scouting at the nest boxes, and detailed individual information on all bees that danced for either nest site. These observations allowed us to relate the collective behaviors at the swarm and nest sites, and to see how the individual behaviors related to the collective patterns.

Figure 7D, like Figure 6, shows the buildup of scouts at the two nest sites (but in a different way, since here we recorded the total number of scouts rather than

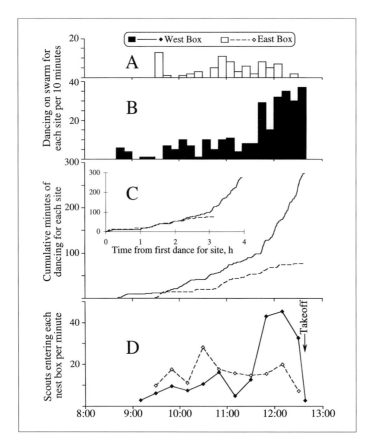

Figure 7 The recruitment dynamics of the Cholla swarm, showing activity at the swarm cluster and the nest boxes

the newly recruited scouts). In parallel with this information, we can see the cumulative buildup of recruitment for each site (Fig. 7C) and the amount of recruitment during any one time interval for each site (Fig. 7A,B). As in previous figures, the cumulative recruitment shows a stairstep pattern. Furthermore, this pattern is strikingly similar for the two sites, but offset in time. The inset in Figure 7C shows the cumulative recruitment lines for the two sites superimposed from their respective beginnings at the time of the first dance for each site.

Individual behaviors

The collective behaviors just described emerge from interactions among individual bees on the swarm. To understand these actions and interactions, we recorded the behaviors of individually marked bees both on the swarm cluster and the nest sites.

Recruitment behavior

Figure 8 presents timelines for four individual bees. In this swarm the homesite search spanned 2 days. The bees exhibited several different patterns of behavior which illustrate patterns we saw repeatedly. The purple-marked bee was marked as a vigorous dancer for the South site on day 2. She danced for a short period of time after being marked, followed dances for the South site, and was observed making several visits there. Shortly before the swarm took off to the North site, she followed dances for the North site, and then left the swarm (presumably for one of sites, though we could not record which one). The white-marked bee, also marked early in day 2, danced for North and then stopped dancing but continued to follow dances for and to visit the North site, and finally danced again just before the swarm took off.

The green-marked bee was one of the early dancers for the North site. She danced for this site, and then spent the rest of day 1 alternately on and off the swarm. Note that she often followed dances for a few minutes, and left the swarm immediately after doing so. On day 2 she began to dance more frequently, and did so, interspersed with dance following and flights off the swarm, for the rest of the day until swarm takeoff.

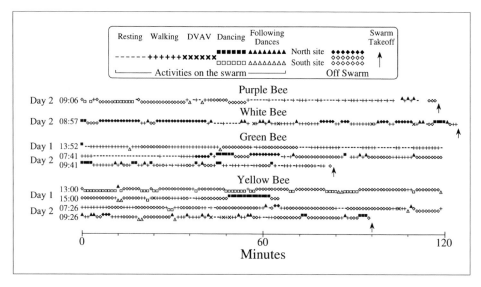

Figure 8 Time lines of bee activity showing the behaviors of four individuals of the Ocotillo swarm

In this figure as well as in Figure 9, the symbols for dancing, dance following, and nest site visiting are filled for the site finally chosen, and open for the alternative site. Also, the symbols for the two sites are displaced above and below the line. "Off swarm" records are gray and on the line when the bee was off the swarm, presumably at one of the nest sites, but not directly observed there (since bees enter and leave the boxes rapidly, and spend most of their time inside the box out of view, their visits are sometimes missed).

The yellow-marked bee, the first vigorous dancer observed on this swarm, initially danced for the South box, then briefly followed a dancer for the North box, then sporadically danced for South, but finally switched to dancing for North late in day 1. On day 2 she again danced for South, but later switched to North. She visited and followed dances for both sites.

To obtain a more complete picture of the individual behavior of scout bees, we marked all dancers on the Cholla swarm and followed their activities from the time they first danced and were marked until the swarm departed. Figure 9 shows these records. This complete transcription of the activity of the scouts allows us to quantify the frequency of the four behavioral patterns seen in the individual bees in Figure 8. Most bees dance for a relatively short period and then stop dancing (type I behavior). A few dance initially, then do not dance, but dance again near the time of swarm takeoff (type II). Many bees dance throughout the process (type III). Only a few visit alternative sites to the one for which they have danced, and may dance for this new site in turn (type IV). The relative frequencies of these behaviors among the 75 marked dancers in the Cholla swarm were 0.32, 0.09, 0.28, and 0.07 for types 1 to IV, respectively. (Note that the frequencies do not sum to 1 since 15 of the dancers could not be categorized because they began dancing shortly before the swarm took off, and an additional three dancers were not categorized because of incomplete records.)

Dance following

To understand how a consensus is reached for a single site, we were especially concerned with the question of whether scouts that danced for one site attended the dances of scouts recruiting for other sites. We found that most bees that dance for nest sites also follow the dances of other scouts, interspersing this following behavior with their own dances (Figs 8 and 9). Furthermore, they attend dances for both their own nest site and for others. In the Cholla swarm, 89% of the 46 dancers that danced more than 30 min before swarm takeoff (and thus had considerable opportunity to follow other dances) followed other dancers. Forty-one percent of these scouts followed dances that were for the same site these scouts had danced for, 13% for the alternative site, and 35% of the scouts followed dances for both sites. These percentages are not significantly different from those expected if scout bees follow dances randomly in proportion to their frequency on the swarm [22].

Scouting nest sites

Bees repeatedly visit the nest sites, even when they no longer perform dances for the sites they have visited (i.e. type I and II behavior). During these visits they spend much of their time inside the nest box, or crawling on the outside of the box (especially near the entrance) or flying around close to the surface of the box.

As noted above, some bees follow dances for both sites, suggesting that they may also visit both sites. By attending dances for other sites, and then visiting the alternative sites, scout bees would be able to compare one site to another, and select the one most suitable. In the Cholla swarm, however, only 6.7% of the marked recruiters visited both sites. Even in the Ocotillo swarm, where we

Figure 9 Time lines of bee activity showing the behaviors of each scout that performed recruitment dances on the Cholla swarm

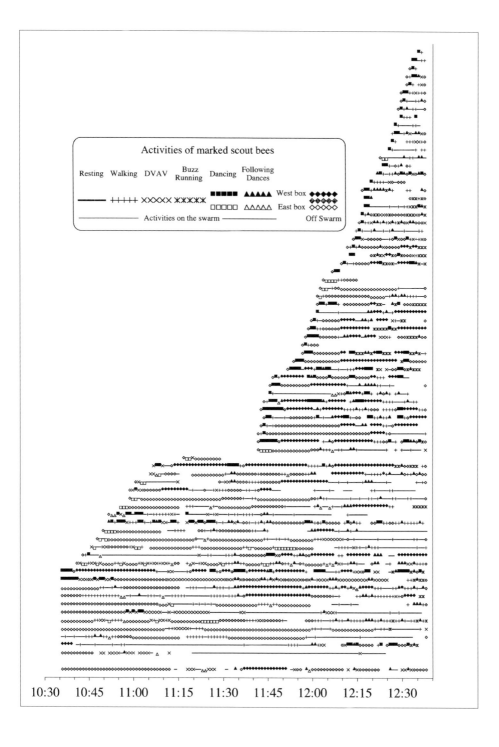

10:30   10:45   11:00   11:15   11:30   11:45   12:00   12:15   12:30

marked all scouts whether they danced or not, we only observed 9.2% to visit both sites. Most of these were bees which had visited the South site, which was finally rejected, and then visited the North site later in the process (when nearly all of the dancing was for North). Of the North site scouts, only 2.9% were observed to also visit the South site.

How do swarms of honey bees decide on a new home?

What we observe during the decision-making process is that a swarm begins with no recruitment dances, then dances build up for several sites; finally the dances for one site begin to predominate at the expense of the others, until all or nearly all dances on the swarm indicate a single site.

Thus, for a unanimous decision to be reached then, two things need to happen. First there must be communication of finds and a mechanism of positive feedback so that the numbers of scouts visiting them and dancing for them increases. Lindauer [16] described the recruitment process as follows: "The better the nesting site the more excited the dances of the scout bees, and the more new scout bees become recruited for this new place; the more bees therefore then fly off to this spot, inspect the home and then likewise announce it to the swarm by the same lively dances."

Second, there must be attrition: somehow scouts must stop dancing for "losing" sites. Lindauer actually presented more than one version of how this attrition comes about. One process of attrition is comparison, described by Lindauer [16] as follows: "Those scout bees which at first had announced the inferior nesting places are won over by the more lively dances of their competitors and as a result themselves inspect this home—so that they can compare the two—then they naturally choose the better one." However, Lindauer also described another process, that of standing down and passing the decision on to the next generation of scouts: "The scout bees do not remain stubborn about their first decision, but after a shorter or longer time, they become silent and leave the further decision to the new scouts" [3].

Two models of how decisions might be reached

These are two very different formulations of the decision-making process: In the first, there is comparison of sites, and switching to the better site; that is, bees change their mind. In the second, bees don't change their mind, they merely drop out after performing a greater or lesser amount of recruitment, and the subsequent dynamics of the process are determined solely through sites competing for the attention of future scouts through gaining an edge in the positive feedback system of recruitment.

Lindauer's description of the househunting process has resulted in subsequent authors seizing upon the comparison aspect of the decision-making process. This may be due in part to our natural inclination to assume that decision-making among other species of animals involves the same mechanisms that we, as humans, would use. Thus, Gould and Gould [17] wrote that a bee will "investigate the competition. If one of the alternatives is clearly better, the scout will switch allegiance and begin dancing for the new site…. Over the course of several days a consensus builds among the scouts, and finally the swarm departs for its new home…. Multiple cues at competing sites are weighed in an ongoing process that resembles more the way humans evaluate real estate than any simple task-related behavior." But Griffin [18] was more hesitant to jump to this conclusion. Even though all three records of the individual behavior of scouts that Lindauer [3] presented showed them to switch allegiance, the records of these few bees were not enough to convince Griffin that a unanimous decision by the bees generally involved switching of allegiance for sites. Thus, in a recent book [18] on animal cognition, he wrote: "Lindauer discovered that bees that had visited a cavity of mediocre quality sometimes became followers of more enthusiastic dances. Then some of them visited the better cavity, returned and danced [for] the superior cavity. Lindauer was only able to observe this in a handful of cases, and it is not clear how large a role this [comparison] plays in the process of reaching a group decision."

Evaluating the two models

Our results suggest that direct comparison of sites is relatively rare in househunting. In the Cholla swarm no bees ever danced for both sites, though at least

5 of the 75 marked bees did visit both, and 16 followed dances for both sites. In the Ocotillo swarm, where we could detect visits to both sites by bees which did not dance, we saw more comparison, with 6.3% of bees which had scouted the losing site switching over to the winning site and 2.9% switching in the other direction. However, the fact that we observed no bees switching dances between sites suggests that comparison plays a minor role if any in the decision-making process. In more recent experiments, we set out to document the extent of dance switching, and have found a few bees which dance for both sites. However, when we experimentally remove these bees that dance for both sites, thus preventing any direct comparison between sites, we find no change in the collective dynamics of decision making [22].

Thus, the alternative model, of positive feedback and attrition, is better supported by our observations. In Figures 6 and 7, the stairstepped exponential rise in recruitment (either as scouting or dancing) reflects this feedback, and the individual records show that by far the most common kind of behavior among scout bees is to recruit for a relatively short period and then fall silent, generally without visiting the alternative site. The stairstepped shape of the buildup curves suggest that scouts recruit a new cohort of scouts for a site, which in turn inspect the site and then recruit the next, larger cohort.

How do alternative sites compete in a positive feedback and attrition model?

For a simple positive feedback and attrition system to function effectively merely requires that the parameters underlying an exponential rise in recruitment at the alternative sites be slightly different, so that one site has a chance to pull ahead onto a steep area of the curve, after which it will automatically come to dominate recruitment on the swarm. There are a number of sources for differences in the recruitment dynamics to two sites, which fall broadly into quality differences and random events at the two sites.

Quality differences

Bees grade their recruitment to sites depending on the perceived quality of the sites. Lindauer [3] mentions differences in both the duration and vigor of dances, and these differences were obvious on our swarms as well. Seeley [5] also showed that the scouts' perception of nest site volume affected their probability of dancing. In nectar foraging, it seems that the duration of dances is the important variable affecting recruitment [19]. Lindauer [16] suggests that scouts might be able read site quality from dances and be "won over by the more lively dances of their competitors." To us, the vigor differences are striking, and it merits further research whether the bees respond in some way to vigor differences aside from duration differences (which are correlated, so a critical test will involve prematurely terminating high-vigor dances).

Random events

Individual bees are quite variable in their recruitment intensity for the same site (as seen in Figs 8 and 9), and thus a site which happens to attract a vigorously recruiting scout will receive many recruits, even if the alternative site is of equal quality. Especially in the early stages of recruitment to our nest boxes, we noted that a single individual could have a dramatic impact on visitation to a site. Another source of random differences is time of discovery. A site that is discovered first may have already had several cycles of positive feedback through recruitment by the time another site of equal quality is discovered, and thus build up to an overwhelming majority before the second site can attract enough recruits to compete. We know from other studies that sources of random variation can have crucial effects on self-organizing processes among social insects. This has been most clearly demonstrated by Deneubourg and colleagues [20, 21] in their studies of food source selection and trail formation in ants.

Distance differences

In addition, distance from the swarm to the nest box could also affect recruitment buildup, even if there the sites were of identical quality and discovered simultaneously. The reason is that differences in travel time could lengthen an individual bee's recruitment and scouting cycle, so that less recruitment would be expected for more distant sites.

Where do we go from here?

In future studies, our primary focus will be to evaluate whether swarm decision making involves the positive feedback and attrition mechanism or that of comparison. One approach will be to experimentally manipulate individual groups of bees and to observe the outcome of the decision-making process. By delicately perturbing the system in specific ways, we should be able to gain a better understanding of the mechanisms involved. As briefly mentioned above, we have already shown that preventing a switch in allegiance has no apparent effect on the dynamics of the decision-making process. We might also be able to test the importance of random events on the outcome of the decision by manipulating the amount of dancing for one site or another. Hand in hand with these experimental studies, we can perform simulations based closely on the individual behaviors of the bees. Building mathematical and computer models in this bottom-up fashion will allow us to make predictions to compare to the results of our field experiments.

As we refine our picture of how the decision is reached, we should be able to more easily address the other mysteries of the swarming process, and generate hypotheses concerning other questions such as how the decision is perceived by the bees of the swarm, and how the takeoff is triggered. We suspect that many of the mechanisms involved will be best understood within the framework of self-organization. The emergence of collective patterns during the process of swarming will likely share many of the features of other self-organized processes among social insects. Among our many goals in all these studies will be the development of better techniques for unobtrusively probing the pathways by which information flows and is amplified within the insect colony. And perhaps of equal importance will be an attempt by us to adopt a mindset that allows us to appreciate the highly efficient ways in which tiny individuals with even tinier brains have evolved extraordinary mechanisms for solving their day-to-day problems in very unusual, nonhuman ways.

Acknowledgments

We wish to thank Karl Crailsheim, Tim Judd, Barrett Klein, Cornelia Koenig, Jennifer Kusznir, Jennifer Rittenhouse, Gavin Sherman, Rick Vetter, and Denise Cope for assistance in the field and laboratory. Special thanks to Jennifer Finley who analyzed the video transcriptions of the Cholla swarm, and to Albert Rozo for his work on the Cranberry Lake swarm. We thank the Cranberry Lake Biological Station for providing excellent research facilities in New York. We also thank the owners of the land in Cactus City for allowing us to work at such a beautiful site.

References

1 Lindauer M (1951) Bienentänze in der Schwarmtraube. *Naturwissenschaften* 38: 509–513

2 Lindauer M (1953) Bienentänze in der Schwarmtraube (II). *Naturwissenschaften* 40: 379–385

3 Lindauer M (1955) Schwarmbienen auf Wohnungssuche. *Z Vergl Physiol* 37: 263–324

4 Seeley TD, Morse R (1976) The nest of the honey bee (*Apis mellifera* L.). *Insect Soc* 23: 495–512

5 Seeley TD (1977) Measurement of nest cavity volume by the honey bee (*Apis mellifera*). *Behav Ecol Sociobiol* 2: 201–227

6 Seeley TD, Morse R, Visscher P (1979) the natural history of the flight of honey bee swarms. *Psyche* 86: 103–113

7 Lindauer M (1961) *Communication among social bees*. Harvard University Press, Cambridge, MA

8 Esch H (1967) The sounds produced by swarming honey bees. *Z Vergl Physiol* 56: 408–411

9 Avitabile A, Morse RA, Boch R (1975) Swarming honey bees guided by phero-mones. *Ann Entomol Soc Amer* 68: 1079–1082

10 Morse RA, Boch R (1971) Pheromone concert in swarming honey bees (Hymenoptera: Apidae). *Ann Entomol Soc Amer* 64: 1414–1417

11 Visscher PK, Morse RA, Seeley TD (1985) Honey bees choosing a home prefer previously occupied cavities. *Insect Soc* 32: 217–220

12 Frisch K von (1967) *The dance language and orientation of bees*. Harvard University Press, Cambridge, MA

13 Visscher PK, Seeley TD (1982) Foraging strategy of honeybee colonies in a temperate deciduous forest. *Ecology* 63: 1790–1801

14 Seeley TD, Towne WF (1992) Tactics of dance choice in honey bees: do foragers compare dances? *Behav Ecol Sociobiol* 30: 59–69

15 Seeley TD (1994) Honey bee foragers as sensory units of their colonies. *Behav Ecol Sociobiol* 34: 51–62

16 Lindauer M (1957) Communication in swarm-bees searching for a new home. *Nature* 179: 63–66

17 Gould JL, Gould CG (1994) *The animal mind*. W. H. Freeman, New York

18 Griffin DR (1992) *Animal minds*. University of Chicago Press, Chicago

19 Seeley TD (1995) *The wisdom of the hive*. Harvard University Press, Cambridge, MA

20 Beckers R, Deneubourg JL, Goss S (1992) Trails, U-turns in the selection of a path by the ant *Lasius niger*. *J Theor Biol* 159: 397–415

21 Deneubourg JL, Pasteels JM, Verhaeghe JC (1983) Probabilistic behaviour in ants: a strategy of errors? *J Theor Biol* 105: 259–271

22 Visscher PK, Camazine S (1999) Collective decisions and cognition in bees. *Nature* 397: 400

Collective behavior in social caterpillars

Terrence D. Fitzgerald and James T. Costa

Summary

The repertoires of social caterpillars are drawn from four categories of behavior centered on shelter building, thermoregulation, colony defense, and trail-based communication. This chapter provides an overview of these collective patterns of behavior and assesses the potential role of individual-level behavior in their overt expression. While few of the 300 or more species of caterpillars that form sib-aggregations have been studied in any detail, our review of the literature indicates that the most promising fronts for the investigation of emergent phenomena in social caterpillars lie in the areas of collective shelter building and trail-marking behavior. Of particular interest is the resemblance of trail-based chemical recruitment communication and collective flexibility in recruitment in caterpillars to similar phenomena in the Hymenoptera. The decision rules underlying recruitment patterns have been explored in ants, and are likely to prove generalizable to the more sophisticated of the recruitment systems found among the lasiocampid caterpillars, but a set of mechanisms dependent on such species-specific factors as resource patchiness, mode of recruitment, and physical properties of silk are likely to uniquely influence collective foraging patterns and shelter-building behavior in social caterpillars.

Introduction

At least 300 species of caterpillars distributed among 27 families of the Lepidoptera aggregate in the larval stage [1]. Although sib-group aggregations of caterpillars were commonly referred to as social species in the older entomological literature (see, e.g. [2, 3]) the use of the term "social" by entomologists became much more restrictive following the publication of Wheeler's definitive study of social behavior in ants [4]. This classic treatise focused the attention of subsequent generations of entomologists on the complex societies of ants, bees, wasps, and termites to the near exclusion of other social forms. In schemes that are currently widely accepted, ants and termites and the societies of bees and wasps that sit at the pinnacle of social evolution are classified as the "eusocial" insects or simply as "the social" insects, whereas less sophisticated cooperative associations of parents and offspring are termed "subsocial". Caterpillar societies, which are aggregations of subimagos, lack the defining attributes of either eusociality or subsociality, and are common referred to as "presocial" insects [5]. As discussed elsewhere [6, 7], these terms are not only etymologically imprecise but, more important, misleading in that they imply a false hierarchy of social evolution and place undue emphasis on the importance of reproduction in the definition of sociality. In referring to caterpillars as social species in this chapter, we are using the term in its broader sense as it applies to familial aggregates in which there is "reciprocal communication of a cooperative nature" [8].

While sibling aggregates of caterpillars are simple societies lacking the complexity of interaction that characterizes the eusocial species, studies conducted over the past 2 decades have revealed that they exhibit a richer diversity of cooperative interactions than previously suspected. Indeed, our survey of the literature indicates that gregarious caterpillars, collectively, exhibit a greater range of social behaviors than are found among those species that have been traditionally classified as subsocial forms, and their social repertoires eclipse even those of species of thrips, beetles, and aphids added recently to the catalog of eusocial insects. Social caterpillars are the only social insects outside of the classically eusocial groups to exhibit sychronized forays away from the communal shelter, trail marking, recruitment communication, synchronized shelter building, and collective thermoregulation.

Though some aggregations of caterpillars, such as those of *Porthetira dispar* during the high-density phase of their outbreak cycle, are nonadaptive aggregates of autonomous individuals, the larvae of the obligatorily social species we consider here are clearly adapted to communal living. Studies conducted over the past 4 decades show that when isolated from their siblings, the larvae of social caterpillars grow more slowly and experience significantly more mortality than caterpillars maintained in groups. This differential survival has been attributed to social facilitation in feeding [9–17], cooperative foraging behavior [18–22], group defense against predators and parasitoids [23–27], collective substrate silking to increase steadfastness [28], and aggregative behavior that facilitates thermoregulation, including shelter building [29–32]. It is the purpose of this chapter to provide an overview of these collective behaviors and to assess, where possible, the role of individual behavior in their expression.

Collective shelter building

Collectively built structures comparable to the nests constructed of wax or paper by the hymenopterans or of carton and clay by the isopterans are not found outside these insect orders. The vast majority of the nests collectively built by insects in other orders, such as the cells constructed in soil by pairs of *Lethrus* beetles [8], or the galleries of wood roaches [33], passalids [34] and scolytid beetles [35], are the passive products of excavation. The only building material in any way comparable to the secreted or processed materials of the bees, wasps, and termites produced by noneusocial insects is silk. Webspinners [36] and social psocids [37] spin roofs of silk over their feeding areas, while social thrips [38] bind phyllodes of the host plant with silk to form a simple enclosure. But the preeminent silk spinners are the lepidopterous caterpillars. Caterpillars spin silk prolifically and in comparison to other noneusocial insects build large and relatively complex structures from the material. Moreover, they are the only insects outside of the Hymenoptera and Isoptera to exhibit true collective building behavior involving colonywide synchronization of activity and periodic shelter expansion. Some social caterpillars such as *Brassolis isthmia* [39] and *Archips cervasivoranus* [40] employ silk to draw the leaves of their host plants into tightly bound shelters in which they rest between foraging bouts. But the most impressive structures collectively built by caterpillars, such as the remarkable "bolsa" of the social pierid *Eucheira socialis* [41] and the tents of the lasiocampid caterpillars [6], are made exclusively of silk.

The communal shelters of caterpillar are multifunctional, serving to facilitate basking and thermoregulation, molting, and antipredator defense. In the tent caterpillars (*Malacosoma*), they may also serve as communication centers where hungry caterpillars are recruited to the discovery of food [6]. The utility of caterpillar shelters in these capacities has been reviewed elsewhere [1, 42] and will be discussed only briefly in the present paper.

In comparison to our knowledge of the adaptive functions of caterpillar shelters, very little is known of the behavioral mechanisms that give rise to the architecturally distinct, collectively built shelters of caterpillars. There are as yet no studies like those for the eusocial insects that show how relatively simple, repetitive motor patterns can lead to the formation of complex structures through an autocatalytic process (see, for example Theraulaz, this volume). Indeed, few of the structures that are communally built by caterpillars have even been described in detail. It is clear, however, that unlike the complex, free-form structures of the eusocial insects, the ultimate shape that the collectively built nests of caterpillars takes is determined to no small extent by exogenous factors. Studies have shown that while colonies may actively select sites prior to the construction of a shelter, or abandon a site that proves inadequate after the shelter-building process has begun [6, 43], all collectively built caterpillar shelters are formed either by pulling together plant parts or by spinning silk about a framework of branches and leaves.

The characteristic form of collectively built caterpillar shelters is also determined by the extent to which the insects spatially focus their activity while spinning. The caterpillars of *M. americanum* and *Hyphantria cunea*, for example, are both prodigious silk spinners, but they create distinctly different shelters. Once an incipient tent is formed by the neonate larvae of *M. americanum*, the caterpillars are strongly drawn to the structure and it becomes the focus all subsequent bouts of collective spinning. The larvae add increments of silk to the tent during distinct daily bouts of spinning activity, extending over a period of a month or more. The final structure is multilayered, angular, and compact, rarely exceeding 15 cm in any one direction, and it stands apart from the colony's foraging arena. In contrast, the larvae of *H. cunea* spin copiously wherever they venture, investing the whole of the foraging arena in silk, and occasionally, in the case of small hosts, enveloping the entire tree. One significant consequence of the difference in the design of these nests is that while the tent of *M. americanum* is a highly effective heat trap, capable of achieving temperatures excesses ($T_{tent} - T_{amb}$) as great as 23 °C [32], the core of the nest of *H. cunea*, where the later instars come to rest, typically achieves much smaller temperature excesses [44]. This difference is consistent with the fact that while *H. cunea* forages in the summer, the larvae of *M. americanum* forage in the spring when mean ambient temperatures are low and colony survival is strongly dependent on the ability of the caterpillars to bask in the tent [6].

Recent studies suggest that subtle differences in the intrinsic properties of the silks of caterpillars, or the way they are spun, may be more important than overt differences in larval motor patterns in accounting for interspecific differences in the form of the web-nest [45, 46]. These studies show that caterpillars stretch silk strands as they spin them, inducing axially retractive forces that strongly affect the emergent form of the web-nest. Both *A. cerasivoranus* and *Yponomeuta cagnagella*, for example, extend strands of silk between one leaf and another during communal bouts of spinning. In *Y. cagnagella*, the activity produces an envelope of silk that loosely envelopes the foraging arena. In *Archips*, spinning causes the leaves of the tree to be drawn tightly together, forming a dense shelter within which the caterpillars feed and rest. Through a collective effort, small groups of *A. cerasivoranus* caterpillars allowed to spin in a tensiometer were shown to generate forces as great as 0.3 N (i.e. a 30-g force), whereas similar-sized groups of *Y. cagnagella* generated forces of little more than 0.1 N [45]. These differences are attributable to both the extent to which the caterpillars stretch the strands while spinning and to the modulus of elasticity (stiffness) of the strand. It has also been shown that strands of the silk of some caterpillars contract instantly when wetted, producing an additional axially retractive force.

Largely unexplored is the possibility that the architecture of collectively built shelters made entirely of silk may also be attributable to differences in the physical properties of silk. The nests of both *M. americanum* and *E. socialis* are constructed over branch frameworks that are similar in form, and the silk spun by both species during successive bouts of "en masse" spinning is deposited directly on the walls of the growing structure. Their efforts, however, result in architecturally distinct structures. Indeed, even if the two species were to employ identical motor patterns while spinning, differences in the modulus of elasticity and wet contractability of their strands, alone, could account for overt differences in the final form of the structures they create.

Collective foraging

The hallmark of caterpillar sociality is communal foraging. While the larvae of some social caterpillars feed in loosely formed subgroups, maintaining proximity to the rest of the aggregate by referencing chemical cues that impregnate the foraging arena, others feed side by side. In the latter instance, bouts of feeding and rest are tightly coordinated among the members of the colony. Synchronized bouts of communal foraging have been reported for a number of social caterpillars including *Thaumetopoea* [47], *Ochrogaster* [48], *Eriogaster* [49], *Eucheira* [50], *Gloveria* [51], and *Malacosoma* [6]. Colonies of these species commonly consist of several hundred individuals, and it is particularly impressive to observe the larvae in tight procession as they move off in search of food. There have as yet been no studies to determine how the activity of colony members becomes temporally synchronized, but tactile cues associated with increased restlessness among the hungry caterpillars, which typically rest side by side, are likely to serve as catalysts.

Studies of *M. americanum* show that the temporal foraging pattern exhibited by colonies of this species is an emergent property of the aggregate [52]. Under both field and laboratory conditions, colonies of this species forage "en masse" three to four times each day [53, 54]. When isolated from their siblings, caterpillars feed at significantly shorter intervals than whole colonies, foraging as many as 12 times per day, whereas caterpillars in groups consisting of 5 to 10 individuals have foraging patterns that increasingly resemble those of whole colonies. Thus, in this species, group living constrains rather than promotes foraging activity. Casey et al. [31] speculated that strong selection pressure for synchronous foraging in colonies of the tent caterpillar has given rise to group-mediated behavioral patterns that override the tendency of individuals to follow foraging schedules dictated by their own hunger levels.

The feeding patterns of social caterpillars fall into three general categories: patch-restricted foraging, nomadic foraging, and central-place foraging. The most common mode of foraging found among the more primitive families of the Lepidoptera is patch-restricted foraging. Colonies of these caterpillars typically confine their feeding activity to the leaves found within a single contiguous patch of leaves or to a series of sequentially exploited patches. The colony's foraging arena is often enveloped in silk, and the caterpillars feed and rest within the envelope. Trail markers play little or no role in constraining patch-restricted foragers, enabling the caterpillars to enjoy an independence of movement not seen in either nomadic or central-place feeders. The characteristic amoebic form of the foraging arena of patch-restricted species arises as a consequence of this independence among individual colony members, allowing individuals or subgroups to simultaneously advance on multiple fronts.

The integrity of colonies of patch-restricted foragers is maintained not by maintenance of close physical contact among the caterpillars, but by boundary markers that defined the limits of the foraging arena. In *Y. cagnagella* [55–57] and *A. cerasivoranus* [40] the boundary markers are pheromonal components of the silk strands the caterpillars deposit wherever they go. Patch-restricted foragers are strongly contained by these markers and venture only short distances beyond the existing envelope of silk to access new leaves. In this way, the arena is gradually expanded to meet the increasing energy demands of the colony.

Patch-restricted foraging is most likely to arise among species that are relatively nonselective in their feeding habits, enabling them to consume vegetation as encountered. The overt form of the host plant must also play some role, since leaves must occur in sufficient density to allow the colony to profitably expand the perimeter of their web in multiple directions. In species such as *H. cunea*, foliage is dense enough to enable the early instars to meet their energy demands within the communal web, but the later instars must range more widely in search of leaves, and they adopt a central-place foraging mode [58].

In contrast to patch-restricted foragers, nomadic foragers wander widely in search of feeding and resting sites. The behavior of nomadic foragers is more tightly integrated than that of the patch-restricted foragers, and periods of activity and rest are typically closely synchronized among all the members of the aggregate. Factors leading to the evolution of nomadicism include the potential for the buildup of disease agents at communal resting sites [59, 60], selective pressure from predators that associate caterpillars with the damage they inflict on the host tree [61], and selectivity in feeding habits that requires caterpillars to range widely in search of leaves. The habit is widespread among caterpillars [1], including *Nymphalis antiopa* [62], *Euphydryas phaeton* [63], *Asterocampa clyton* [64], *Chlosyne lacinia* [65], *Hemileuca lucina* [66], and *Pieris brassicae* [67].

Despite the commonness of the nomadic habit, the only nomadic species for which we have detailed knowledge in *Malacosoma disstria*. Here, nomadicism is, at least in part, driven by feeding selectivity [68], but the caterpillars are forest dwelling insects, and their need to bask in this shady habitat also requires that they have sufficient mobility to track the sun. Like that of other nomadic foragers, the integrity of colonies of this caterpillar is maintained by trail pheromones and subserved by a strong reluctance on the part of the caterpillars to forage independently. In *M. disstria,* each caterpillar lays down a chemical trail by pressing the tip of its abdomen against the substrate as it advances along a branch, much in the manner of ants and termites [68]. Individuals proceed hesitantly when moving over previously unexplored terrain, and those in the van turn back shortly if they lose physical contact with other members of the cohort. But once the pheromone trail is established, the caterpillars move along it with little or no hesitancy, and they do not require physical contact with others.

Studies show that the caterpillars of *M. disstria* are sensitive to minute differences in trail strength and prefer stronger over weaker trails at choice points [69]. Thus, though colonies in transit may initially fragment at branch junctures, a bias toward the branch selected by the greater number of caterpillars soon develops, and the majority of the colony goes off in one direction. The high mobility of the caterpillars, the long-term persistence of their trails, and a strong aggregative tendency leads to the eventual regrouping of fragments [68]. During forays the caterpillars settle "en masse" at feeding sites. Immediately after feeding to repletion, the caterpillars set off in search of a suitable bivouac. The integrity of the fed colony is maintained because the caterpillars mark with a recruitment pheromone the branches they follow as they move away from the food-find, and reassemble at the new resting site [68].

Central-place foraging is the least common mode of foraging found among the social caterpillars, and it is largely limited in occurrence to the most recently evolved families of the Lepidoptera [1]. Central-place foragers establish permanent or semipermanent resting sites from which they launch intermittent forays in search of food. The caterpillars may rest in the open, but many species such as *Eriogaster* [70], *Gloveria* [71], *Malacosoma* [6], *Thaumetopoea* [72], *Ochrogaster* [47], and *E. socialis* [41] build large, communal shelters of silk. In addition, some gregarious lycaenids, particularly in the genera *Paralucia* [73] and *Ogyris* [74], are not only central-place foragers but are also ant tended, but the extent to which their foraging patterns are influenced by ants is largely unknown [1].

The advent of central place foraging in the Lepidoptera set the stage for the evolution of cooperative interactions that go beyond those seen in colonies of patch-restricted and nomadic foragers. At its greatest development, the nest, like those of the ants and termites, serves as a communication center where hungry caterpillars are alerted to the discovery of food. But the full potential of this mode of foraging has not been realized in all central place foragers. The foraging systems of *M. americanum*, *E. socialis*, and an as yet unidentified species of *Gloveria* are compared in Table 1. These three species are the only central-place foragers studied in sufficient detail to allow us to relate the trail-marking behavior of the individual to the colony's overt pattern of foraging. All three species mark new terrain with a nonvolatile pheromone secreted from the tip of their

Table 1 Comparison of the foraging and trail marking behavior of Gloveria *sp.,* Malacosoma americanum, *and* Eucheira socialis

Trait	Gloveria[*]	Malacosoma[†]	Eucheira[‡]
Trails marked with tip of abdomen:	yes	yes	yes
marking motor pattern	continuous	continuous	cyclic
marking new branches	yes	yes	yes
marking on turn-back when searching for food	no	no	yes
marking after feeding	yes	yes	no
marking onto nest after feeding	no	yes	no
recruitment to trails	yes	yes	yes
bilevel trail system recruitment	no	yes	no
recruitment to food	weak	strong	none
individuals recruit to food	no	yes	no
sensitivity to differences in trail strength	high	high	high
Role of silk in trail following	minor	minor	minor
Patchiness of food supply	low/occasional	high/chronic	low/occasional
Selective foraging	no	yes	no
Collective flexibility	low	high	low

Refererences: [51], [6, and references therein], [50]

abdomen and detect the marker with contact chemoreceptors of the head [75]. Each also shows a high degree of sensitivity to trail strength as determined by the number of caterpillars that have previously marked a trail. Thus, all three can be considered to recruit to trails. *Gloveria* and *M. americanum*, but not *E. socialis* reinforce the pathways they follow after feeding so that these two species also recruit to food-finds.

Despite similarities in their trail-marking behavior, the foraging systems of *M. americanum* and *Gloveria* differ in significant ways. The fed, returning caterpillars of *M. americanum* lay down trails that extend without interruption from a food-find to the surface of the tent, thus rendering the structure a communication center [6]. In contrast, the fed, returning larvae of *Gloveria* stop marking soon after they reestablish contact with a major trunk trail and consequently do not recruit from the tent surface. In addition, the recruitment system of *Gloveria* is less efficient than that of *M. americanum*, enabling colonies of the latter to abandon a trail to an exhausted feeding site more rapidly than the former [51]. The evolutionary refinement of trail-based communication leading to a colony's ability to redirect foraging activity from one site to another has been previously described in ants and termed "collective flexibility" [76]. Once a trail to a food-find is established, species of ants that have low collective flexibility continue to forage at the site even if a richer food source is discovered elsewhere [76, 77]. In contrast, species of ants that are collectively flexible override the attraction of an established trail by varying the amount of trail marker they lay down, by using different chemicals, or by directly leading small groups of colony mates to the food-find [76–78]. The trail deposited by a larva of *Gloveria* as it returns to the nest after feeding does not appear any more effective in eliciting trail following than a trail deposited by a hungry caterpillar searching for food, and there is no evidence that colonies of *Gloveria* exhibit the collective flexibility necessary to efficiently abandon an established feeding site in favor of a newer and richer food-find. In contrast, trails deposited by *M. americanum* caterpillars as they return to the tent after feeding are followed much more readily than the trails deposited by caterpillars searching for food, and it is this bilevel trial system that allows colonies to quickly abandon a trail to an exhausted site in favor of one leading to a new and profitable food-find. It is unknown whether the ability of the caterpillars of *M. americanum* to discriminate among trails is due to differences in the quantity of marker deposited or to the employment of qualitatively different marker chemicals.

Differences in the relative sophistication of the communication systems of *Gloveria*, *E. socialis*, and *M. americanum* may have evolved in response to differences in their feeding selectivity. Both *Gloveria* and *Eucheira* are nonselective feeders and utilize a patch-restricted mode of foraging in the early stadia, switching to central-place foraging when it becomes necessary to move greater distances to obtain food. In contrast, *M. americanum* is a highly selective feeder that from the earliest instars prefers a tree's youngest leaves, and the caterpillars search extensively to find them [6]. The trail-marking system of *M. americanum* appears fine-tuned relative to those of *Gloveria* and *E. socialis*, allowing the caterpillars to deal with food as a chronically scarce resource. Studies show that a caterpillar of *M. americanum* does not recruit to sites that fail to provide food in quantities sufficient to sate it; thus foragers are channeled to uncrowded feeding sites. In addition, the caterpillars assess differences in food quality and recruit preferentially to the leaves of favored host species and to the young leaves that occur at the tips of branches [20].

Collective defense in caterpillars

Aggregation allows caterpillars to collectively defend themselves against the attacks of predators and parasitoids. Such defense may be passive or active. Passive modes of collective defense involve dilution effects since the mathematical probability that any one individual will be randomly singled out by a predator decreases with group size. Individuals also gain protection from predators by surrounding themselves with others. Hamilton's "selfish herd" models [79] have been widely applied to both vertebrate and invertebrate groups [27, 80–84] and appear broadly applicable to social caterpillars, but exceptions have been reported. Medium-sized groups of larval *Euphydryas phaeton*, for example, had lower rates of parasitism than either smaller or larger groups [26, 85], whereas medium-sized egg masses of *Laspeyresia pomonella* showed the highest survivorship rates [86]. In addition to the effects of dilution on individual survival, the overall rate of mortality suffered by colonies of central place foragers such as those of *M. americanum* and *E. socialis* may be reduced when caterpillars make tightly synchronized "en masse" forays away from their shelters during brief and discrete intervals, crowding predators that wait outside the nest in time as well as space [50, 53, 54].

In addition to its direct effects in reducing predation, aggregation in caterpillars may also enhance larval growth rates due to the joint benefits of group thermoregulation, cooperative foraging, and reduced predator harassment. Rapid growth, in itself, also constitutes a defensive benefit because larval vulnerability to predators and parasites is often greatest in the early instars. Predator- and parasitoid-induced mortality may decrease as larval size increases because (i) larger caterpillars more readily rebuff small predators such as ants or stinkbugs, (ii) chemical and structural defenses often do not constitute an effective defense until later instars, and (iii) the small incipient shelters of young colonies are more easily penetrated than large shelters [1].

Shelters collectively built by caterpillars play an important role in antipredator defense. The tough silk shell of the bolsa formed by a colony of *E. socialis* caterpillars is virtually impregnable to both birds and invertebrate predators. Like many other shelter builders, the resident caterpillars venture from the bolsa only under the cover of darkness [41] when birds and predatory wasps are inactive. Bound-leaf shelters, and the more loosely spun shelters of the tent caterpillars and the fall webworm, *Hyphantria cunea*, cannot exclude predators completely [6, 24, 87], but all are likely to be deterred to some degree [88], providing the residents with greater protection than they would enjoy resting in the open.

It is generally thought that aposematic coloration is most effective in deterring predators when insects group together [89–92, but see 93]. Indeed, the most common mode of active defense among social caterpillars is aposematic display, often combined with synchronous body rearing, flicking, and "en masse" regurgitation of toxic or unpalatable chemicals [12, 64, 89, 94–96; reviewed in 1 and 6]. Simple displays include those of the many social notodontid caterpillars, such as *Datana* spp., which synchronously raise the anterior and posterior ends of their bodies into a distinctive U-shape when disturbed [97]. The social sawfly *Croesus latitarsus* [97] also performs simple displays by rearing the posterior portion of its body up over its head in sinuous S-shaped fashion, but some such as *Neodiprion* spp. rear the anterior portion of the body while simultaneously regurgitating sequestered compounds.

Synchronous flicking of the body has been described from many social caterpillars, and some, such as the eastern tent caterpillar (*Malacosoma americanum*), also combine these displays with defensive regurgitation [96]. Multicomponent displays increase the signal value of the display and may have evolved in response to selective pressures from different types of predators that favor different sensory modalities. Entomophagous insects, for example, are likely to be deterred by synchronized body movements and defensive regurgitation, whereas birds may be more effectively deterred by aposematic displays. The effectiveness of aposematism and associated chemical defenses also depends critically on predator conditioning. Here, gregariousness itself plays a role in conditioning by increasing the contact rate of predator with aposematic prey, facilitating the association of warning coloration with unpalatability [98, 99, but see 100].

What is of particular interest here is how antipredator displays become synchronized among colony members, leading to the emergence of group signals. Group transmission of predator avoidance signals have been reported in a number of gregarious invertebrates [93, 101–103], but little is known of the phenomenon in social caterpillars. It is not surprising that alarm pheromones, like those of the eusocial insects that draw defenders to the site of a disturbance, are not found among the social caterpillars. Although there is at least one anecdotal report of caterpillars enveloping a would-be attacker in their silk [104], caterpillars lack offensive weapons in any way comparable to those of the eusocial insects, and there are no credible reports of the larvae of any species behaving altruistically in a defensive context. It is perhaps more surprising that social caterpillars are not known to employ dispersal pheromones of the sort used by aggregates of aphids and membracids [105, 106], since caterpillar colonies are typically sib groups within which this sort of kin-selected defensive behavior could be expected to evolve.

Studies to date indicate that the spread of alarm through colonies of social caterpillars are mediated largely by tactile and, possibly, visual cues. Studies have shown that caterpillars can detect the airborne signals generated by the beating wings of flying wasps and flies and respond with rapidly jerking movements [94]. Vibrational signals set up by the agitated caterpillars and propagated by the communal web would appear the most likely means of alerting the colony to danger. Although there are as yet no definitive studies of the influence of discrete vibrational signals on group display, it is known that the defensive body thrashing initiated by caterpillars that detect a predator radiates rapidly through colonies, and that physical contact is not necessary for such radiation to occur [42]. Less is known of the role of vision in the spread of defensive behavior among colony members, but studies show that caterpillars are able to detect stationary objects, such as the stems of trees, at distances exceeding 1 m [107, 106]. So it would appear well within the sensory capacity of caterpillars to detect the frenetic movements of nearby siblings.

Basking and collective thermoregulation

Although species such as the larvae of the gypsy moth and the many cryptic species of caterpillars are thermal conformers, many social caterpillars are thermoregulators [109, 110]. The caterpillars of *Hemileuca oliviae* regulate their body temperature by ascending grass stems under hot conditions to avoid excessive heat radiating from the ground [111]. These insects respond independently to temperature excesses, and their behavior does not directly influence the thermal ecology of colony mates. In contrast, the thermoregulatory behavior of many species of social caterpillars is markedly enhanced by the presence of siblings.

At low air temperatures, the dark-colored larvae of the Australian pergid sawfly *Perga dorsalis* experience reduced lateral heat loss by insulating one another in basking groups. At about 30 °C, the larvae begin to overheat, and they raise their abdomens above the substrate to facilitate convective heat loss. Under more extreme conditions (>37 °C) the larvae secret from the anus a semiliquid material which when smeared onto the surfaces of their bodies induces evaporative cooling. Seymour [112] observed that the caterpillars had difficulty directing the material onto their own body, leading to waste and inefficient cooling when caterpillars were isolated. When grouped, misdirected coolant was spread onto nearby siblings, facilitating the cooling of the whole aggregate.

Caterpillars that feed at times of the year when air temperatures are low are particularly likely to benefit from aggregative basking. The spring feeding larvae of the nymphalid butterfly *Euphydryas aurinia*, a nonshelter building species, bask en masse in the open, packing their bodies tightly together to minimize convective heat loss [30]. Under high levels of solar radiation on cold days, gregariousness and the darkness of their cuticle enable the larvae to gain temperature excesses ($T_{body} - T_{ambient}$) as great as 30 °C, allowing them to readily achieve the 35 °C body temperature optimal for the assimilation of food.

The construction of heat-trapping shelters enables social species to gain even more control over their body temperature. The tentlike shelters of *Eriogaster lanestris* and *M. americanum* provide a surface extensive enough to enable the colony to bask en masse, and the catepillars oriented their nests to take full advantage of the sun [43, 70]. The silk walls of the structures are dense enough to serve as barriers to convective heat loss, allowing them to function as miniature glasshouses [32]. Casey et al. [31] showed that when the tents of *M. americanum* are shielded from the sun, the caterpillars are unable to raise their body temperature above the cool ambient temperatures that prevail in the spring, and they fail to grow. When tents are exposed to sunlight, their layered structure creates a thermally heterogeneous microhabitat within which the caterpillars can thermoregulate by moving from compartment to compartment [32, 44]. In addition, studies by Joos et al. [32] show that caterpillars basking side by side in groups on the surface of the tent are able to achieve significantly higher body temperatures, due to boundary layer effects and convective shielding, than solitary caterpillars basking in the open on nearby branches.

The limitations of the social repertoire

Lacking from this brief synopsis of the collective behaviors of social caterpillars are unequivocal instances of altruism and division of labor. In addition, although colonies of social caterpillars begin as family units, there is as yet no evidence that the insects practice any form of kin discrimination. Indeed, in the eastern tent caterpillar, a species which stands near the pinnacle of social evolution among caterpillars, colonies that are the products of genetically different parents commonly combine, and megacolonies consisting of as many as 10 or more colonies are not uncommon [43, 113]. However, fewer than 5% of the species of caterpillars known to form familiar aggregates have been studied in any detail, and future studies may well reveal instances of these phenomena in some species. Moreover, there were early reports of a weak division of labor in nest and trail construction among tent caterpillars [114], but subsequent studies [115, 116] failed to substantiate them. More recently, Porter et al. [117] reported that the male members of colonies of *E. socialis* spin more silk than the females, but the full significance of this behavior remains to be evaluated. Wellington [114, 118] also suggested that colonies of tent caterpillars may be genetically predisposed to produce a class of caterpillars, which because of their sluggish behavior, act as sinks for parasitoids [118]. But in the only rigorous experimental study to investigate the potential for this sort of kin-selected behavior, Stamp [119] tested and rejected the hypothesis that caterpillars of *Euphydryas phaeton* doomed by parasitoid attack behave in a manner that favors their intact colony mates.

Despite the lack of evidence for the occurrence of altruistic behavior, division of labor, and kin recognition in colonies of social caterpillars, the potential for the evolution of at least simple forms of these phenomena would appear to be within the reach of caterpillars. Furthermore, it is not difficult to envision circumstances under which these phenomena might act to enhance the fitness of colonies. It is of some interest, therefore, to know why these phenomena are not a more common feature of caterpillar societies, and future studies might be directed profitably at furthering our understanding of ecological factors that constrain as well as promote the behavioral repertoire of social caterpillars.

Conclusions

With the exception of the trail systems of the tent caterpillars, the collective structures of the species of colonial caterpillars studied so far are relatively simple and as such may be ideally suited to the study of emergent phenomena. Our review indicates that the "simple rules" model of emergent, collective behavior applied to eusocial species (e.g. [120, 121], see Detrain et al., this volume) may also be profitably applied to caterpillar societies. Elements of trail-making, recruitment, and nest-building behavior by social caterpillars present the clearest examples of simple rules followed by individuals collectively resulting in autocatalytic group phenomena. The sensitivity of caterpillars to minute differences in trail pheromone concentration coupled with the positive correlation of marking intensity and patch quality permits collective flexibility: recruitment pathways readily shift in response to stronger signals, allowing larvae to abandon patches of low profitability in favor of more profitable patches [20, 51]. The activities of individuals in the colony are thus structured in that response to and reinforcement of the messages of initially successful foragers quickly guides the remainder of the colony in exploiting a patchy resource. In this respect, the channeling of information directing the colony represents autocatalytic behavior similar to that discussed by Deneubourg and Goss [122] in relation to army ants.

The simple rules model also applies to collective shelter-building behavior in that simple, repetitive, motor patterns associated with silk spinning underlie the formation of distinctive, emergent structures. Studies indicate, however, that species-specific differences in the physical properties of the building material may have a significant influence on the nature of the emergent structure. Furthermore, the collective building behavior of caterpillars differs from that of eusocial insects such as wasps [120, 123] and termites [121] in that the stigmergic, structure-derived cues that direct the construction of complex structures in eusocial insects appear to play no role in the formation of the relative simple shelters of caterpillars.

Acknowledgments

This work was supported in part by NSF grant BNS93-17897.

References

1 Costa JT, Pierce NE (1997) Social evolution in the Lepidoptera: ecological context and communication in larval societies. *In*: JC Choe, BJ Crespi (eds): *The evolution of social behavior in insects and arachnids.* Cambridge University Press, Cambridge 407–422

2 Scudder SH (1889) *The butterflies of the Eastern United States and Canada with special reference to New England 1*, Cambridge

3 Balfour-Browne F (1925) The evolution of social life among caterpillars. *III Inter Cong Entomol* Zurich 334–339

4 Wheeler WM (1928) *The social insects: their origin and evolution*. Harcourt Brace, New York

5 Eickwort GC (1981) Presocial Insects. *In*: HR Hermann (ed): *Social insects,* vol 2. Academic Press, New York, 199–279

6 Fitzgerald TD (1995) *The tent caterpillars.* Cornell University Press, Ithaca, NY

7 Costa JT, Fitzgerald TD (1996) Developments in social terminology: semantic battles in a conceptual war. *Trends Ecol Evol* 11: 285–289

8 Wilson EO (1971) *The insect societies.* The Belknap Press of Harvard University Press, Cambridge, MA

9 Long DB (1953) Effects of population density on larvae of Lepidoptera. *Trans R Entomol Soc Lond* 104: 533–585

10 Hosoya J (1956) Notes on the biology of the tea-tussock-moth *Euproctis pseudo-conspersa* Strand. *Jpn J Sanit Zool* 7: 77–82

11 Ghent AW (1960) A study of the group-feeding behaviour of the jack pine sawfly *Neodiprion pratti banksianae* Roh. *Behaviour* 16: 110–148

12 Lyons LA (1962) The effect of aggregation on egg and larval survival in *Neodiprion swainei* Midd (Hymenoptera: Diprionidae)

Can Entomol 94: 49–58

13 Sugimoto T (1962) Influences of individuals in aggregation or isolation on the development and survival of larvae of *Artona funeralis*. *Jpn J Appl Entomol Zool* 6: 196–199

15 Watanabe N, Umeya K (1968) Biology of *Hyphantria cunea* Drury (Lepidoptera: Arctiidae) in Japan. IV. Effects of group size on survival and growth of larvae. *Jpn Plant Prot Serv Res Bull* 6: 1–6

16 Shiga M (1976) Effect of group size on the survival and development of young larvae of *Malacosoma neustria testacea* Motschulsky (Lepidoptera: Lasiocampidae) and its role in the natural population. *Kontyu* 44: 537–553

17 Shiga M (1979) Population dynamics of *Malacosoma neustria testacea* (Lepidoptera, Lasiocampidae). *Bull Fruit Tree Res Stn* 6: 59–168

18 Fitzgerald TD (1976) Trail marking by larvae of the eastern tent caterpillar. *Science* 94: 961–993

19 Tsubaki Y (1981) Some beneficial effects of aggregation in young larvae of *Pryeria sinica* Moore (Lepidoptera: Zygaenidae). *Res Pop Ecol* 23: 156–167

20 Fitzgerald TD, Peterson SC (1983) Elective recruitment communication by the eastern tent caterpillar (*Malacosoma americanum*). *Anim Behav* 31: 417–42

21 Fitzgerald TD, Peterson SC (1988) Cooperative foraging and communication in social caterpillars. *Bioscience* 38: 20–25

22 Peterson SC (1987) Communication of leaf suitability by gregarious eastern tent caterpillars (*Malacosoma americanum*). *Ecol Entomol* 12: 283–289

23 Tostowaryk W (1971) The effect of prey defense on the functional response of *Podisus modestus* (Hemiptera: Pentatomidae) to densities of the sawflies

Neodiprion swainei and *N pratti banksianae* (Hymenoptera: Neodiprionidae) *Can Entomol* 104: 61–69

24 Morris RF (1976) Relation of parasite attack to the colonial habit of *Hyphantria cunea. Can Entomol* 108: 833–836

25 Stamp NE (1980) Egg deposition patterns in butterflies: why do some species cluster their eggs rather than deposit them singly? *Amer Naturalist* 115: 367–380

26 Stamp NE (1981) Effect of group size on parasitism in a natural population of the Baltimore checkerspot *Euphydryas phaeton. Oecologia* 49: 201–206

27 Lawrence WS (1990) The effects of group size and host species on development and survivorship of a gregarious caterpillar *Halisidota caryae* (Lepidoptera: Arctiidae) *Ecol Entomol* 15: 53–62

28 Robison DJ (1993) *The feeding ecology of the forest tent caterpillar,* Malacosoma disstria *Hübner, among hybrid poplar clones,* Populus *spp.* PhD. dissertation, University of Wisconsin, Madison

29 Evans EW (1982) Influence of weather on predator/prey relations: stinkbugs and tent caterpillars. *J New York Entomol Soc* 90: 241–246

30 Porter K (1982) Basking behaviour in larvae of the butterfly *Euphydryas aurina Oikos* 38: 308–312

31 Casey TM, Joos B, Fitzgerald TD, Yurlina ME, Young PA (1988) Synchronized group foraging, thermoregulation, and growth of eastern tent caterpillars in relation to microclimate. *Physiol Zool* 61: 372–377

32 Joos B, Casey TM, Fitzgerald TD, Buttemer WA (1988) Roles of the tent in behavioral thermoregulation of eastern tent caterpillars. *Ecology* 69: 2004–2011

33 Seelinger G, Seelinger U (1983) On the social organization, alarm, and fighting in the primitive cockroach *Cryptocercus punctulatus* Scudder. *Z Tierpsychol* 61: 315–333

34 Schuster JC, Schuster LB (1985) Social behavior in passalid beetles (Coleoptera: Passalidae): Cooperative brood care. *Fla Entomol* 68: 266–272

35 Kirkendall LR, Kent DS, Raffa KA (1997) Interactions among males, females and offspring in bark and ambrosia beetles: the significance of living in tunnels for the evolution of social behavior. *In*: JC Choe, BJ Crespi (eds): *The evolution of social behaviour in insects and arachnids.* Cambridge University Press, Cambridge, 181–215

36 Edgerly JA (1997) Life beneath silk walls: a review of the primitively Embiidina. *In*: JC Choe, BJ Crespi (eds): *The evolution of social behaviour in insects and arachnids.* Cambridge University Press, Cambridge, 14–25

37 Mockford EL (1957) Life history studies on some Florida insects of the genus *Archipsocus* (Psocoptera). *Bull Florida St Museum, Biol Sci* 1: 253–274

38 Crespi BJ, Mound LA (1997) Ecology and evolution of social behavior among Australian gall thrips and their allies. In JC Choe, BJ Crespi (eds): *The Evolution of Social Behaviour in Insects and Arachnids.* Cambridge Univ. Press, Cambridge, 166–180

39 Dunn LH (1917) The coconut tree caterpillar (*Brassolis isthmia*) of Panama. *J Econ Entomol* 10: 473–488

40 Fitzgerald TD (1993) Trail and arena marking by caterpillars of *Archips cerasivoranus* (Lepidoptera: Tortricidae). *J Chem Ecol* 19: 1479–1489

41 Kevan PG, Bye RA (1991) The natural history, sociobiology, and ethnobiology of *Eucheira socialis* Westwood (Lepidoptera: Pieridae), a unique and little-known butterfly from Mexico. *Entomologist* 110: 146–165

42 Fitzgerald TD (1993) Social caterpillars. *In*: NE Stamp, TM Casey (eds):

Caterpillars: ecological and evolutionary constraints on foraging. Chapman and Hall, New York, 372–403

43 Fitzgerald TD, Willer DE (1983) Tent-building behavior of the eastern tent caterpillar *Malacosoma americanum* (Lepidoptera: Lasiocampidae). *J Kans Entomol Soc* 56: 20–31

44 Wellington WG (1950) Effects of radiation on the temperatures of insects and habitats. *Sci Agric* 30: 209–234

45 Fitzgerald TD, Clark KL, Vanderpool R, Phillips C (1991) Leaf shelter-building caterpillars harness forces generated by axial retraction of stretched and wetted silk. *J Insect Behav* 4: 21–32

46 Fitzgerald TD, Clark K (1994) Analysis of leaf-rolling behavior of *Caloptilia serotinella* (Lepidoptera: Gracillariidae). *J Insect Behav* 7: 859–872

47 Floater GJ (1996) Estimating movements of the processionary caterpillar *Ochrogaster lunifer* Herrich-Schaffer (Lepidoptera: Thaumetopoeidae) between discrete resource patches. *Aust J Entomol* 35: 279–283

48 Mills MB (1951) Bag-shelter caterpillars and their habits. *W Austr Natur* 3: 61–67

49 Talhouk AS (1975) Contributions to the knowledge of almond pests in east Mediterranean countries. I. Notes on *Eriogaster amygdali* Wilts. (Lepid., Lasiocampidae) with a description of a new subspecies by E.P. Wiltshire. *Z Ang. Ent* 78: 306–312

50 Fitzgerald TD, Underwood DLA. Trail marking by the larva of the Madrone butterfly *Eucheira socialis* and the role of the trail pheromone in communal foraging behavior. *J Insect Behav* 11: 247–263

51 Fitzgerald TD, Underwood DLA. Communal foraging behavior and recruitment communication in *Gloveria* sp. (Lepidoptera: Lasiocampidae). *J Chem Ecol* 24: 1381–1396

52 Fitzgerald TD, Visscher CR (1996) Foraging behavior and growth of isolated larvae of the social caterpillar *Malacosoma americanum*. *Entomol Exp Appl* 81: 293–299

53 Fitzgerald TD (1980) An analysis of daily foraging patterns of laboratory colonies of the eastern tent caterpillar, *Malacosoma americanum* (Lepidoptera: Lasiocampidae), recorded photoelectronically. *Can Entomol* 112: 731–738

54 Fitzgerald TD, Casey T, Joos B (1988) Daily foraging schedule of field colonies of the eastern tent caterpillar *Malacosoma americanum*. *Oecologia* 76: 574–578

55 Hoebeke B (1987) *Yponomeuta cagnagella* (Lepidoptera: Yponomeutidae): a palearctic ermine moth in the United States, with notes on its recognition, seasonal history, and habits. *Ann Entomol Soc Amer* 80: 462–467

56 Roessingh P (1989) The trail following behaviour of *Yponomeuta cagnagellus*. *Entomol Exp Appl* 51: 49–57

57 Roessingh P (1990) Chemical marker from silk of *Yponomeuta cagnagellus*. *J Chem Ecol* 16: 2203–2216

58 Cannon WN (1985) Social feeding behavior of *Hyphantria cunea* larvae (Lepidoptera: Arctiidae) in multiple choice experiments. *Great Lakes Entomol* 18: 79–80

59 MacLeod DM, Tyrrell D (1979) *Entomophthora crustosa* n. sp. as a pathogen of the forest tent caterpillar, *Malacosoma disstria* (Lepidoptera: Lasiocampidae). *Can Entomol* 111: 1137–1144

60 Bucher GE (1957) Diseases of the larvae of tent caterpillars caused by a sporeforming bacterium. *Can J Microbiol* 3: 695–709

61 Heinrich B (1979) Foraging strategies of caterpillars. *Oecologia (Berl.)* 3325–3337

62 Sharplin J (1964) The mourning cloak butterfly. University of Alberta, Department of Entomology leaflet 475

63 Stamp NE (1982) Behavioral interactions of parasitoids and Baltimore checkerspot caterpillars *Euphydryas phaeton*. *Environ Entomol* 11: 100–104

64 Stamp NE (1984) Foraging behavior of tawny emperor caterpillars (Nymphalidae: *Asterocampa clyton*). *J Lepid Soc* 38: 186–191

65 Stamp NE (1977) Aggregation behavior of *Chlosyne lacinia* (Nymphalidae) *J Lepid Soc* 31: 35–40

66 Capinera JL (1980) A trail pheromone from the silk produced by larvae of the range caterpillar *Hemileuca olivae* (Lepidoptera: Saturnidae) and observations on aggregation behavior. *J Chem Ecol* 3: 655–644

67 Long DB (1955) Observations on subsocial behaviour in two species of larvae, *Pieris brassicae* L. and *Plusia gamma* L. *Trans R Entomol Soc Lond* 106: 421–437

68 Fitzgerald TD, Costa JT (1986) Trail-based communication and foraging behavior of young colonies of the forest tent caterpillar *Malacosoma disstria* Hubn. (Lepidoptera: Lasiocampidae). *Ann Entomol Soc Amer* 79: 999–1007

69 Fitzgerald TD, Webster FX (1993) Identification and behavioural assays of the trail pheromone of the forest tent caterpilllar *Malacosoma disstria* Hübner (Lepidoptera: Lasiocampidae). *Can J Zool* 71: 1511–1515

70 Balfour-Browne F (1933) The life-history of the "small eggar moth," *Eriogaster lanestris*. L *Proc Zool Soc Lond*: 161–180

71 Franclemont JG (1973) Mimallonoidea and Bombycoidea. *In*: RB Dominick et al., *The moths of America north of Mexico*, vol 20.1. EW Classey and RBD Publications, London

72 Breuer M, Devoka B (1990) Studies on the importance of nest temperature of *Thaumetopoea pityocampa* (Den. and Schiff.) (Lep., Thaumetopoeidae). *J Appl Entomol* 109: 331–335

73 Cushman JH, Rashbrook VK, Beattie AJ (1994) Assessing benefits to both participants in a lycaenid-ant association. *Ecology* 75: 1031–1041

74 Common IFB, D F Waterhouse (1981) *Butterflies of Australia*. Angus and Robertson, London

75 Roessingh P, Peterson SC, Fitzgerald TD (1988) The sensory basis of trail following in some lepidopterous larave: contact chemoreception. *Physiol Entomol* 13: 219–224

76 de Biseau JC, Deneubourg JL, Pasteels JM (1992) Collective flexibility during mass recruitment in the ant *Myrmica sabuleti* (Hymenoptera: Formicidae). *Psyche* 98: 323–336

77 Beckers R, Deneubourg JL, Goss S, Pasteels JM (1990) Collective decision making through food recruitment. *Insect Soc* 37: 258–267

78 de Biseau JC, Schuiten M, Pasteels JM, Deneubourg JL (1994) Respective contributions of leader and trail during recruitment to food in *Tetramorium bicarinatum* (Hymenoptera: Formicidae). *Insect Soc* 41: 241–254

79 Hamilton WD (1971) Geometry for the selfish herd. *J Theor Biol* 31: 295–311

80 Taylor RJ (1977) The value of clumping to prey. *Oecologia* 30: 285–294

81 Taylor RJ (1977) The value of clumping to prey when detectability increases with group size. *Amer Naturalist* 111: 229–301

82 Turner GF, Pitcher TJ (1986) Attack abatement: a model for group protection by combined avoidance and dilution. *Amer Naturalist* 128: 228–240

83 Sill(n-Tullberg B, Leimar O (1988) The evolution of gregariousness in distasteful insects as a defense against predators. *Amer Naturalist* 132: 723–734

84 Wrona FJ, Jamieson Dixon RW (1991) Group size and predation risk: a field analysis of encounter and dilution effects.

Amer Naturalist 137: 186–201

85 Stamp NE (1981) Parasitism of single and multiple egg clusters of *Euphydryas phaeton* (Nymphalidae). *J New York Entomol S* 89: 89–97

86 Subinprasert S, Svensson BW (1988) Effects of predation on clutch size and egg dispersion in the codling moth *Laspeyresia pomonella*. *Ecol Entomol* 13: 87–94

87 Evans EW (1983) Niche relations of predatory stinkbugs (*Podisus* spp, Pentatomidae) attacking tent caterpillars (*Malacosoma americanum*, Lasiocampidae). *Amer Midland Naturalist* 109: 316–323

88 Damman H (1987) Leaf quality and enemy avoidance by the larvae of a pyralid moth. *Ecology* 68: 88–97

89 Prop N (1960) Protection against birds and parasites in some species of tenthredinid larvae. *Arch Neerl Zool* 13: 380–447

90 Eisner T, Kafatkos FC (1962) Defense mechanisms of arthropods. X. A pheromone promoting aggregation in an aposematic distasteful insect. *Psyche* 69: 53–61

91 Young AM (1978) A communal roost of the butterfly *Heliconius charitonius* L. in Costa Rican premontane tropical wet forest (Lepidoptera: Nymphalidae). *Entomol News* 89: 235–243

92 Pasteels JM, Gregoire JC, Rowell-Rahier M (1983) The chemical ecology of defense in arthropods. *Annu Rev Entomol* 28: 263–289

93 Vulinec K (1990) Collective security: aggregation by insects in defense. *In*: DL Evans, JO Schmidt (eds): *Insect defenses.* State University of New York Press, Albany, NY, 251–288

94 Myers JH, Smith JNM (1978) Head flicking by tent caterpillars: a defensive response to parasite sounds. *Can J Zool* 56: 1628–1631

95 Cornell JC, Stamp NE, Bowers MD (1987) Developmental change in aggregation, defense and escape behavior of buckmoth caterpillars, *Hemileuca lucina* (Saturniidae). *Behav Ecol Sociobiol* 20: 383–388

96 Peterson SC, Johnson ND, LeGuyader JL (1987) Defensive regurgitation of allelochemicals derived from host cyanogenesis by eastern tent caterpillars. *Ecology* 68: 1268–1272

97 Johnson WT, Lyon HH (1988) *Insects that feed on trees and shrubs*, 2nd ed. Cornell University Press, Ithaca, NY

98 Tinbergen N, Impekoven M, Franck D (1967) An experiment on spacing-out as a defence against predation. *Behaviour* 28: 307–327

99 Smith JNM (1974) The food searching behaviour of two European thrushes II: the adaptiveness of the search patterns. *Behaviour* 49: 1–61

100 Wiklund C, Jarvi T (1982) Survival of distasteful insects after being attacked by naive birds: a reappraisal of the theory of aposematic coloration evolving through individual selection. *Evolution* 36: 998–1002

101 Treherne JE, Foster WA (1980) The effects of group size on predator avoidance in a marine insect. *Anim Behav* 28: 1119–1122

102 Treherne JE, Foster WA (1981) Group transmission of predator avoidance in a marine insect: the Trafalgar effect. *Anim Behav* 29: 911–917

103 Treherne JE, Foster WA (1982) Group size and anti-predator strategies in a marine insect. *Anim Behav* 30: 536–542

104 Sullivan CR, Green GW (1950) Reactions of the eastern tent caterpillar *Malacosoma americanum* (F.), and the spotless fall webworm *Hyphantria textor* Harr., to pentatomid predators. *Can Entomol* 82: 52–53

105 Dixon AFG (1985) *Aphid ecology.* Blackie and Son Limited, Glasgow

106 Nault LR, Wood TK, Goff AM (1974) Tree hopper (Membracidae) alarm pheromones. *Nature* 149: 387–388

107 Doane CC, Leonard DE (1975) Orientation and dispersal of late-stage larvae of Porthetria dispar (Lepidoptera: Lymanthriidae). *Can Entomol* 107: 1333–1338

108 Roden DB, Miller JR, Simmons GA (1992) Visual stimuli influencing orientation by larval. psy moth, *Lymantria dispar* (L.). *Can Entomol* 124: 287–304

109 Knapp R, Casey TM (1986) Activity patterns, behavior, and growth in gypsy moth and eastern tent caterpillars. *Ecology* 67: 598–608

110 Casey TM, Knapp R (1987) Caterpillar thermal adaptation: behavioral differences reflect metabolic thermal sensitivities. *Comp Biochem Physiol* 86a: 679–682

111 Capinera JL, Weiner LF, Anamosa PR (1980) Behavioral thermoregulation by late-instar range caterpillar larvae *Hemileuca oliviae* Cockerell (Lepidoptera: Saturniidae). *J Kans Entomol Soc* 53: 631–638

112 Seymour R (1974) Convective and evaporative cooling in sawfly larvae. *J Insect Physiol* 20: 2447–2457

113 Costa JT, Ross KG (1993) Seasonal decline in intracolony genetic relatedness in eastern tent caterpillars: implications for social evolution. *Behav Ecol Sociobiol* 32: 47–54

114 Wellington WG (1957) Individual differences as a factor in population dynamics: the development of a problem *Can J Zool* 38: 289–314

115 Edgerly JS, Fitzgerald TD (1982) An investigation of behavioral variability in within colonies of the eastern tent caterpillar *Malacosoma americanum* (Lepidoptera: Lasiocampidae) *J Kans Entomol Soc* 55: 145–155

116 Papaj DR, Rausher MD (1983) Individual variation in host location by phytophagous insects *In*: S Ahmad (ed): *Herbivorous insects: host seeking behavior and mechanisms.* Academic Press, New York

117 Porter AH, Geiger H, Underwood DLA, Llorente-Bousquets J, Shapiro AM (1997) Relatedness and population differentiation in a colony of the butterfly *Eucheria socialis* (Lepidoptera: Pieridae). *Ann Entomol Soc Amer* 90: 230–236

118 Wellington WG (1960) Qualitative changes in natural populations during periods of abundance. *Can J Zool* 38: 289–314

119 Stamp NE (1981) Behavior of parasitized aposematic caterpillars: advantages to the parasite or host? *Amer Naturalist* 118: 715–725

120 Karsai I, Pénzes Z, Wenzel JW (1996) Dynamics of colony development in *Polistes dominulus*: a modeling approach. *Behav Ecol Sociobiol* 39: 97–105

121 Bonabeau E, Theraulaz G, Deneubourg JL, Aron S, Camazine S (1997) Self-organization in social insects. *Trends Ecol Evol* 12: 188–193

122 Deneubourg JL, Goss S (1989) Collective patterns and decision-making. *Ethol Ecol Evol* 1: 295–311

123 Karsai I, Theraulaz G (1995) Nest building in a social wasp: postures and constraints (Hymenoptera: Vespidae). *Sociobiology* 26: 83–114

Self-organization or individual complexity: a false dilemma or a true complementarity?

Jean-Louis Deneubourg, Scott Camazine and Claire Detrain

The collective decisions and patterns discussed in this section are examples of processes based upon behavioral rules of thumb [1] executed by individuals which have only limited access to global information. Most of these behavioral rules can be expressed as relatively simple "if-then" statements which correspond to stimulus-response pairs based upon a change in the internal or motivational state of the individual. These processes involve multiple interactions (the individuals or events are numerous) in the form of positive or negative feedback loops. These interactions modify the characteristics of the system and provide new stimuli for further interactions. In these systems, forms of positive feedback, such as recruitment, often involve specific behaviors by individuals; in contrast, the negative feedback often arises "automatically " as a result of the limits or constraints in the system (e.g. the depletion of building materials or the consumption of food).

The coupling between such feedbacks can lead to some surprisingly non-intuitive dynamics such as the oscillations discussed by Cole and Trampus in this volume. In this case, very simple interactions such as collisions and changes in activity levels are enough to produce organized structures. This contribution suggests that ant colonies might provide useful and interesting models for the study of oscillations in biology [2].

The environment itself plays a key role in the emergence of patterns. The chapter of Theraulaz et al., which concentrates on building, illustrates how a model involving positive feedback in the form of chemotatic orientation of termites to cement pheromone, coupled with an air flux, can explain the shift from regular pillars to walls [3].

Food recruitment also provides numerous examples of such coupling between positive and negative feedback. Changing patterns of crowding at a food source [4], or variations in the food characteristics due to changes in its exploitation by foragers can lead to foraging dynamics which are far more complicated than the

C. Detrain et al. (eds) Information Processing in Social Insects

simple choice of one food source over another. For example, the coupling of food exhaustion due to foraging activity with trail-laying behavior can lead to a homogeneous exploitation of the environment or to foraging trails that rotate systematically around a nest [5] depending upon the dynamics of available food. Similarly, the plasticity in the pattern of recruitment trails may be driven simply by an increase in the number of food sources, without any requirement for changes in the communication system or the individual behavior of the ants. A colony may select only one source if the number of sources is small or if its forager population is large. Alternatively, a colony may exploit all sources equally if the number of food sources is large or if the colony population is small.

In the above examples, collective patterns emerge without any modulation of positive feedbacks. However, in insect societies, random searching of scouts for food or home sites can lead to discoveries and recruitments which can be modulated. The collective decisions may arise as a result of competition between different sources of information flow which are conveyed to nestmates and amplified in a number of different ways. For example, an individual recruiting scout bee or ant may modulate its amount of dancing or trail laying in relation to its perception of the profitability of a particular sugar source [6, 7]. In bees the flow of recruited insects is proportionnal to the number of dancing bees, whereas in ants no such relationship is found between the numbers of recruiting and recruited workers. This seemingly small difference in the recruitment systems is responsible for different properties of collective foraging in honeybees and many ant species. For example, bees are always able to select the most rewarding nectar source even if many weaker food sources have been discovered previously. This is not the case for the ants, which may not be able to break free from a previously selected food source and to switch to a more rewarding source. These different strategies are not primarily due to differences in the sensory or physiological capacities of the individuals involved. They are rather a by-product of small differences between the recruitment "devices" used by the ants or the bees [8].

As discussed above, an individual scout bee or ant may modulate its recruitment behavior in relation to its local perception of food profitability. In addition, other factors unrelated to individual behaviors, such as characteristics of the environment (Detrain et al., this volume) can alter the amplification of recruitment to food sources. Thus, efficient decision making can arise without any modulation of the individual's behavior. In such cases, any constraint which reduces the rate of recruitment (such as the distance between the nest and the food source or the time spent to discover food) can lead to that food source losing out in its competition with other sources.

The simplicity of the rules involved in the production of a global foraging response appears even more clearly if other colony dynamics, such as changes in the internal demand of workers or larvae, are also taken into account. Such colony dynamics can play a role that is of equal importance to the specifics of the recruitment behaviors. For example, in the case of ants when there is a choice between sources of different quality (e.g. proteins versus sugars), workers will selectively forage to meet colony requirements. In this case, non-foraging workers within the colony can provide important information to foragers by means of trophallactic and behavioral interactions [9–10] (see also Cassill and Tschinkel, this volume). This exchange of information between foragers and domestics is indirect, since it is based on time delays in the unloading of foragers (see also Anderson and Ratnieks, this volume).

With this capacity to produce efficient responses or patterns, it is not surprising to find a large number of such self-organized social systems. Recruitment is used in ants for other purposes such as nest defense [11], exploration [12–14] or trail networks between nests [15]. It is also involved in nest-site selection in honeybees described by Visscher and Camazine in their chapter as well as nest-moving in wasps and ants [16, 17]. The example of social caterpillars discussed by Fitzgerald and Costa, this volume, and [18] also provides a beautiful instance of spatial organization involving recruitment and trails formation. The problem of selection of information can be easily solved by the colony of caterpillars through competition between simple recruitments. Trail recruitment has also been found in very different animals such as mollusks [19], Norway rats [20] and naked mole rats [21].

Besides recruitment, activities such as clustering of workers, brood or food within the nest [22–24] also involve amplification processes. The ability of ants to discriminate between different items (e.g. larval instars) coupled with amplifying processes can explain the spatial organization of these items within the nest. Similarly, mathematical models and intuition suggest that the same basic logic could be extended to task regulation and could be one of the touchstones of social organization [25, 26].

Moreover, self-organization is not limited only to these questions of social organisation but can also be extended to studies on the behavioral specialisation of individuals as well as the emergence of hierarchy and division of labour ([27],

Huang and Robinson, this volume). All these examples address the fundamental question of how the most rewarding site, activity and so on can be selected by a group of individuals or by one individual.

In their approach to social life based on self-organization, all the contributions of this section stress the behavioral simplicity of the individual. This begs the question of how much more diverse and efficient the responses could be with small increases in the complexity of the insects' responses to the stimuli. In the case of trail recruitment, mathematical models show that an increase in individual complexity may not lead to a large increase in collective efficiency [28].

But in other cases in which workers are likely to be sensitive to a great diversity of cues, both from nestmates and the external environment, and to be able to modulate their behavior in response to these cues, we expect to find a greater diversity of colony-level responses. Cooperative defense of *Myrmecocystus* ants is an example that illustrates the alternative: Is complexity of global behavior a by-product of complex individual assessments or the expression of simpler behavioral rules and amplification processes? *Myrmecocystus* tournaments are seen as sites for global harvesting of information and integrative sampling to tally colony strength. Several decision-making processes have been suggested to account for these ritualized fights [29, 30]. On the one hand, the first two models hypothesize behavioral complexity of the individual ant able to census the number of aggressive encounters and the caste of the opponents. The ant "measures", by some analogical counting, the colony size and even gets an image of the opponent colony's caste system. On the other hand, in the third model, recruitment for information storage and processing by individual ants is less demanding. In this case, the mere detection of a supply of unengaged nestmates is in itself sufficient to release recruitment behavior and to result in a raid on the weaker colony. This implies that ants are not engaged in some accurate sampling procedure but are locally reacting to simple behavioral rules.

Odors of nestmates or signals arising from the genetic diversity of a colony are potential sources of diverse cues for individuals. An insect can modulate its response to signals emitted by the conspecifics. This can be achieved in different ways. For example genetic relatedness can favor mutual recognition: cockroaches discriminate their odor from that of other lineages, and alter their gregarious behavior [31]. Some ant species are also able to distinguish their trail from those produced by their nestmates [32]. The ability to learn the characteristics of a pheromonal blend is a means of facilitating mutual recognition in ants

(Lenoir et al., this volume). The degree to which individuals can recognize one another affects the dynamics of the amplification involved in the clustering and leads to a diversity of patterns such as mixed or segregated clusters [33].

Researchers can expect to obtain a great deal of information about insect social life by examing behavioral rules of the individual. Our conviction is that self-organized "scripts" are numerous in group living organisms and that noneusocial species provide important material for exploring the link between self-organization and evolution of sociality. Aggregation is a rather simple social behavior and could be a first step towards more sophisticated social interactions. Most aggregations result from the simple gathering of individuals at the same place. The selection of a valuable site and the spatial organization of the populations are very often by-products of an "amplification-competition" script. The resulting increase in density is a favorable ground for the emergence of new social behaviors such as a cooperative defense or a new function for the cluster such as an information center for food collection (see, for example, Fitzgerald and Costa, this volume).

Put in an evolutionary perspective, an increase in the cooperation among individuals would be difficult to achieve if we assume it to be governed by a plethora of behavioral mechanisms and feedbacks. In this respect, self-organization scripts allow some parsimony in the coding and management of information. Our intuition about the importance of self-organized patterns in social organisms is partially due to our fascination with the impressive structures emerging from simple models of behavior. However, one of the current weaknesses in the approach of self-organization is the few instances in which it can be clearly identified. The paucity of data at both individual and colony levels prevents us from making any general assessment of the contribution of self-organization to the working of biological systems. For this reason, we suggest that more would be achieved at the present time through greater emphasis on experiments enabling us to discriminate between self-organization and other hypotheses on pattern formation and decision making, rather than further speculations and facile computer models [8].

References

1 Krebs JR, Davies NB (1987) *Behavioural ecology*. Blackwell, Oxford

2 Goldbeter A (1996) *Biochemical oscillations and cellular rhythms*. Cambridge University Press, Cambridge

3 Bonabeau E, Theraulaz G, Deneubourg JL, Franks N, Rafelsberger O, Joly JL, Blanco S (1998) The emergence of pillars, walls and royal chambre in termite nests. *Phil Trans R Soc Lond B* 353: 1561–1576

4 Wilson EO (1971) *The insect societies*. Harvard University Press, Cambridge, MA

5 Goss S, Deneubourg JL (1989) The self-organising clock pattern of *Messor pergandei*. (Formicidae, Myrmicinae). *Insect Soc* 36: 339–346

6 Camazine S, Sneyd J (1991) A model of collective nectar source selection by honey bees: self-organization through simple rules. *J Theor Biol* 149: 547–571

7 Seeley T D, Camazine S, Sneyd J (1991) Collective decision-making in honey bees: how colonies choose among nectar sources. *Behav Ecol Sociobiol* 28: 277–290

8 Camazine S, Deneubourg JL, Franks N, Sneyd J, Bonabeau E, Theraulaz G (1999) *Self-organization in biological systems*. Princeton University Press; *in press*

9 Howard DF, Tschinkel WR (1980) The effect of colony size and starvation on food flow in the fire ants, *Solenopsis invicta* (Hymenoptera: Formicidae). *Behav Ecol Sociobiol* 7: 293–300

10 Sorensen AA, Busch TM, Bradleigh Vinson S (1985) Control of food influx by temporal subcastes in the fire ant, *Solenopsis invicta*. *Behav Ecol Sociobiol* 17: 191–198

11 Hölldobler B, Wilson EO (1990) *The ants*. Harvard University Press, Cambridge, MA

12 Deneubourg JL, Aron S, Goss S, Pasteels JM (1990) The self-organizing exploratory pattern of the Argentine ant *Iridomyrmex humilis*. *J Insect Behav* 3: 159–168

13 Detrain C, Deneubourg JL, Goss S, Quinet Y (1991) Dynamics of collective exploration in the ant *Pheidole pallidula*. *Psyche* 98: 21–31

14 Fourcassié V, Deneubourg JL (1994) The dynamics of collective exploration and trail-formation in *Monomorium pharaonis*: experiments and model. *Physiol Entomol* 19: 291–300

15 Aron S, Deneubourg JL, Goss S, Pasteels JM (1990) Functional self-organisation illustrated by inter-nest traffic in the argentine ant *Iridomyrmex humilis*. *In*: W Alt, G Hoffman (eds): *Biological motion*. Springer, Berlin, 533–547

16 Jeanne RL (1991) The swarm-founding Polistinae. *In*: KG Ross, RW Matthews (eds): *The social behavior of Wasps*. Cornell University Press, Ithaca, NY, 191–231

17 JC Verhaeghe JC, Selicaers N, Deneubourg JL (1992) Nest-moving and food-location in *Tapinoma erraticum*. (Hymenoptera, Formicidae). *In*: J Billen (ed): *Biology and evolution of social insects*. Leuven University Press, Leuven, 335–342

18 Fitzgerald T D (1995) *The tent caterpillars*. Cornell University Press, Ithaca, NY

19 Focardi S, Deneubourg JL, Chelazzi G (1985) How shore morphology and orientation mechanisms can affect the spatial organization of intertidal molluscs. *J Theor Biol* 112: 771–782

20 Galef BGJr Buckley LL (1996) Use of foraging trails by Norway rats. *Anim Behav* 52: 765–771

21 Judd T, Sherman P (1996) Naked mole rats recruit colony mates to food source. *Anim*

Behav 52: 957–969

22 Camazine S (1991) Self-organizing pattern formation on the combs of honey bee colonies. *Behav Ecol Sociobiol* 28: 61–76

23 Camazine S, Sneyd J, Jenkins MJ, Murray JD (1990) A mathematical model of self-organized pattern formation on the combs of honeybee colonies. *J Theor Biol* 147: 553–571

24 Franks NR, Sendova-Franks AB (1992) Brood sorting by ants: distributing the workload over the work-surface. *Behav Ecol Sociobiol* 30: 109–123

25 Pacala SW, Gordon DM, Godfray HCJ (1996) Effects of social group size on information transfer and task allocation. *Evol Ecol* 10: 127–165

26 Adler FR, Gordon DM (1992) Information collection and spread by networks of patrolling ants. *Amer Naturalist* 40: 373–400

27 Bonabeau E, Theraulaz G, Deneubourg JL (1996) Mathematical model of self-organizing hierarchies in animal societies. *Bull Math Biol* 58: 661–717

28 Beckers R, Deneubourg JL, Goss S (1992) Trails and U-turns in the selection of a path by the ant *Lasius niger. J Theor Biol* 159: 397–415

29 Hölldobler B (1982) Foraging and spatiotemporal territories in the honey ant *Myrmecocystus mimicus* Wheeler. *Behav Ecol Sociobiol* 9:301–314

30 Lumsden CJ, Hölldobler B (1983) Ritualized combats and intercolony communication in ants. *J Theor Biol* 100:81–98

31 Rivault C, Cloarec A (1998) Cockroach aggregation: discrimination between strain odours in *Blattella germanica. Anim Behav* 55: 177–184

32 Aron S Deneubourg JL, Pasteels J M (1988) Visual cues and trail-following idiosynchrasy in *Leptothorax unifasciatus*: an orientation process during foraging. *Insect Soc* 35: 355–366

33 Deneubourg JL, Goss S, Franks N Sendova-Franks A Detrain C, Chrétien L (1991) The dynamics of collective sorting robot-like ants and ant-like robots. *In*: J-A Meyer, S Wilson (eds): *From animals to animats*. MIT Press, Cambridge, MA, 356–365

Subject index